PRACTICAL ENGINEERING MANAGEMENT OF OFFSHORE OIL AND GAS PLATFORMS

PRACTICAL ENGINEERING MANAGEMENT OF OFFSHORE OIL AND GAS PLATFORMS

NAEIM NOURI SAMIE
MSc Hydraulic Structures

Amsterdam • Boston • Heidelberg • London
New York • Oxford • Paris • San Diego
San Francisco • Singapore • Sydney • Tokyo
Gulf Professional Publishing is an imprint of Elsevier

Gulf Professional Publishing is an imprint of Elsevier
50 Hampshire Street, 5th Floor, Cambridge, MA 02139, USA
The Boulevard, Langford Lane, Kidlington, Oxford, OX5 1GB, UK

Notices
Knowledge and best practice in this field are constantly changing. As new research and experience broaden our understanding, changes in research methods, professional practices, or medical treatment may become necessary.

Practitioners and researchers must always rely on their own experience and knowledge in evaluating and using any information, methods, compounds, or experiments described herein. In using such information or methods they should be mindful of their own safety and the safety of others, including parties for whom they have a professional responsibility.

British Library Cataloguing-in-Publication Data
A catalogue record for this book is available from the British Library

Library of Congress Cataloging-in-Publication Data
A catalog record for this book is available from the Library of Congress

ISBN: 978-0-12-809331-3

For information on all Gulf Professional Publishing publications
visit our website at https://www.elsevier.com/

  Working together
to grow libraries in
Book Aid International developing countries

www.elsevier.com • www.bookaid.org

Publisher: Joe Hayton
Acquisition Editor: Katie Hammon
Editorial Project Manager: Kattie Washington
Production Project Manager: Sruthi Satheesh
Designer: Mark Roger

Typeset by TNQ Books and Journals

To my daughter Mitra
(Goddess of Light)

I wish she also performs her contribution to human civilization.

CONTENTS

PREFACE

Human civilization relies for its survival on energy sources. Up to now with the explosion of technological developments in using renewable energies like wind, water, solar, and other sources like atomic energy, the main energy source worldwide has been the hydrocarbons in oil and gas reservoirs. These are found both onshore and offshore in specific parts of the world.

This book concentrates on the engineering management role for design of oil and gas offshore platforms. The book's title shows it is related to the oil and gas sector (Part 1). It is known that other than wind turbine platforms in northern Europe, like in Denmark, and a few surveillance platforms [1] all other platforms are fabricated for oil and gas production. Part 2 of this book is applicable to all engineering sectors.

I started my career as a structural engineer and gradually shifted to offshore structures and covered all steps one by one up to the level of engineering manager. I worked during university years and have experienced all positions including drafter, surveyor, workshop technician, estimator, junior/senior/lead engineer, and finally engineering manager. This book is the outcome of more than 30 years of experience in different projects.

Since offshore platform design is a multidisciplinary task, I have faced many questions related to other disciplines. Maybe some of these questions were clear to even a junior engineer in those disciplines. But for me, coming from another discipline, the concepts needed further study. I strongly believe that no engineering manager can succeed just by concentrating on general items and project schedule issues. He shall actively participate in technical discussions. It doesn't mean that the engineering manager shall interfere with the lead engineer's role, but he needs to have an idea of what the discussion is about and what solutions have been found for similar problems in other projects. In addition, he shall specifically understand the consequences of each decision on other disciplines. In an exaggerated sense, he shall know whether the undergoing discussion is about a single valve or a large vessel or a complicated package that may impact other disciplines' design and activities. This will enable him to differentiate between minor and major issues and foresee consequences of a decision on other disciplines and total design.

If engineering disciplines involved in the design of an offshore platform were separate islands in an ocean with no impact on each other, the task would be easy. But in the small area of a platform (maybe less than 30 m × 30 m for each level) every single decision in a discipline may affect another discipline's design. Some of these impacts may be small but there are also some major impacts that require the engineering manager to understand and find suitable solutions for them. Therefore, he has to have some knowledge of all involved disciplines' activities. This is not possible unless the engineering manager clearly knows the following:

- What documents are expected from each discipline?
- What main topics shall be addressed in each document?
- What are the prerequisites for generation of each document or revising it to the next stage?
- How does the information in each discipline impact and change other disciplines' design?
- What interdisciplinary information shall be transferred?
- What tasks are expected from each discipline to enable other disciplines to perform their duties?
- What is the engineering impact on other project departments like procurement, construction, installation, and commissioning?

Interdisciplinary tasks may range from handing over data or filling out part of a document or participating in a joint meeting with a client, vendor, etc. Its delay or malfunction may lead to considerable change in schedule and cost.

During my career I couldn't find a simple introductory book to explain what systems and equipment are installed in an offshore platform. Some useful books covered portions of the question. References [2] and [3] provide very good explanations.

In this book I have tried to provide brief answers to several basic questions including:

- What are offshore platforms?
- What functions do they perform?
- What engineering disciplines are involved in their design?
- What systems and equipment are installed on a platform?
- How to generate necessary documents and distribute them?
- How to prepare an engineering proposal?
- How to monitor and control engineering progress (physical/invoicing)?

The main problem is that a book of this nature needs to cover a wide range of subjects in several disciplines and at the same time be concise and simple. Technical information as well as project control procedures shall be discussed. For technical data in each discipline, several books have been written. I have to emphasize that it is neither the intention nor the function of this book to explain theoretical aspects in detail. Only a brief introduction to some concepts is given here.

In some cases simplified methods have been explained to calculate the order of magnitude of variables. I have to emphasize that these simplifications in no way replace accurate calculations. Their sole intention is to give the engineering manager confidence that large mistakes do not happen. In many cases he may use information from similar projects instead of these "rule-of-thumb" calculations. Reference [4] gives some useful information for the mechanical engineering discipline.

In addition, the book will try to explain the author's experience in dealing with engineers involved in the design of platforms. After that, preparing bid proposals, organizing an engineering team, and managing to the end of a specific project are briefly discussed.

This book is divided to two parts. Chapters " Introduction to Offshore Platforms," "Disciplines Involved in Offshore Platforms Design," and "Systems and Equipment for Offshore Platforms Design" comprise Part 1, which concentrates on engineering and design aspects of offshore platforms. Chapters "Balancing Between Client and Task Force Engineers," "Handling Design Documentation," "Proposal Preparation," and "Planning the Project Budget" are included in Part 2, which covers management, budgeting, and scheduling of engineering projects in a general manner.

Chapter "Introduction to Offshore Platforms" gives a general description of offshore platforms and why engineering and, consequently, its management are important. Chapter "Disciplines Involved in Offshore Platforms Design" describes the disciplines that are involved in the engineering design of a platform. It also discusses the main documents that they need to produce, including the main topics discussed/explained in each document and some of the major items that have to be addressed in each document. Chapter "Systems and Equipment for Offshore Platforms Design" describes the systems that must be provided in an offshore platform, including the functions of each of them.

The information given in chapters "Disciplines Involved in Offshore Platforms Design" and "Systems and Equipment for Offshore Platforms Design" on many occasions refers to certain editions of standards. The content of standards and recommended practices undergo continuous change.

Every effort is made to ensure that the information given in this book is accurate and conforms to international standards. In spite of this, readers are recommended to refer to the latest edition of all standards. Any human activity may have some errors. This book is also not an exception.

It is understood that a detailed discussion of all disciplines and systems is beyond the scope of this book. In fact, if I borrow a sentence from one of my university teachers, this book may provide "1 cm depth of an ocean-wide knowledge." The extent of required information is very widespread but the degree of information given in this book is very preliminary and suffices only to give a list. Once more it is important to emphasize that the engineering manager need not and shall not undertake/ interfere with the responsibility of lead engineers. But when two lead engineers are talking, then the engineering manager has to distinguish whether they are talking about the fiction movie shown last night on the TV or a system that is required for the function of the intended platform, which may impact the environment, safety of the operators on the platform, and comfort of the people living inland.

The second part of this book is broad enough for all consultants. Chapter "Balancing Between Client and Task Force Engineers" describes how the engineering team shall be selected and how their working environment, salaries, etc. defined. A competent engineer is not a commodity that can be easily procured off the shelf of the first store on the street. In addition, the smartest engineers working on "separate islands" without links to one another may not be able to design a platform. A team of engineers working together, identifying other disciplines' requirements and clients' interests is needed to produce the desired result.

Chapter "Balancing Between Client and Task Force Engineers" also explains how this team shall work as individual disciplines and together as an integrated engineering team. Performing a sophisticated analysis may be easier than matching two interface lines. Changing/impacting the design of other disciplines due to decisions made in another discipline is inevitable. The important point is that critical interface information shall be transferred accurately and on time to avoid heavy rework and, more importantly, before the point of no return.

Chapter "Handling Design Documentation" describes how project documents shall be handled. This covers both engineering documents produced by the team and those generated by vendors. The procedure to control smooth flow of documents is described.

Chapter "Proposal Preparation" defines how to prepare a proposal for an engineering study. What factors shall be considered? The external and internal factors affecting an engineering project are explained. This chapter also describes how to deal with change requests encountered during a project.

Finally, chapter "Planning the Project Budget" explains how to run a project. Dividing the project budget into subsections and allocating budget sources to different parts need careful consideration, which varies from consultant to consultant.

Although the book's name refers to offshore oil and gas platforms, many of the concepts described herein can be used effectively in onshore projects.

Men and women alike can serve in this industry (especially on the engineering team). Therefore in all the chapters, when referring to engineering team I have tried to use either he/she or remove reference to gender. If anywhere only one gender is mentioned (like in this preface) it means both men and women without any discrimination.

I have tried my best to explain whatever seems to be practicable when encountering different issues. However, my experience is limited to a specific geographic location. Cultural differences and engineering practices in each region may require different approaches to the same issue. The common basis is dealing with oil and gas projects and highly trained/qualified engineers.

It is important to note that examples given in this book are typical of a wide range of offshore platforms. However, in spite of all similarities each offshore platform is a unique design. As the saying goes, there are many ways to skin a cat. Some of the platforms may have chosen different options. It is a fact that oil and gas offshore platforms have many similarities considering their layout, equipment size and type, production operation, crude composition, etc. In spite of this, they have many differences in reservoir capacity, daily production, production profile, crude composition, gas—oil ratio, water cut, reservoir shut in pressure, flowing pressure, temperature, etc. These differences impose different designs.

Geographical location is also a major factor. A design that is economically feasible in one location may not be so in another region. The author has encountered cases where in one location production started from 40,000 BOPD and ended at 5000 BOPD. While in another location it started from only 7000 BOPD.

Having said all of the above, this book is not intended to be a manual. It only describes my experience in offshore oil and gas projects.

It is important to note that information given in this book is gathered from a number of different projects. I am not sure, but it seems Albert Einstein has said: "That which remains in the brain after studying everything and forgetting everything is knowledge."

A major portion of the information is gathered from available literature. Since the information was gathered over the course of several years, in some cases I may have forgotten the references. If in any case proper references are not given, I sincerely apologize. A large part of the information is also selected from project documents. No engineering activity starts from discovery of fire. Each project builds on go-by documents available from previous projects.

My colleagues and friends have had an important share in teaching me the unknowns. With some of them I have worked more than 15 years in the same organization. My work experience with the Iran Marine Industrial Company (SADRA) is one of the best parts of my life. All of my colleagues in the SADRA structural group are now working in high-level positions in reputable companies around the globe. If I start to list their names, I would have to give a very long list of engineers in other departments as well and still might forget some names. In addition to colleagues, I have enjoyed working with very highly qualified managers. These managers and engineers have finished many projects and have overcome countless unknowns. Therefore, I prefer to thank all of them and wish them happiness and success in any part of the world in which they are now and any job or role that they have or will have.

After SADRA I spent long years with Snake Lay Technology Engineering (SLTE). Some of the sketches in this book refer to SLTE projects. It was my privilege to work with several highly qualified managers and engineers at this company.

I will be more than glad to receive any criticisms, comments, technical advice and further explanations from valued readers. The role of the engineering manager is not a rigid concept. It deals with technical facts and human relations. In some cases in both of these areas two multiplied by two may not always be equal to four.

Challenges in each project are an indispensable part of the process. Each project has its own challenges. In some you work only to receive salary. In some projects you actually enjoy working! Challenging technical aspects, knowledgeable clients and colleagues, cooperating teams, sense of responsibility to world environment, your country, and company all boost morale for better efficiency. Half of our useful time in a day is spent at work. We have to try to get best benefit of what we do.

I had many photos (taken by myself or my colleagues) from offshore projects and equipment. Since they were clients' property I couldn't insert them in this book. Therefore, to avoid copyright issues I have drawn AutoCAD sketches from them. In some figures I have written numbers on the sketch to identify the described subject in the text. In some cases they may not be as good as original photo, but most of them I have found to be quite satisfactory. My colleague Ms. Rezvani did a great job of drawing these sketches from photos. I have tried to explain major items of offshore complexes, platforms, and equipment in these figures.

Throughout the book I have tried to use an informal language style. This is not a preaching book. It is not a textbook either. In this book experiences from different projects are shared with the reader.

I would like to thank the following for their help. Mr. Mehdi Zamani read parts of this book and made valuable comments. Mr. Arne Skaven gave advice about the preface. Mr. Maldonado gave comments on chapter "Introduction to Offshore Platforms." Dr. James Speight, Mr. Kevin Juniel, Mr. Ramesh Bapat and Dr. Mohamed El-Reedy reviewed the full text and made important comments. I am grateful for their contributions. I have also to thank Ms. Katie Hammon (Elsevier Senior Acquisitions Editor) for her continuous support during manuscript preparation.

This book is dedicated to my daughter Mitra (Goddess of Light in Iranian mythology). I hope that she will make her own meaningful contributions to our world.

Naeim Nouri Samie

ABBREVIATIONS

ACH	Air Change per Hour
AFC	Approved For Construction
AHU	Air Handling Unit
AISS	Anti Intrusion and Surveillance System
AP	Advance Payment
API	American Petroleum Institute
ASD	Abandon Platform Shut Down
ASME	The American Society Of Mechanical Engineers
AVL	Approved Vendor List
AVR	Automatic Voltage Regulator
BDV	Blow Down Valve
BOD	Biochemical Oxygen Demand
BOPD	Barrel Oil Per Day
BWPD	Barrel of Water Per Day
CA	Certifying Authority
CAA	Civil Aviation Authority
CAPEX	Capital Expenditure
CBE	Commercial Bid Evaluation
CCPS	Center for Chemical Process Safety
CCR	Central Control Room
CCTV	Closed Circuit Television
CD	Chart Datum
CEV	Carbon Equivalent Value
CFD	Computational Fluid Dynamics
CGD	Combustible Gas Detector
COG	Center Of Gravity
CRA	Corrosion Resistant Alloy
CS	Carbon Steel
CTICM	Centre Technique Industriel de la Construction Métallique
CV	Curriculum Vitae
DAF	Dynamic Amplification Factor
DCC	Document Control Center
DCS	Distributed Control System
DG	Diesel Generator
DHSV	Down Hole Safety Valve
DNV	Det Norske Veritas

DS	Data Sheet
DVC	Design Verification Certificate
EIA	Environmental Impact Assessment
EOR	Enhanced Oil Recovery
EPCIC	Engineering, Procurement, Construction, Installation and Commissioning
ESD	Emergency Shut Down
ESDV	Emergency Shut Down Valve
ESP	Electrically Submerged Pump
FAT	Factory Acceptance Test
FDB	Final Data Book
FEED	Front End Engineering Design
FGS	Fire and Gas System
FWKO	Free Water Knock Out
GOR	Gas Oil Ratio
GPG	Good Performance Guarantee
GPM	Gallon Per Minute
GRP	Glass Fiber Reinforced Plastic
GTG	Gas Turbine Generator
HAT	Highest Astronomical Tide
HAZID	Hazard Identification
HAZOP	Hazard and Operability
HC	Hydrocarbon
HMB	Heat and Material Balance
HMI	Human Machine Interface
HOD	Head Of Department
HP	High Pressure
HPHT	High Pressure High Temperature
HPU	Hydraulic Power Unit
HRS	Hollow Rectangular Section
HVAC	Heat, Ventilation and Air Conditioning
ICSS	Integrated Control and Safety System
IDC	Inter Disciplinary Check
IDLH	Immediately Dangerous to Life and Health
IEC	International Electrotechnical Commission
IFA	Issued For Approval
IFC	Issued For Comment
IP	Ingress Protection
IR	Infra Red
IS	Intrinsically Safe

ITB	Invitation To Bid
ITP	Inspection and Test Plan
JB	Junction Box
KOM	Kick Off Meeting
LAT	Lowest Astronomical Tide
LCP	Local Control Panel
LLI	Long Lead Item
LOI	Letter Of Intent
LP	Low Pressure
LPCD	Liter Per Capita Daily
LQ	Living Quarter
LSD	Local Shut Down
LSF	Loadout Support Frame
LV	Low Voltage
MCT	Multi Cable Transit
MDP	Master Development Plan
MDR	Master Document Register
MEG	Mono Ethylene Glycol
MOM	Minutes Of Meeting
MOV	Motor Operated Valve
MPA	Mega Pascal
MR	Material Requisition
MSDS	Material Safety Data Sheet
MSL	Mean Sea Level
MTO	Material Take Off
MV	Master Valve
MV	Medium Voltage
MW	Mega Watts
MWS	Marine Warranty Surveyor
NDB	Non Directional Beacon
NDT	Non Destructive Testing
NFPA	National Fire Protection Association
NORSOK	Norsk Sokkels Konkurranseposisjo (Norwegian Petroleum Standard)
NPSH	Net Positive Suction Head
NRV	Non Return Valve
OIC	Offshore Installation Contractor
OPEX	Operating Expenditure
OS	Operating System
OP	Operating Procedure

PABX	Private Automatic Branch Exchange
PAGA	Public Address and General Alarm
PCV	Pressure Control Valve
PDS	Plant Design System
PDMS	Plant Design Management System
PFD	Process Flow Diagram
PID	Piping and Instrumentation Diagram
PLC	Programmable Logic Controller
PM	Project Manager
PO	Purchase Order
PP	Production Platform
PPM	Part Per Million
PPMW	Part Per Million Weight
PSD	Process Shut Down
PSI	Pound per Square Inch
PSV	Pressure Safety Valve
PWHT	Post Weld Heat Treatment
QA	Quality Assurance
QC	Quality Control
RH	Relative Humidity
ROS	Ready On Site
RT	Radiographic Test
SACS	Structural Analysis Computer Systems
SCFD	Standard Cubic Feet per Day
SDRS	Supplier Documentation Requirements Schedule
SLD	Single Line Diagram
SOW	Scope Of Work
SPC	Specification
SS	Stainless Steel
SSV	Surface Safety Valve
SSSV	Subsurface Safety Valve
STEL	Short Time Exposure Limit
SWL	Safe Working Load
TBE	Technical Bid Evaluation
TC	Technical Clarification
TEG	Tri Ethylene Glycol
TGD	Toxic Gas Detector
TPA	Third Party Authority
TTP	Through Thickness Property
UCP	Unit Control Panel

UFD	Utility Flow Diagram
ULCC	Ultra Large Crude Carrier
UPS	Uninterrupted Power Supply
USGS	United States Geological Survey
UV	Ultra Violet
VDC	Vendor Data Controller
VIV	Vortex Induced Vibrations
VLCC	Very Large Crude Carrier
WHCP	Wellhead Control Panel
WHP	Wellhead Platform
WPA	Work Progress Appraisal
WV	Wing Valve

PART 1

Engineering Design of Offshore Platforms

CHAPTER 1

Introduction to Offshore Platforms

Offshore platforms have been fabricated for a variety of purposes. Army support units, surveillance activities, navigational purposes, etc. are all kinds of offshore platform usage. The study by Dawson [1] provides several examples of usage other than oil and gas sector. Many studies [5−14] have introduced different types of offshore platforms with an emphasis on structural behavior and configuration.

This book concentrates only on offshore platforms that are fabricated for drilling, transferring crude oil and gas to the surface, production enhancing, preliminary processing and transferring to onshore facilities, or exporting. In addition it concentrates on fixed-type jacket platforms. The studies by Patel [6] and Wilson [7] have given sketches showing platform type versus water depth. Above ∼400 m depth the fixed jacket platform becomes uneconomical. In ultra-deep waters, using this concept is not feasible.

Other types of platforms like guyed towers, tension legs, semi-submersibles, floating structures, etc. are out of this book discussion. Several issues have to be pointed out:

- The structural behavior of fixed jacket platforms is different compared with others. This applies to all aspects like analysis, approval, construction, transportation, and installation. Discussions in chapter "Disciplines Involved in Offshore Platforms Design," Section 2.10 are related to fixed jacket platforms.
- In deep waters, due to higher cost of the supporting structure, it is not economical to dedicate each platform to a specific task, like accommodating operators, production, processing, etc. For these platforms normally all functions are performed on the same unit.
- Process design, power generation/distribution, control systems, piping, and installed equipment more or less follow the same concepts as explained in this book.
- Document handling and procedures applicable to engineering office are also the same.

Practical Engineering Management of Offshore Oil and Gas Platforms
ISBN 978-0-12-809331-3
http://dx.doi.org/10.1016/B978-0-12-809331-3.00001-6

It is not intended that we give a complete history of developments in this field. For a better description please refer to listed references.

Platforms can be divided based on their material of construction (ie, steel, concrete) and/or structural behavior (ie, fixed, floating, gravity, tension leg, etc.). These divisions serve very well in a structural design book. This book intends to describe the equipment/systems, design process, and design team interactions. The emphasis is on the performance. Therefore platforms are divided based on their function (ie, production, living quarters, etc.).

1.1 OFFSHORE PLATFORMS HISTORY

This book does not intend to give a history of offshore platforms. It only lists the main achievements. Probably the first oil wells were drilled on spots where oil had surfaced. Under heavy rock pressures oil had gradually seeped out to surface. People had been using them for a long time. Probably early Zoroastrians in Iran built their temples of worship over these locations, which had nonextinguishable fires. The tar collected from these pots was used for sealing ships.

It is said the first wells were drilled in California over these seeping points. Gradually operators understood that in some locations wells nearer to shore produce more oil. At that time and much later seismological investigations had not been developed. Therefore oil explorers were guided by actual production rates. Perhaps the first offshore platform was driven in 1887 when H. L. Williams proposed the idea of building a wharf and erecting a drilling rig on top of it. This first offshore well extended about 90 m in the ocean.

Oil exploration pioneers progressing into the ocean encountered many problems. If the supports were wooden, living organisms attacked it, and if they were made of steel, sea water would corrode it. Heavy storms hit the platforms. It was very costly to fabricate very huge structures. Therefore, rational evaluation of environmental forces became necessary. Costs were high and some of the failures were catastrophic. It was necessary to develop environmental data acquisition/forecasting techniques. As offshore platforms moved away from the beach to deeper waters, gaining geotechnical data became more important.

Before starting development on a location and fixing drilling point, studies are needed to ensure hydrocarbon exists in this location. This necessity has led to seismological studies.

In 1947 the first steel offshore platform was installed and drilled far away from shoreline. Companies continuously progressed to deeper water. The first offshore platform was installed in 6-m water depth. About 30 years later, in 1978 jackets were installed up to a depth of 311 m [6]. Cognac Platform has been fabricated in 386-m (1265-ft) water depth [7]. Beyond such depth using conventional jacket-supported platforms may not be economical. Other offshore platform types such as guyed tower and tension leg platforms were developed.

Fixed jacket structures were fabricated in different designs like monopods, tripods, four-, six-, or eight-legged jackets, lifted jackets, launched jackets, gravity-based structures, etc. to suit design requirements.

In deep waters fixed jackets may not be economical. Instead, floating structures like semisubmersibles and drill ships have been used.

1.2 OIL AND GAS OFFSHORE PLATFORM TYPES

Oil and gas may be found in large reservoirs either offshore or onshore. In some cases these reservoirs are located several thousand meters below the surface. Long wells must be drilled to access them. Offshore wells can be drilled from a fixed jacket, drillship, or template installed at the seabed. Based on oil and gas composition, reservoir pressure, and temperature, some processing needs be done to enable their safe/economical transfer to onshore or exporting facilities. Necessary equipment shall be installed on a platform. In addition to process equipment, operating personnel need living space. For moderate depths these requirements are served on different platforms. For deep waters it is more economical to build all facilities on a single platform. This imposes additional safety measures and requires a compromise between different criteria.

Offshore complexes use several types of platforms. Figs. 1.1 and 1.2 show two oil complexes. An offshore complex is comprised of several platforms each serving a particular intention connected to each other via bridges or infield subsea pipelines. Normally the bridge length does not exceed 80–100 m. An offshore complex may include the production platform, living quarter, wellhead platform, riser platform, satellite platform, and flare platform. Satellite platforms can be connected via subsea cables (power and control signals), infield pipelines (crude oil or gas transfer) to the complex, and from there to the onshore processing unit. Their distance from the complex may be in the range of several kilometers.

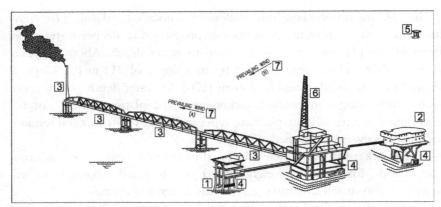

Figure 1.1 Offshore oil complex with living quarter (LQ), production platform (PP), wellhead platform (WHP), and flare.

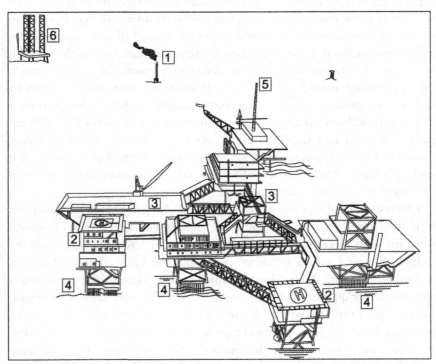

Figure 1.2 Offshore oil complex with living quarter (LQ), compression/production platform, and flare.

On bridge-connected platforms, electrical cables, instrument wires, and various piping are run across the bridges to link the systems together.

An offshore complex configuration is influenced by financial studies. In some areas it may become viable to perform oil processing onshore. In this case minimal processing (may be only first-stage separation) is done offshore and crude product is transferred onshore via pipelines. In Persian Gulf, water depth is moderate and coastline may be a maximum 100–150 km away from installed platforms. Therefore pipeline cost is acceptable. In cases where pipelines are not economical, offshore oilfields use floating production, storage, and offloading (FPSO) units. Crude export may be done with very large crude carrier/ultra large crude carrier tankers.

Fig. 1.1 shows three platforms and one flare plus two intermediate support platforms (ISPs).

1. At the left end a wellhead platform, bridge connected to the production platform (PP), is shown.
2. At the right end a living quarter (LQ) platform with helideck is installed, which is again bridge connected to the PP.
3. The flare platform is located far away from the production platform in a line perpendicular to the wellhead platform (WHP)-LQ connecting line. Three bridge spans and two intermediate support jackets separate it from the PP. This relatively long distance indicates considerable flaring volume.
4. A boat landing is installed in the north side of all three: LQ, PP, and WHP. Platform north may not necessarily coincide with geographical north.
5. A satellite wellhead platform is visible at a distance from the main complex. It is connected with an infield pipeline to the production platform. Its size shows a minimum facility platform without any processing.
6. The long tower installed on the PP is most probably a radio antenna. To ensure line-of-sight access with the next radio station, a tall antenna is needed. This is needed to overcome earth curvature plus noises due to radio wave refraction in contact with air layers. Satellite connections need a much smaller dish-type antenna on the platform. In some cases, for security/redundancy reasons in addition to satellite or LOS radio contact, subsea fiber-optic cables are also installed.
7. In this layout one of two approaches can be followed:
   a. Line connecting production platform to flare shows prevailing wind direction. This ensures heat and radiation from flare is blown away from platforms.

b. Line connecting the LQ to the PP shows prevailing wind directions. This ensures toxic or explosive gases are away from the LQ.

The author proposes a third approach as explained in Section 2.7.1.

Normally main power and compressor utilities are installed on the production platform. Bridges not only provide crew passage but also support different piping/cabling. The flare platform is installed far away. In this case three long span bridges connect the PP to the flare platform. The figure clearly shows that wind is taking flared gases (heat and smoke) away from the complex. In some cases it may be more economical to flow surplus gases via subsea pipelines to the flare platform. Pipeline to flare may consist of several lines including high pressure (HP)/Low Pressure (LP) flare lines, ignition gas, and cabling. LQ has a helipad. In the figure the wellhead platform also has a helipad.

Fig. 1.2 shows a much more complex configuration. Here two living quarters, three wellhead platforms, one manifold platform, one gas compression, and one production platform have been constructed.

1. In this complex the line to flare is via subsea pipeline. Again it is seen that wind is blowing flare gases away from the complex.
2. The two platforms located at the most faraway point compared to flare, have a helipad. This way a 210-degree obstacle-free sector for chopper approach and departure can be provided.
3. The sequence for offshore installation of these platforms shall be carefully planned. Outer jackets restrict access to internal platforms. The main limitation in this design was available crane capacity in the region. Instead of only two or three levels, constructing a LQ up to five levels is not uncommon. With suitable lifting capacity it would have been more economical to construct fewer platforms with bigger topsides. In this specific case development was also done in several stages.
4. Boat landings are installed at the opposite side of the flare. Berthing of a tugboat to the middle platform shall be very carefully done. In the middle boat landing, if (due to any reason) the workboat drifts uncontrollably, a possibility of the boat's mast clashing with the bridge exists.
5. In this complex also a long antenna is installed to ensure continuous radio link.
6. Jack-up shown at top left point may be working on a nearby subsea well. It is standing on its legs. Considering this jack-up to have typical 110-m legs and observing the fact that about half of the legs are above seawater, it may be concluded that water depth at this region is about 40–50 m.

Platforms in a complex may include:

a. Living quarter: This platform combines the roles of hotel, preliminary medical treatment facilities, laboratories, workshops, offices, and recreation facilities. Specific safety regulations shall be strictly followed to provide a suitable comfort level and working environment condition for personnel. Limiting noise level, safe guards against presence of smoke, hazardous gases, fire hazard, and rescue/evacuation means shall be provided based on safety standards. Telecommunication means not only for official purposes but also for family matters are necessary to keep productivity of the personnel.

Fig. 1.3 shows a living quarter capable of accommodating 94 personnel on a 12-working-hour per day and 2-week rotation basis. This is still in the fabrication yard before loadout for installation. It will be connected with a bridge to the production platform shown in Fig. 1.4.

1. Four level layouts of this LQ are described in section 2.1.2. It is seen that cabins, living and working area walls are white colored. White color has maximum light reflection. This reduces absorbed heat from the environment and consequently reduces consumed HVAC power. At the same time the structural color is yellow paint. This has maximum light reflection in foggy conditions. To distinguish colors we have only changed hatches in the sketch.

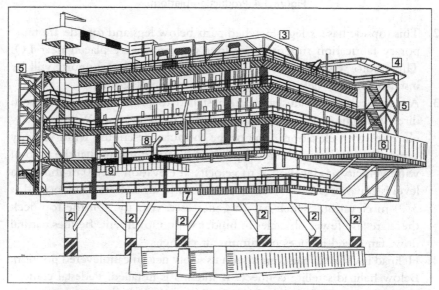

Figure 1.3 Living quarter platform before weighting and loadout.

Figure 1.4 Production platform.

2. This topside has six legs. Hatched parts below legs and topside are temporary fabrication supports. Loadout skid beams are placed near LQ. This photo was taken before weighting. Temporary supports will be installed directly below offshore transportation elastomers.

3. At top deck some sunshades are installed to protect equipment from direct sunlight. Air coolers are installed at this elevation. Wind velocity increases with elevation. One of the formulas describing wind velocity variation versus elevation is the power 1/8 law. It means that the ratio of wind velocity at elevation Z to velocity at elevation +10.0 m above sea level is equal to $(Z/10)^{0.125}$. On this platform air coolers were located at 31.0-m elevation, which has 15% increased velocity. On the top deck there are also fewer obstacles to hinder wind movement. Besides natural flow, fan speed induces maximum air change.

4. Helipad is visible at the corner with its safety nets in cantilevered position. Below helipad satellite connection antenna is installed. Pedestal crane is installed opposite to helideck to avoid obstacle in helicopter approach.

5. Two stair towers are constructed at opposite ends of the platform. This ensures that if one side is blocked for any reason the other approach is available for escape. Access to the boat landing is from the pedestal crane side adjacent to the column connected to the stair tower.

6. Overhang area covered with acoustic protection walls contains two survival crafts. Each one has 70-persons capacity. They are installed at the same level with offices and working area. Cabins are at higher elevations.

7. Joint strengthening is not visible on the sketch but is similar to the production platform in Fig. 1.4. This LQ was installed by floatover operation. This can be understood from large free span between two side legs and braces supporting long span beams at approximately quarter span points. This large span allows floatover barge entrance between legs.

8. Installation guides for connecting bridge to PP are shown at side. Bridge support is installed below deck to enable bridge walkway at the same elevation with deck. This allows trolley passage to PP.

9. Generators' exhausts are at bridge walkway level. Since prevailing wind is in LQ to PP direction, operators walking at this bridge may face hot air. This subject may be considered as a flaw in this design. It cannot be installed in the opposite direction because wind will blow hot gases to cabins. The two perpendicular directions also have the boat landing and lifeboats. Considering complex layout installing at this point is okay. However, hot gas release point could have been elevated.

b. Production platform: This platform normally takes in crude oil and gas from satellite wellhead platforms for processing to enable economic transmission to onshore facilities or to offshore export point. Normally processing crude oil and gas offshore is limited to a degree to enable it being safely and with minimal costs transferred onshore or exported. Providing process facilities on offshore platforms is very expensive. However, if minimal processing is not performed, then pipeline shall be fabricated from special alloy material, which is very expensive and normally is only used in short distances or as platform piping. Chemical injection to prevent corrosion or hydrate inhibition is another means to enable safe transmission to onshore.

Fig. 1.4 shows a six-legged oil production platform that is bridge connected to LQ in Fig. 1.3. Numbers on the figure refer to explanations that follow. In the original photo a wellhead platform was visible at the end that is removed in the sketch.

1. Water being discharged to sea is visible at side. Treated water separated from crude shall be directed to open drain caisson and not to sea. Part of service water used for pressurizing fire water ring via jockey pump may be directly discharged to sea. It may be better to change water discharge location away from boat landing. Splashes may reach personnel boarding or leaving via supply boat.

2. At the opposite side, external risers with their structural guard are visible. For safety reasons internal risers are more preferred. If a riser is added after platform installation, external configuration shall be used. For this platform, topside installation method (floatover) has dictated external riser.

3. Heavy stiffened joints at quarter span points in first elevation plus brace connection at the same location shows that during transportation these points have been used as temporary supporting points. These stiffened joints plus a long, unobstructed span between jacket legs and the fact that drain deck is not constructed shows this platform is installed by floatover method. To avoid clash with installation barge, external risers have been constructed. Below support joint shear keys as guide to elastomer pads are visible.

4. Attached to each leg, barge bumper can be seen.

5. Portable tanks containing different chemicals are stored at edge of first elevation.

6. Bridge support is installed below second elevation. This enables direct cargo transfer using trolley from platform via bridge to the next platform. If they didn't need to move transport trolley over the bridge, then support could have been fabricated at the same elevation with platform level and two or three steps could be used to access bridge.

7. Bridge installation padeye with supporting rings is also visible at one side. Similar padeyes at opposite side and next to support at nearby platform are also installed.

8. Pedestal crane and its boom rest are visible at top elevation. It has full access over boat landing.

   c. Service platform: In some offshore complexes specific functions may be located in a separate platform. Normally these are utility functions like power generation, gas compression, water/gas injection pumps, etc.

d. Wellhead platform: Includes wells drilled to required depth to bring crude oil and gas to the surface. In areas where reservoir pressure is not sufficient, artificial lifting methods may be used. These methods may vary from injecting treated/highly pressurized water/gas to the reservoir to installing special pumps below the well inside the reservoir. Wellhead platforms may be simple satellite tripods with one or two wells or relatively big platforms performing some sort of preliminary process. From time to time well maintenance is required. Therefore, wellhead platforms shall have suitable facility/configuration either to support tender-assisted drilling rigs or jack-up rigs.

Fig. 1.5 shows a wellhead platform with some process facilities.

1. At front of the figure boat landing with ladder access to two different elevations and stair access to drain deck level is visible.
2. Two caissons are shown. One is terminated in drain deck and the other is extended to cellar deck. Drain deck level shall provide access to all risers, J tubes, caissons, etc.
3. Internal riser clamped to jacket leg is installed. This configuration is used for lift installed topside. Riser clamps shall provide sufficient stability in all transportation, installation, and operation stages.

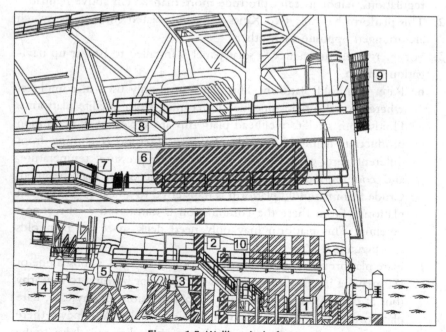

Figure 1.5 Wellhead platform.

4. Barge bumpers are connected with elastomer to jacket legs.
5. Crown pieces connect piles to jacket legs in an artistic manner. Half-circle cutting has a considerable waste of thick through thickness property high-quality steel plates.
6. Flare knock-out drum is immediately installed above drain deck. This is to ensure all drains can be easily directed to it. It is at the same level as bridge to flare.
7. At the same level, a laydown area is shown. Several cylinders are shown on a transport bucket. They may be empty N_2 or CO_2 cylinders. Intelligent pig handling trolley is also shown near cylinder bucket.
8. Immediately above cellar deck another laydown area is shown. They are arranged to simplify pedestal crane access to each elevation without hindering access to other elevations.
9. Helipad safety net is also visible at top deck edge. They are installed only to the edge of 210-degree obstacle-free zone.
10. Conductors are visible extending to cellar deck above drain deck.
Fig. 1.6 shows a four-legged minimum-facility satellite platform.
1. On top of this platform a helideck is installed. Safety net extends all around helipad other than stair access. As per Civil Aviation Authority regulations, handrails can't protrude more than 25 cm above helipad.
2. This platform has only two Xmas Tree and related conductors, which are arranged opposite each other.
3. For energy self-sufficiency, a solar panel is installed to power up navigation lantern.
   e. Riser/manifold platform: This platform is only used in complexes where several infield pipelines are coming from satellite platforms. Fluids from satellite wellhead platforms are mixed in the common production manifold on this platform. If crude is coming from different parts of the reservoir with major pressure, temperature, and composition differences, two different trains may be used. Crude from manifold platform is transferred to a nearby production platform. From there the combination is transferred to the export pipeline. This platform may only need deck space for manifolds and pigging.
   f. Flare platform: If large flaring volumes are expected, a subsea line or ISP may be installed to transfer gas or fluid for burning to a nearby (but at the same time sufficiently away) location. In any case, flare platform is the rescue means of the main complex. Whenever there is a danger, all combustible fluids/gases are directed there to be burned in a sufficiently faraway location.

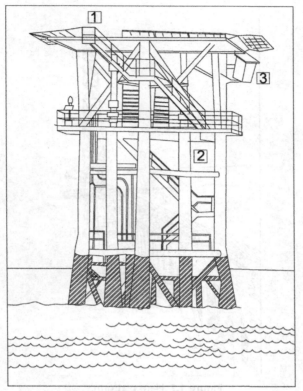

Figure 1.6 Satellite platform.

g. ISP and bridges: Normally burning liquid/gas in emergency cases produces very high temperature and radiation. Therefore, the flare platform shall be sufficiently away from others. This distance may be connected via subsea lines or intermediate platforms linked via bridges.

Fig. 1.7 shows such a condition.

1. One intermediate support and one flare platform are shown with their connecting bridges. For safety reasons and to prevent unwanted intrusion, ISP and flare support platform don't have boat landing. This will allow restricted access only from wellhead platform to check the time personnel have entered flare area and the equipment they carry.

2. Flare line expansion loop is clearly visible on top of the bridge. During emergency burning, temperature on the flare tower may reach 200°C and on the bridge to 100°C. Considering flare line length, a large

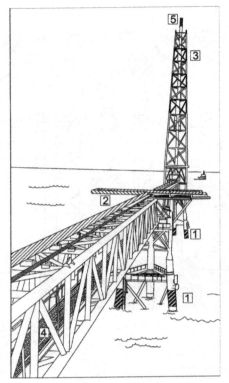

Figure 1.7 Flare platform.

expansion loop is needed to compensate flare line expansion. One large (HP) and one smaller (LP) flare line transfer crude for burning.

3. Paint material in this region has to withstand very high temperatures. On the flare platform, zebra white-and-red color shows specific feature applied on flare towers.
4. Walkway inside bridge allows maintenance visits to flare.
5. Flare tip is visible at top of tower. It shall be smokeless and under sonic.

Function of a platform defines its dimensions, layout, equipment, and in a word, its basic configuration and design basis.

Many types of offshore platforms are fabricated. For material they may be from concrete or steel. For resistance against environmental loads, they may be fixed-type piled to seabed or gravity based or floating and kept in location via wires, anchors, or dynamic positioning thrusters. For structural behavior, they may be fixed, compliant, or mobile. The authors' main experience is in fixed-steel platforms and one mobile drilling unit project.

A majority of the world's offshore platforms are pile-supported fixed-steel type. The study by Dawson [1] and many such studies [5–14] describe different types.

1.3 IMPORTANCE OF ENGINEERING

In an offshore project, several groups are involved. The client, Engineering, Procurement, Construction, Installation & Commissioning (EPCIC)/ drilling and pipeline contractors, consultant(s), third-party authority, marine warranty surveyor, and vendors each has a certain role. For mobile units, classification societies are also involved. Although third-party authority (TPA) and marine warranty surveyor are also selected from classification societies, their roles are somehow different.

A single platform project may be divided to several activities like engineering, procurement, construction, installation, and commissioning. As a rough estimate, engineering costs may be 5%, procurement 50%, construction 25%, installation 10%, and commissioning 10% of total project cost. Since engineering is the first portion of the job, EPCIC contractors may assign higher value to engineering. This is to get a better invoice at the start of the project to improve financial status.

It is understood that in different projects these figures may vary. For example, if several similar platforms are fabricated under one contract, the procurement cost of similar equipment may be less. The same may be applicable for their engineering. Or if several platforms consecutively become ready for installation, offshore fleet mobilization costs will be much less.

Regardless of its share in total project cost, engineering will impact not only all other phases but also platform performance. This means that the important aspect of saving human lives in a potentially dangerous environment and preventing damage to the environment and huge financial losses are mainly governed by good engineering. This book doesn't intend to neglect the important roles of fabrication team, equipment manufacturers, and other groups on final safety of the operators. It only wants to emphasize that if in a project total engineering cost is 5%, it doesn't mean 5% importance. Engineering not only affects all other stages like procurement, construction, and installation but also affects platform operation within its design life.

1.3.1 Importance of Engineering in Procurement

Offshore projects are normally given to EPCIC contractors. It is impossible to define all minute details in the BID documents of each individual package. Therefore, only several main specifications are listed and the rest are referred to standard requirements and general industry practice. Some Clients have developed their own extensive specifications. Standards only define the minimum requirements. The engineering team may define strict, tailor-made requirements. This will considerably increase a package's manufacturing cost. Manufacturers are very familiar with their normal production line equipment. Any change in package specifications increases their cost. In addition, some manufacturers have specific requirements for their brand. For example, Caterpillar is not changing the top coat of their equipment. Almost all manufacturers have a painting line with certain equipment. It will not be economically feasible to change their standard painting procedures and adapt it to a certain requirement of a single client.

This is different from internal design of a package. Certain items may be industry cutting-edge technology, and manufacturers may not even explain its features. Besides, manufacturers have to guarantee their product can operate under intended conditions. Therefore, they shall be able to design with their own responsibility.

Each platform has several packages with numerous interconnections. These interfaces shall be properly managed by the engineering team. This shall be done during the vendor data review stage.

It is possible to transfer all responsibility to manufacturers for internal design. But interfaces between equipment shall be designed by the engineering team. Otherwise, general costs and project duration may increase.

Effective management of interfaces will reduce total project cost and facilitate trouble-free platform commissioning, handover, and future operation.

1.3.2 Importance of Engineering in Fabrication

Other than the equipment that is fabricated by manufacturers and installed by contractors, other items shall be fabricated by the EPCIC contractor in the yard. These may be:

- Structural steel: like primary, secondary, and tertiary members, handrails, stairs, walls, etc.
- Platform piping: like carbon steel, stainless steel, copper nickel, galvanized steel, Inconel pipes, flanges, valves, etc.
- Electrical, instrument, and telecommunication cable trays, cable ladders, and their supports. Lighting, field instruments, loudspeakers, beacons, etc.
- Power and signal cabling of all field instruments and packages.
- Installation of all packages and vessels.

Fabrication of these items requires strict QA/QC systems. A properly engineered design reduces clashes, allows sequential construction, and fabrication procedures. This will considerably reduce costs. It has to be emphasized that machinery and equipment involved in construction are very costly. Their daily renting, maintenance, and operational costs are very high. If they are left idle or not used efficiently, this will increase fabrication costs.

Many construction details are left to the construction engineering scope of the work. Many fabrication yards have their own developed and familiar construction details. For them mandatory details proposed by the detail design team may be more costly. Normally the detail design team proposes some typical details for later usage. Each yard has their own details matching their machinery and routine practice.

1.3.3 Importance of Engineering in Operations

After a platform starts operation it shall continuously produce oil and gas. Any shut down due to any reason will have considerable financial impacts. Assume a platform is designed for 40,000 BOPD. When this book was being written, the average oil price was 105 $/B. When I was doing last reviews, it dropped to less than 50 $/B. Who knows how much an oil barrel will cost when this book is published. A single day shut down for this platform with 105 $/B would cost US$4,200,000.

The offshore environment is very hazardous. A malfunction may lead to disastrous results. It is a fact that industry is learning from these accidents. But at the same time, they impose heavy costs and may lead to loss of life and damage to the environment. Several topsides have been capsized during transportation. The Piper Alpha platform accident is very well known to all engineers in this field. An Internet search shows photos from this platform before and after explosion and fire.

The Piper Alpha disaster happened in 1988 in the Scottish part of the North Sea. Gas leak led to explosion, which resulted in the death of 167 platform crew. The platform was totally destroyed. The investigating committee concluded that the accident was caused by a series of human errors due to lack of safety procedures.

This and similar accidents have established Murphy's law as an important principle in the oil and gas industry. The origin of Murphy's law is not quite clear. In an Internet search I found the following information but am not sure about its authenticity. It seems that the first reference to it started from a US Air Force base in 1949 during tests measuring human response to very high decelerations. At first test strain gauges didn't measure anything. Capitan Murphy who was responsible to install them wasn't in cooperative mode with the team and refused to calibrate gauges before the actual test. Either he himself or his assistants had wired gauges in the wrong direction. Therefore nothing was measured. He put the responsibility of the failure on his assistants by stating: "if there is one way of doing a thing wrong he will find it."

Murphy's law states that "Anything that may go wrong will go wrong." At first glance this may seem a pessimistic point of view. But it is connected with a very important "fail safe" concept in oil and gas engineering.

For example, in a flow line the crude shall pass. As long as the controlling device works properly, it is possible to both let it flow or stop it. If the controlling device fails, then if shutdown valve closes there won't be any safety hazard because ignitable, combustible, or toxic material will not be released. But if the shutdown valve remains open allowing the crude to flow, there may or may not be a safety hazard. In case there is no ignition source or leakage, still there may be no hazard. Therefore it is concluded that for the flow line valve it is safer to close the flow whenever the controller fails. This "fail close" concept may either be achieved by a mechanical spring, which pushes the valve to close position, or the flow to be directed on a plate pushing it to close valve passage or a combination of them. To open the valve, a hydraulic system (or any other mechanism) pushes the spring or closure plate to open the flow passage. As far as this system is working, it is possible to close or open the passage. Whenever it fails the flow is closed. In this condition only loss of production may occur and dangerous situations (loss of life/property and damage to environment) are avoided. Better safe than sorry. This has led to the "fail close" concept.

In a fire water line the setup is contrary. During normal operation the valve is in the closed position. But in case of a fire, it shall open to let water pass and extinguish the fire. Valves in this line shall act opposite to flow line valves. They shall open whenever the controlling system fails. In this condition only some water is discharged and possibility of dangerous fire spreading has been prevented. This has led to the "fail open" concept.

1.4 ENGINEERING PRACTICE

For basic design services, engineering companies may be directly employed by oil companies. This stage will define client basic needs. The results of this study will be used in invitation to bid documents. EPCIC contractors' pricing will be based on the extent of equipment and facilities defined in basic design documents.

For detailed design services, consultant companies may get a subcontract from EPCIC contractors. Although this definition is more or less widely accepted, however, the author has seen cases in which some of the design basis, analysis, and calculation notes had been classified as basic design activities and the drawings and miscellaneous designs were classified as detail design.

As already stated, based on cost-wise evaluation, engineering may only be about 5—8% of an EPCIC project. However, this small portion affects all remaining, about 95%. A majority of EPCIC contractor managers may have started their career from construction group. Some of them may not appreciate how much good engineering can reduce their problems. Although there are few of this type of manager, the author has encountered one manager who thought the yard fabricators were actually building the platform. Besides, since money was paid by the financial department his main concern/favorite was the financial department. He thought engineering is a sort of paper work necessary to satisfy the owner and TPA that everything has been foreseen. This is a very dangerous attitude, which may lead to huge losses to a company.

Although this statement may also be applicable to other managers, I find it more suitable for the engineering manager role. I would like to state: "The engineering manager shall combine science, technique and art."

It is science because all disciplines are dealing with the latest technological developments. They shall continuously upgrade their knowledge and be aware of the new trends and developments in science. Each discipline may use the latest software developed and shall be aware of the industry trends. Technological achievements in process, mechanical, and instruments design, progress in material property, automation, etc. all impact new designs compared to old ones even if crude composition and production rate are similar.

Technique is important because methods and procedures to coordinate diverse interests and orientations in a cooperative mode can only be accomplished with properly established and practiced methods. Matching the different interests of owner, EPCIC contractor, consultant team, and environmental preservation NGOs together, and at the same time following safety regulations and minimum standard requirements, may be impossible if certain guidelines are not established. Simple procedures followed by every team member may produce better results than sophisticated procedures that nobody knows about.

The engineering manager's role is art because human factor is the most important item in an engineering team. Cutting-edge technology computers and software are not able to design a platform unless run by qualified and competent engineers. In many cases human relations does not follow the rigid bureaucratic rules set up in some companies. Although contractual terms govern the contract, they are ultimately decided upon in court. Before that, team members from all parties should be able to communicate with each other.

Fabricators and installation contractors have expensive crawler cranes, CNC cutting machines, yard facilities, offshore vessels, etc. as their main assets. An engineering company has the individuals as the main asset. Consultant profit comes from the results of the work of employees and employees' income is the salary received from the consultant.

Clients may adapt different policies toward engineering design of their platforms.

Some may want the cutting edge technology. Consultant shall notify them that from one hand latest design equipment may require specific training and operating procedures. This may lead to increased cost in operation. From the other hand cutting edge technology may still need time to prove itself in the field for removing small design faults to be adapted to the specific usage and environments.

Some clients may adapt a conservative approach and want their accustomed proven methods and equipment. They may be reluctant to change even in small ways. The author has seen a client change a multiphase flow meter to test separator just because their operation people were reluctant to use it.

Some clients have their own specific standards and guidelines. They do not allow deviations. In some cases standardization and using similar equipment may reduce their procurement and maintenance costs.

Some clients may want their platform to perform its intended function as specified. They have given a lump sum contract to produce a specific amount of hydrocarbon within a design life. It shall be others' headache how to design, fabricate, and operate it.

It is a fact that a competent engineering consultant when working with a knowledgeable and experienced client provides the best result. Bureaucratic measures from all sides including client, EPCIC contractor, and consultant will hinder engineering design considerably. The author has encountered a project manager who after 30 years of experience had made a motto for himself to avoid as much as possible involvement in the decision-making process. His motto was: "never write down everything which you say, never sign everything which you write and never do everything which you sign." This is totally wrong. Only a bureaucratic and inefficient organization can accept such a philosophy. This approach inevitably leads first to increased cost to the client and then to stoppage and delay in projects. Every manager has to decide and accept the responsibilities of his decision. In some cases it may be said that to some extent a wrong decision is better than no decision.

Client project managers shall always consider the huge revenue loss they face due to each single day delay in production. Production loss may be due to delay in platform installation and commissioning or to compulsory shutdowns due to bad engineering or faulty equipment during operation. Consultant project managers should consider that meticulous insistence on specifications will increase project duration, which means delay in final platform handover.

Some bureaucratic organizations impose so many restrictions on their engineers that they are afraid to approve any document. The author encountered a case in which client engineers required material requisitions to be issued only with final approved for construction (AFC) documents. When issuing data sheets and specifications in AFC stage, they required piping and instrumentation diagrams (PIDs) to be issued AFC first. When upgrading PIDs to AFC, they required hazard and operability (HAZOP) to be performed first and its results implemented in PIDs. Finally, when performing HAZOP, they required vendors to participate in the HAZOP meeting and vendor data to be incorporated. It cost the consultant a considerable sum of work hours, meetings, and schedule delays to comprehend this illogical egg and chicken sequence. Subsequently parts of the structure were procured, fabricated, and even installed without having the stamp of approval. The reason wasn't that fabricated parts were not according to standards or contractual requirements. The reason was that client engineers were waiting to see if anything might go wrong. In case of any mishap they could have rejected the related documents to save their neck in the bureaucratic system.

A good design foresees the problems that might be encountered during fabrication, installation, and operation. Regardless of what prices the manufacturers provide for their equipment, a good design will always be cost-effective.

Definition of clear interface lines between the involved parties and engineering team members is a *must* in having a successful project.

CHAPTER 2

Disciplines Involved in Offshore Platform Design

A fixed platform consists of jacket as supporting structure and topside as crude handling part that houses all equipment. Several engineering disciplines are involved in a platform design. As an example consider the structural discipline, which is responsible for design of jacket structures. It needs cooperation of structural, geotechnical, corrosion, and naval architectural specialists in addition to drafting and modeling engineers. In a broad sense topside design requires interface between several engineering disciplines. Within each discipline several specialists shall be engaged. For example, in the mechanical group, a rotary machinery specialist normally does not interfere with pressure vessels or stationary packages. In alphabetical order, the following disciplines are involved in design of offshore oil and gas platforms:

- Architectural
- Control
- Electrical
- Heat, Ventilation and Air Conditioning (HVAC)
- Instrument
- Mechanical
- Piping
- Process
- Safety
- Structure
- Telecommunication

In onshore projects the civil engineering discipline is also involved, but in offshore projects it is not engaged. In university courses structural discipline is part of the civil engineering major, but here due to its importance structure is considered independently.

In this book the emphasis is on conceptual, basic, and detail engineering design phases. For a complete offshore project some other activities like vendor data review, construction engineering, shop drawing preparation, installation engineering, precommissioning/commissioning engineering, and as-built documentation are also needed but are not referred to or dealt

Practical Engineering Management of Offshore Oil and Gas Platforms
ISBN 978-0-12-809331-3
http://dx.doi.org/10.1016/B978-0-12-809331-3.00002-8

with in this book. Some of the discussions are common for all consultants, engineering, procurement, construction, installation, and commissioning (EPCIC) contractors, or even onshore projects. However, the author only emphasizes these three engineering fields. The major emphasis is also limited to basic and detail engineering.

Before basic/front-end engineering design (FEED), a conceptual study should have been performed, which is based on a master development plan for the specific field under consideration. In the conceptual phase decision is made on what to do on the platform. The decision may range from fabricating a minimum facility wellhead platform to constructing a complicated offshore complex with production, processing, and living quarter platforms.

Based on consultant practices, some disciplines may be integrated within each other. For example, the architectural discipline may be included inside the structural department or HVAC may be included in the mechanical department. Or control and telecommunication may be included in the instrument department. The author has even encountered cases where piping was included as part of mechanical department, although I do not like it.

Discipline division may not be so detrimental. The main issue is the correct interface between different disciplines. As an example, electrical discipline is responsible for providing power for all equipment on the platform. Assume a compressor package so big that it may trigger generator shutdown during start-up. This may happen due to high current demand during start-up. The mechanical team shall circulate this necessary information to electrical to ensure compressor package start-up requirements have been foreseen. For this package electrical team shall provide suitable power generation capacity plus necessary number and sizing of cables.

As another example the electrical discipline shall give generator fuel consumption to mechanical. They have to consider necessary fuel tank and sufficient pumping capacity. This is in case it is a diesel generator. If it is turbogenerator, again mechanical discipline shall check fuel gas requirements. This includes both volume of supplied gas and its condition considering moisture/H_2S contents, etc. Besides, for start-up of turbogenerators and before platform gas becomes available, diesel power shall be available.

Piping shall also know dimensions and layout of each package to accommodate them in the platform layout and provide sufficient maintenance area. Almost always in offshore platforms there are space limitations. Increasing platform dimensions increases its weight and consequently cost.

Client always asks for new control systems for equipment. They increase package dimensions. Minimum safety requirements for escape route and operational requirements for maintenance area shall also be followed. They all necessitate an accurate piping check in layout. Since platform construction starts long before final vendor data becomes available, estimating and assigning suitable plan dimensions is vital for a successful design.

One of the items that the instrument discipline shall check is the signal type via transmitters. This shall be adaptable to input signal type to platform central control system.

The previous sample explanations show some examples of how data in a discipline may impact other disciplines' design. This shows that no discipline is free to do whatever they want.

All of the previous discussion does not mean that until all data in any single package has been finalized other disciplines shall not order their packages or fabrication shall not start. On the contrary, all disciplines shall move simultaneously. They shall start with preliminary assumptions based on previous projects' information and improve their design step by step. New information either received from selected vendors' quotation or received vendor data during package manufacturing shall be reviewed and implemented in design.

The main issue is correct interface between different disciplines. A correctly defined interface will reduce reworks/delays due to clashes and ensures that designed equipment will work in harmony. In order to define a clear picture of interfaces, the engineering manager has to first understand what is each discipline doing.

In this chapter some of the main documents that are prepared in each discipline, their main discussion subjects, and interfaces are explained. This list and explanations in no way may replace a comprehensive document register. The project document register shall be prepared based on platform type/function and owner/yard practice. Design shall be comprehensive enough to enable safe and full construction of the platform.

Documents that are prerequisites of others appear first. Once more it is emphasized these are only a selected list of important documents.

2.1 ARCHITECTURAL

An offshore complex may have several platforms including the living quarter (LQ). In addition, each wellhead or production platform may have rooms dedicated to specific tasks like control room, battery room, generator room, telecommunication or radio room, electrical (switchgear) room, mechanical workshop, temporary refuge, etc.

Definition of the wall ratings, furniture type, ergonomic requirements, noise level, access, and other related issues to provide a suitable environment for operators working and/or living on these platforms shall be done by architectural discipline.

Documents generated by this discipline may vary from architectural specifications to interior finish specifications to layout/general arrangement of each room and isolation/partitioning details, simulations of interior design, list of facilities, utensils, furniture, etc.

Several years ago and especially for living quarters, clients used to request a model replica (mock-up) of the living cabin. This was a full-scale model of a typical cabin. Before approving architectural documents and start of living quarter fabrication, it had to be fabricated by the EPCIC contractor and examined by the client. It was intended to see in full-scale reality if operators would feel comfortable during their offshore stay. It had both time and cost impact and delayed issuing engineering documents and start of fabrication. Nowadays with 3D simulation software and the ability to "walk through" the model, this is no longer required or requested.

In addition to general standards, many operators have their own minimum specifications for architectural design of living quarters. These specifications cover all aspects like number of cabins, their size, common toilets and baths, services like laundry, messroom/kitchen, recreation facilities, etc. Chakrabarti [5] and TOTAL [15] give detailed information.

2.1.1 Architectural Specifications

Architectural specifications is the main document in the architectural discipline. It shall clearly define the areas and rooms in a platform, the requirements of each area, furniture/equipment inside each area, characteristics of the partitions separating different areas, flooring/ceiling type, doors/windows, noise/fire rating of each partition (based on safety requirements), ergonomic requirements of each area, etc.

For living quarters, design of cabins, their furniture, safe access, suitable calmness, and clean and cozy environment are in the architectural discipline scope of work.

Sufficient air conditioning, noise-free environment, clean and hygienic cabins enable crew to rest completely and become ready for next-day challenges. An exhausted body may not react properly in critical situations.

Material for food shall be transferred from onshore to platform. This is normally done via supply boats. Platforms may be more than 100 km away from shore and therefore not easily accessible in rough weather. Considering the difficulty in access to such platforms, sufficient material for several weeks' consumption shall be stored in the platform in suitable quality and quantity. The freezer and cold room storage shall be designed for this purpose.

Offshore work is in 12-h shifts. Long, exhausting working hours away from family members increase the importance of having recreational and sport facilities. The author agrees that sunrise/sunset at sea are very beautiful. But imagine that every day you see it for 10 years in a rough environment full of pipes, vessels, pumps, noise, vibrations, foul smells, and hazards. Then it will not be so charming.

Entertainment and exercise are important aspects of living on platforms. Telephones to communicate with family, video televisions or satellite facilities and gymnasiums are considered as basic requirements of an LQ platform. I have myself lived long times away from my family. I am sure that for a father nothing is comparable to hearing his child's voice or seeing his or her loving face.

Water and wastewater handling is another important topic. Based on the geographical area in which the platform is located, providing hot and cold water for usage and suitable transfer/disposal of the produced wastewater is very important. Normally gray and black sewage lines are separated and after treatment are disposed to sea. The solid waste is burned in an incinerator or stored to be transferred via supply boats to the shore treatment/disposal facilities.

Platform kitchen and messroom shall have sufficient facilities and space to prepare and serve hygienic, delicious food for the operators. Simply moving up and down stairs and moving heavy tools and equipment consumes a lot of energy. Therefore food calorie content shall be plentiful. Since platform work is round the clock and night shifts also need proper catering services, kitchen and messroom shall also be able to work round the clock and serve five to six meals. From working on oil and gas platforms, splashes of oil, contact with grease, etc. require continuous laundry service for dirty clothes. Heavy duty/large capacity machines are required to wash rough clothes and support big living quarters.

2.1.2 Architectural Layouts

The major intention of architectural layout is for the living quarter. But some of the explanation may also be applicable for rooms in production or wellhead platforms. LQ contains different spaces each with specific intended usage as listed later. Some of the listed areas may not be present in all living quarters. Some of them may be combined into one area or a multipurpose area dedicated to several operations. Separation of area facilities, arranging necessary utility for equipment and their future maintenance while keeping a quiet and comfortable atmosphere for cabins is vital.

A. Equipment Area

Several packages may be installed in each LQ. They shall be away from the living area. Based on their function some are placed at the lowest elevation and some at the highest elevation. The equipment is further explained in chapter "Systems and Equipment for Offshore Platform Design."

- Generator: Normally the living quarter gets power from the main generators installed in the production platform plus an emergency generator. This generator, which may be installed in a dedicated room, will only provide power for specifically defined functions.
- Firewater pump: As per National Fire Protection Association (NFPA) each living quarter, although placed at a suitable distance from production platform, shall have dedicated firewater pump system, which shall be redundant. Firewater pump shall have its own dedicated power supply.
- Potable water maker including desalination, mineralization, ultraviolet (UV) sterilization, and storage tanks shall be provided with sufficient capacity. In a small shelter, a freshwater system may only include a tank that will be filled periodically by supply boat plus sterilization unit. This tank will be placed in the highest elevation to ensure gravity flow. For a big LQ it is more cost-effective to generate freshwater on the offshore platform.
- Gray/black water disposal system: This system consists of two independent piping systems that collect gray (like kitchen, shower, etc.) and black (like toilets) water from different areas with gravity flow and end at the sewage treatment system.
- Sewage water treatment: This package separates the solid waste and water and disinfects them. Water may be disposed to sea after treatment.
- Incinerator package: Solid waste need not always be transferred to onshore. They can also be burned. Specific regulations apply to the items that can be burned.
- Transformer area: To avoid voltage drop in subsea cable, electrical power is transferred in high voltages. Besides, electrical power supplied from gas turbogenerators will be in higher voltages compared to normal

equipment used in LQ. High voltage shall be reduced to enable using normal-sized cables and reduce high electric voltage potential hazards. At entrance to LQ voltage difference shall be lowered to be distributed to platform medium voltage (MV)/low voltage (LV) switchboards. To prevent hazards to human life due to high voltages, access to transformer area is restricted via fencing.

- Equipment lift: This is installed in case pedestal crane and laydown areas cannot transfer the equipment easily to all parts of LQ. This may be due to layout restrictions. It is clear that full package transfer is not intended. Only capital spares, food, tools, and similar items are transferred within platform levels.
- Personnel lift is only installed in living quarters with several elevations.
- Aviation refueling station/tank is only installed at highest elevation for quick access to helicopter. This is in case platform is installed at remote locations.
- Helideck firefighting monitors/equipment are directly installed below helipad with very quick access to helideck. As per LQ operating procedures, helicopter landing is not permitted until firefighting and safety personnel are fully mobilized at their intended positions.

B. Working Area/Offices

A major part of operators' work is in the field to maintain packages, replace damaged gaskets/flanges/bolts, overhaul rotary equipment like pumps/generators/compressors, clean vessels, etc. Other parts of routine work are performed in dedicated areas. Some of the areas listed following can be combined with each other.

- Central control room (CCR): This is the brain of the platform. Crude production, flow volume/pressure/temperature measured at key points and all packages' main signals are controlled here. Thanks to new technologies practically all information can be reproduced on the operating system (OS) screens. In addition, for emergency conditions overriding alarms can be generated. In fact full piping and instrumentation diagrams (PIDs) are modeled in the controlling software. Operator can select any item and get its online information. This is in addition to pop-up alarms that are built in the software. Predetermined cause-and-effect logics enable automatic decisions. Platform control system programmable logic controller issues necessary commands to various package control panels. Operator interference and override is also permitted based on safety requirements with a predefined hierarchy.

- Switchgear room: This room contains all electrical switchboards to distribute generated power to consumers. In small facilities it may be part of control room. It is customary to dedicate specific boards to some functions.
- Battery room: This room contains battery racks for the uninterrupted power supply (UPS) system. It shall be ventilated to avoid hydrogen concentration. Electrolyte can be either acid or hydroxide. Both require corrosion protection measures. Battery room air conditioning may be instructed by certain clients. Some battery manufacturers may also recommend a specific temperature range for maximum battery life. In some cases it may be more economical to change batteries every 10 years instead of installing an air conditioning package.
- UPS room: Dedicated switchboards for systems fed by UPS are located in this room. In some platforms they may be located in the same switchgear room.
- Telecommunication room: This may contain telecom console and video/radio transmission facilities. In small platforms it may be only a desk with OS located in CCR.
- Mechanical workshop: A complete workshop may include drilling machines, lathe, bending tools, handling equipment, workbenches, etc. For main overhauls, a specialist engineer with special tools may be assigned for a specific period to offshore platform.
- HVAC workshop: used for replacing/maintenance of filters, motors, fans, ducts, louvers, manual and automatic dampers, etc. It may be combined with mechanical workshop.
- Instrument workshop: Used for calibrating field instruments and maintenance, repair, and replacing activities.
- Electrical workshop: Used for winding electromotors, repairing termination points in junction boxes (JBs), replacing lightbulbs etc. It may be combined with instrument workshop.
- Wireline and well maintenance workshop is installed in case frequent drilling rig operation is anticipated. Otherwise any jack-up rig will have its own workshop.
- Meeting room for group of personnel to decide on a plan or course of action for a specific task or briefing visitors from other organizations. Other than routine activities that are clearly defined in platform Ops, all new operations (including alternative plans and risk mitigation measures) shall be explained to involved personnel.

- Offices may be divided into managerial and staff offices. Strict operational procedures shall be followed. This is required to prevent any mishap. Paperwork for all work permits in production area shall be recorded. Any activity shall be traceable. Due to space limitations, hard copy archives will be transferred onshore periodically.
- Library: Normally contains all packages' final data books (FDBs), complex final design dossier, platform's operation and control procedures/manual, and general information books.
- Transit lounge is located with direct access to helipad. In this room crew to be transferred by chopper are listed, leave permits are checked, and their luggage/themselves are weighed to avoid excessive loading. Security checks of personnel and luggage for incoming and outgoing crew are also performed here.
- Hospital: Medical staff in LQ may treat operators exposed to overheating, skin surface burning/cutting, contact/inhaling/swallowing hazardous chemicals, normal fevers and diseases like catching colds, preliminary operations (more than first aid) when a limb is broken or cut, etc. It may even contain facilities for some emergency surgeries.

C. Catering Area
- Messroom: Shall have sufficient area to serve all personnel within 1 or 2 h. In some cases a small area may be dedicated to managers or guests.
- Laundry: May be divided to rough work suits and normal bed sheets, shirts. Each operator may need several laundry services per week. Heavy-duty machinery shall be used.
- Kitchen: Shall have possibility to prepare food round the clock. Bakery to prepare fresh breads may also be installed.
- Cold storage: For storage of items like eggs, jam, honey, fruits, vegetables, milk, etc. Its capacity depends on accessibility by boat in different environmental conditions. Normally storage capacity for a 2-week interval is selected.
- Freezer room: For long-term storage of meat, butter, and other foods.
- Storage: May be used for different material.
- Cleaning room may be used for food preparation purposes.

D. Recreation and Fitness Area
- Recreation room: May include TV, video, VCD/DVD player, play stations, music player, satellite receivers, etc.

- Gymnasium: May include table tennis, fixed bicycle, treadmill, lifting apparatus, etc.
- Cinema: With advent of large panel LED/LCD televisions, its usage may be reduced. But still, some cinema halls are constructed.

E. Living Area
This area is usually installed at higher decks.
- Bedrooms/Cabins: Single-bed cabins may be used for managers. Double-bed cabins may be given to engineers and crew. In some cases four-bed cabins may also be constructed for crew. The author has never seen eight-bed cabins in a platform design. All single-bed and some of the double-bed cabins have their own toilet and shower. New regulations require all cabins to have their own toilet and shower.
- Toilets: Only some of the double-bed and all four-bed cabins may use public toilets. Their location shall be such that while providing easy/immediate access, cleaning and ventilation is possible without disturbance to cabin's privacy. Number of toilets shall be such that in rush hours crew can attend their work location without delay.
- Shower: Similar to toilets, only some of double-bed and almost all four-bed cabins may use public shower. Its location shall be separate from toilets to follow hygienic considerations.
- Door: Shall always open in escape direction and be equipped with panic handle. If doors open to outside safe area may have fire rating, otherwise may only be comfort rated. Doors shall be gas tight.
- Window: Provides natural light but at the same time induces a weak point in wall anti-explosion/fire resistance property. As much as possible windows in process area side shall be avoided. Similar to doors they shall be gas tight. Window fire or comfort rating shall match the wall that is installed in it.

Fig. 2.1 shows a level that contains four workshops (wireline, electrical/HVAC, instrument, and mechanical), store for tools and consumables, switchgear room, generator and firewater room, transformer area, equipment lift, sewage treatment, laundry, freshwater-maker system, and HVAC package. Wireline operation introduces tools inside drill pipe, which is under high pressure.

This area is located in LQ, which is separated from production platform with a bridge ~52.0 m long, therefore is considered to be safe from presence of hazardous gases. As shown in the sketches, workshops and diesel generator room are not gas tight. To facilitate transfer of maintenance items,

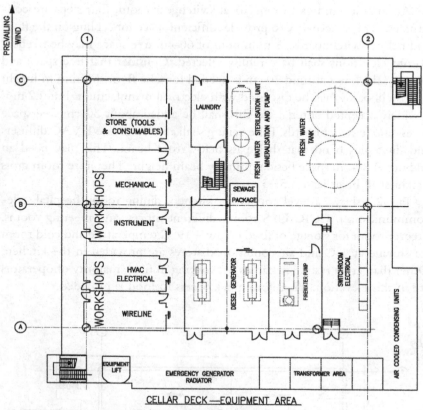

Figure 2.1 Living quarter cellar deck (equipment area).

sliding doors are installed for these spaces. Although only spaces with a length more than 12.0 m need to have two exits, in this design all spaces follow the same requirement.

Two stair towers are located at diagonal ends compared to each other. This configuration ensures that if passage from one side is blocked (due to any reason) operators can reach muster area and escape to safety. Corridors at this level allow transfer of equipment and are much wider than corridors in living area.

For all tanks and vessels, ladder or access platform to top was foreseen. But for clarity they have been removed from this sketch. Monorails are installed both in workshops and operating spaces like generator room, fire water room, etc. They shall extend from above equipment to laydown area to place lifted items on the handling trolley. In some cases, two perpendicular monorails may be used for transportation.

At entrance and exit to electrical switchgear room, four steps are con-structed. This is necessary to provide sufficient space for cabling on the floor and below switchboards. A minimum of 60-cm free space may be needed. Room elevations shall be carefully selected. Consider that free space and flooring thickness around 70 cm is needed for false floor. Minimum height of switchboards shall be discussed with electrical manufacturer, but 2.2 m is a widely accepted criteria. To this shall be added about 30-cm free space access above switchboards. False ceiling shall accommodate HVAC diffusers and duct plus bend that shall pass below roof beam. This may need an additional 80-cm space besides primary beam height. Therefore room gross height shall be at least 4.5–5 m.

Fig. 2.2 shows a level containing kitchen, dining room, hospital, tele-communication, CCR, UPS room, different offices, and meeting rooms. Freezer room for storage of food below −19°C temperature and cold room for storage at 4°C and air-conditioned stores are provided in the kitchen. Other than crew resting in cabins or working in field, majority of operators are in this elevation. Therefore quick access to food is provided.

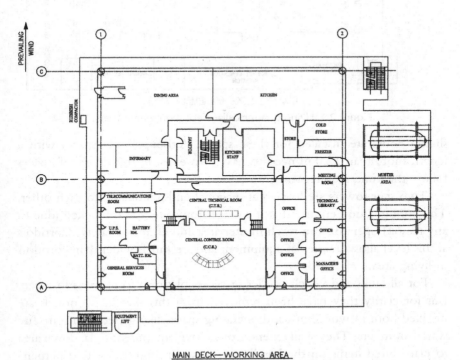

MAIN DECK—WORKING AREA

Figure 2.2 Living quarter main deck (working area).

Although this elevation is considered a safe area, around control room a fire-rated wall is constructed.

Since most of the platform crew work in this level, muster area and lifeboats are stationed in this level. They are protected with acoustic wall and provide sufficient escape time for personnel. Lifeboats' arrangement shall be such that if their motor fails, prevailing wind carries them away from production platform.

CCR also has height limitation criteria similar to switchgear room. Considering larger number of cables, height requirement for false floor may be even more. Five step access to CCR shows larger height below false floor. This may be due to thicker cables that require longer bending radius or large number of incoming cables.

Access to outer space is provided with an airlock. This includes a space in a corridor closed with two doors. Each time it is opened only a small amount of internal air is displaced. This reduces air conditioning load.

The level in Fig. 2.3 contains recreation area, cinema, gymnasium, and some of the cabins. A large area is dedicated to official large meetings and presentations to future visitors.

To prevent personnel wearing dirty working suits and boots from entering clean living area, a changing room is provided in this level. Immediately connected to this area common toilets and showers are provided. Normally field crew must change/clean up here before proceeding to their cabins or messroom. Equipment in fitness and recreation area may be selected based on regional/cultural preferences. Alcoholic drinks and drugs are prohibited in all offshore platforms regardless of region.

Cabin dimensions in this LQ are more or less uniform. External dimension of 6.5 × 3.0 m is chosen for single- and double-bed cabins including toilet and corridor area. This includes bunker, desk/chair, locker, cupboard, corridor, etc. For bed 900-mm width and 2100-mm length and for desk 600 mm depth times 1300-mm width are sufficient. For easy access to chair, additional space equal to chair width may be left in front or behind chair. Toilet area may include shower, toilet, and washbasin. Around shower a curtain can be installed to prevent water splash all around. Some regional standards (like the Norwegian Petroleum Standard; NORSOK) may require larger area for cabins or additional facilities.

Airlocks at access points to outdoor environment prevent wasting air-conditioned atmosphere. At airlocks in working area, which are the first entrance from dirty space, some footwear may be provided. This may not be required at recreation area airlocks.

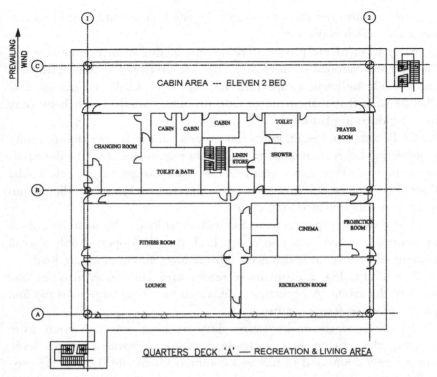

Figure 2.3 Living quarter deck A (catering, recreation, and living areas).

The level in Fig. 2.4 contains only cabins. This platform had only four one-bed cabins intended for managers. All other cabins had two beds. All cabins in this elevation had dedicated shower and toilet, but in other elevations some used common toilet/shower between two adjacent cabins. Although in accommodations installed in ships four-bed cabins are constructed, in platforms it is avoided as much as possible. In ships four-bed cabins may use double bunk beds to save plan area. Each elevation may have specific housekeeping crew and store for clean linens, blankets, pillows, etc.

Cabins are used for sleeping and after a heavy working day shall provide suitable resting atmosphere. Brightness and sound level in cabin area shall be limited.

Smoking is normally prohibited in offshore platforms. In addition to health-related concerns, one of the reasons is that detectors may mistake heavy smoke as fire. Besides, if fire detector is disabled or not installed in an area, to dedicate it as smoking room, there is no guarantee that an actual fire may not happen in that area.

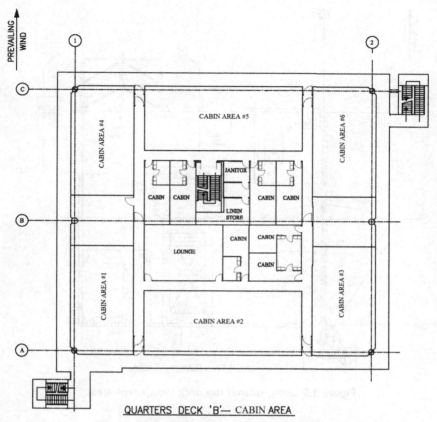

PREVAILING WIND

QUARTERS DECK 'B'— CABIN AREA

Figure 2.4 Living quarter deck B (living area).

Fig. 2.5 shows a level containing HVAC, aviation fuel, firefighting, and pedestal crane. Helideck is connected by a cantilever truss to this elevation away from pedestal crane access.

Normally crew change is with supply boats. But managers and in emergency cases injured personnel or low-weight, urgently needed equipment are transferred by choppers. Choppers do not stay for a long time. Immediately after offloading cargo, disembarking passengers and boarding the new ones they take off. In rare cases weather may become too rough. In that case helicopters are tied down to platform and its crew stay overnight on the platform. Therefore living quarter platforms shall always have refueling and firefighting facilities for choppers. Helipad design follows certain rules for 210 degree obstacle-free access zone, structural strength, tie-down points arrangement and structural shape, safety net and

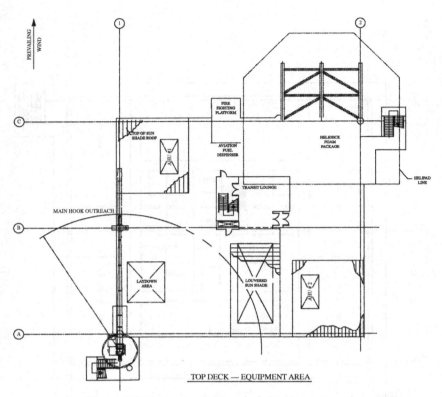

Figure 2.5 Living quarter top deck (equipment area).

discharge gutter around pad, landing net on the pad, special painting form and material, personnel approach, refueling/firefighting equipment, and drainage gutter around pad. The civil aviation agency in each area will examine all requirements and issue landing certificate.

Safety personnel have permission to enter helideck firefighting platforms or refueling skids. Access to helideck for other personnel is only through transit lounge.

One important aspect of architectural layout is that it shows the location, configuration, and rating of fire/comfort-rated walls. This is an important item that requires proper interface checking and involves several disciplines. Location and rating of fire-rated walls are defined by safety discipline. Dimension/configuration of blast/fire-rated walls are determined by structure. Dimension and location of comfort-rated walls are determined by architectural. The outcome is the layout that is used by mechanical, piping, electrical, and instrument.

For a description of fire ratings refer to safety discipline and for blast walls refer to structure.

Refer to Fig. 2.6 for a blast-rated wall with internal thermal insulation and outside jet fire-rated plate or coating. Comfort-rated walls may consist of some vertical posts in less than 2-m spans. These may be hollow rectangular or normal profile sections accompanied by similar horizontal members. Corrugated steel plates cover the span between two posts. Only wind load shall be used in designing these vertical posts and the corrugated plates. Suitable wind velocity averaging time and return period shall be selected for the design gust velocity. In many cases, 3-s gust and 100-year return period have been used. Insulation material is placed adjacent to corrugated plate.

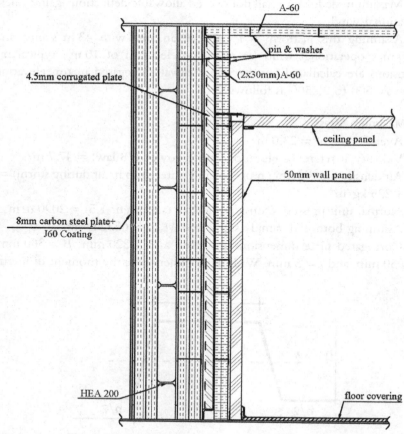

Figure 2.6 J60 Fire-rated wall and internal insulation.

In between vertical posts normally 4—6-mm-thick (depending on the span) corrugated plates are placed to support the insulation panels. Insulation consists of one or two layers of insulating material (like rock wool 30—50 mm thick). The intention is to reduce noise level inside premises and to reduce heat conductivity. The internal surface of the rooms depends on its usage. Where required, an architectural wall panel can be used. This mainly serves for esthetic purposes but has some thermal insulation property as well. In other locations like workshops, only a galvanized plate may be installed on top of insulating panels. Flat bars can be installed to support galvanized plates or they may be corrugated to get structural strength.

A typical section for a corrugated wall is shown in Fig. 2.7. Two design criteria shall be followed:

- Maximum stress shall not exceed allowable stress against gust wind
- Maximum deflection shall not exceed allowable deflection against operational wind

Assuming this platform is located in an area with 43 m/s gust and 26.5 m/s operational wind (at reference elevation of 10 m), typical dimensions are calculated for the upper elevation located in top elevation from 20,500 to 25,500 as follows:

A. Maximum Stress

Average height $= 23.0$ m

Velocity at reference elevation (using power 1/8 law) $= 47.7$ m/s

Air density (considering moisture and water drop in air during storm) $= 1.225$ kg/m^3

Normal unit pressure (considering drag coefficient 1.5) $= 2090$ n/m

Assuming both ends simply supported $M_{max} = 6532$ n—m

Corrugated plate dimensions are taken as $b = 220$ mm, $B = 360$ mm, $h = 50$ mm, and $t = 5$ mm. With these dimensions, the moment of inertia

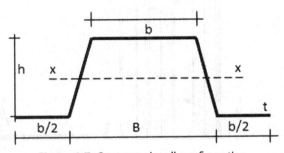

Figure 2.7 Corrugated wall configuration.

is calculated to be equal to 147.9 cm^4. Using the bending stress formula, the critical normal stress is calculated equal to 71 MPa. The allowable bending stress is taken to be equal to 0.6 yield stress. If S275 steel is used, the allowable stress becomes 165 MPa, which is much above the applied stresses. In this case even a 4-mm plate would have been sufficient. However, we keep the higher thickness for deflection criteria. Allowable stresses against rare environmental conditions can be increased by one-third. This is to cater to the low possibility of their occurrence during platform design life. In the above calculation this condition has not been used yet.

B. Maximum Deflection

Deflections shall be checked to ensure that during operational conditions architectural panels like sound and heat insulation are not damaged. Operational storms are normally taken as storms with a 1-year return period. For these calculations allowable stresses are not increased. In our example case operation wind velocity at middle of wall becomes 29.4 m/s.

Normal unit pressure $(C_d = 1.5) = 794$ n/m

For this cross-section the moment of inertia is calculated to be 147.9 cm^4.

Maximum deflection at center with both ends simply supported is 1.2 cm. This is compared to the allowable deflection, which is dictated by distance between wall panels/corrugated plate and conditions of attachments (like windows). Normally a value of L/360 is used. For our design conditions the allowable deflection is calculated to be 1.39 cm, which is acceptable.

It is to be noted that if corrugated plate thickness was taken equal to 4 mm, then most probably deflection criteria could not have been satisfied.

Preventing heat transfer from both sides is the next function of these walls. This is achieved by placing thermal isolating panels between walls. From basic thermodynamics it is understood that two objects with different temperatures (T_1, T_2) placed nearby each other for a sufficiently long time will change to an equilibrium temperature. The value of this equilibrium temperature depends on ΔT (temperature difference of the two bodies) and mass of each body. But the duration to attain equilibrium depends on the heat transfer coefficient. Outside environment is considered infinity. Therefore thickness of thermal insulation governs the HVAC equipment power.

Steel material has very high conductivity. Air gap and insulation material have very low conductivity. Carbon steel thermal conductivity varies with carbon content. At 20°C temperature for 0.1% carbon it is 64 W/(m K) and for 0.5% carbon it is 54 [16]. Normal structural steel has about 0.2% carbon content. Therefore its thermal conductivity is assumed to be 61.5 W/(m K). Rock wool thermal conductivity is 0.04 W/(m K). Comparing a 10-mm carbon steel plate and two 50-mm rock wool panels (all 4-m length by 4-m height) show a drastic difference. Assume outside temperature is 45°C and inside is 20°C.

$$Q = qA = -k\frac{\Delta T}{L}A$$

Q is in watt, q is in watt per square meter, and A is in m². Thermal conductivity is defined by k in W/(m K). Temperature difference between the two sides of a media is shown by ΔT. Media thickness in meters is shown by L.

For carbon steel: $Q = 61.5 \times 25 \times 4 \times 4/0.01 = 2,460,000$ W

For rock wool panels: $Q = 0.04 \times 25 \times 4 \times 4/(2 \times 0.05) = 160$ W

Comparison shows carbon steel plates require more than 15,000 times higher energy. It has to be noted that these values are without considering the air boundary layer near the media inside and outside. Without wind the air boundary layer will act as a cushioning effect to reduce this tremendous ratio. However, this is used as an indication of the order of magnitude difference.

2.1.3 Architectural Interfaces

There are many architectural interfaces with other members of the engineering team like safety, structure, piping, electrical, instrument, and mechanical. Necessary information may be exchanged between several groups. For example, a package procured by ME may receive power supply from EL, transfer signals to IN, receive utility from piping, follow safety ratings, and comply with architectural limitations.

Safety shall define wall, floor, and ceiling rating for all architectural spaces. This is done based on usage of that space, occupancy, equipment and type of furniture present, its proximity to hazard sources, presence of ignitable and explosive materials, etc. Architecture shall coordinate with vendors to ensure procured materials comply with specified rating.

Structure shall support walls, doors, and windows against applied loads. Hollow section members and corrugated plates can be used for this purpose. Structural load-bearing members increase total wall thickness and occupy part of available space. Architecture shall ensure remaining space is adequate for planned usage of that area and that structural elements do not obstruct movement or placing of equipment.

Piping outside architectural premises will be designed by piping discipline while inside this area has to be taken care of by architectural discipline. Proper tie-in points to connect potable water, cold and hot water, black and grey sewage, etc. shall be carefully defined. In process area access for maintenance, installation sequence, piping size, operating pressure/temperature, etc. govern piping general arrangement. In architectural area pipes containing hazardous or high-pressure, high-temperature material are not present. Esthetic considerations shall be considered. To enable access for future repair or replacement, ducts with removable panels are used.

Electrical discipline shall provide power for all utilities including lighting, HVAC, cooking and refrigeration facilities, etc. Power cables and outlets shall be both accessible and at the same time sufficiently away from inadvertent touches.

Instrument discipline shall install all necessary detectors and provide for required signals/alarms. While detectors shall have full coverage of the whole area, they shall be away from accidental impacts or other damages.

Mechanical discipline shall procure some of the required items like calorifier, HVAC units, incinerator, potable water, etc. All of these interfaces shall be carefully foreseen during design and later in vendor data review stage.

2.2 CONTROL

All equipment in the platform have instruments that measure required information and transfer data to the CCR. In CCR all package information is entered as input to the control software. This software displays the required info for control or monitoring by operator. In addition, it has some built-in functions that automatically decide for some of the encountered cases.

There are three main control systems: distributed control system (DCS), emergency shutdown system (ESD), and fire and gas system (FGS). The signals for each one are transmitted to a separate system. (The three systems will be further discussed in the next sections.) Control discipline defines the architecture of these control systems, their interfaces with each other, and the hierarchy.

The list of signals in each platform depends on the existing packages. Those that are needed for normal platform operation are directed to DCS. Those signals whose malfunction may cause a safety hazard are directed to ESD, and those that shall trigger an action in response to a fire or gas detection event are directed to the FGS system. A sample brief list of each signal type is given below. Based on each platform's number of wells, installed process trains, and available equipment, the actual list may be several times more extensive. These three systems can be combined in the integrated control and safety system (ICSS).

A. Typical Signals in DCS

Signals transferred to the DCS are needed for normal platform operation. They may include pressure, temperature, and flow measurements in different lines, vessels, and equipment both in process and utility areas. Some of them may be directed to the related package unit control panel (UCP). In that case software/logic implemented in the UCP will operate the system and only a few running/stop, general fault, or only status/monitoring information signals are sent to DCS.

Decisions selecting packages that need to be controlled by UCP or DCS shall be made before starting procurement process. This will impact package manufacturer and DCS vendor scope of work (SOW). Decision may be based on several factors like:

- Package complexity: A single package may have several pumps, motors, vessels, filters, etc. Command signal for individual or simultaneous operation of main/standby motors shall be based on specific logic. Operating parameters in one part of the package may impact others. For example, an analyzer may start or stop a recycle pump. Very complex packages preferably shall be controlled by their dedicated UCP.
- Vendor guarantee: If parts of package control are governed by logic prepared by another vendor and a mishap occurs (for example, during commissioning), package may not work as intended. In this case vendor X (package manufacturer) will blame vendor Y (DCS manufacturer) and vice versa. Even if finally it is possible to find the problem and rectify it, elapsed time cannot be recovered. Packages that have an important role in platform operation shall preferably be controlled by manufacturer UCP.
- Platform operation: In vessels in which improper control of liquid level may lead to gas entrance in liquid line or draining hydrocarbon to water line, control is transferred to DCS.

Large and complex packages like turbo/diesel generators, sweetening plant, dehydration plant, compressed air system, and HVAC are normally controlled by their dedicated UCP. One of the reasons is their manufacturer shall accept responsibility for package proper functioning and issue guarantee. Other packages have an LCP that only has some switches to interface with operator with no control logic. These packages are controlled by DCS.

Packages whose start requires coordination between operator and other crew (like supply boat captain) shall get control commands from DCS. For example, filling diesel oil system needs coordination with platform crew to connect transfer hoses to tie-in point and open inlet flow valves. Chemical injection package or closed drain pump also may be controlled by DCS.

Local shutdown (LSD) and process shutdown (PSD) can be initiated automatically or manually from DCS.

B. Typical Signals in ESD Package

Signals transferred to ESD may come from process and/or utility systems. Based on severity and importance of a signal, it may trigger only a warning or an action. The action may vary from local shutdown of a valve, pump, vessel, or package to total platform shutdown. These are carefully categorized to each section. Any mistake in identifying the required action may trigger an action that may be either too much or too low for confronting the event. In the first case, it may lead to loss of production and consequently have commercial impact. In the second case, it may lead to a series of events that may finally become catastrophic both for loss of human life and economical costs. Some of the signals that may cause ESD action are listed here:

- Subsurface safety valve (SSSV) pressurization condition.
- Surface safety valve (SSV) pressurization/position (low/high) condition
- Master valve (MV) position (low/high)
- Automatic to manual ESD or PSD activation
- Each well local shutdown
- Production lines' pressure
- Production trains' blowdown valves commands/position
- Water/condensate level in production/test separators
- Gas/oil operating pressure in production/test separators
- Open/close command of blowdown and shutdown valves for each production/test separator
- If applicable, same signals for liquid pressure and level in different oily water treatment vessels in addition to position and open/close commands for blowdown valve (BDV) and emergency shutdown valve (ESDV) in this package

- Open/close condition of main emergency shutdown valve in export line to riser
- Chemical injection system signals including their pumps, chemical level, pressure, etc. Lack of injection of different liquids has different consequences. For example, if methanol is not injected during start-up, the crude may freeze and cause long delay. If corrosion inhibitor is not injected, in the long term the pipeline will corrode. Normally platform piping is from corrosion resistance material like Inconel. If antifoam is not injected in test separator, the measuring may be wrong. For each case, proper action shall be foreseen.
- Liquid level (low/high) in flare vessel
- Liquid level (low/high) in open/closed–drain systems
- Pump start/stop signal in open/closed–drain systems
- Wellhead control panel hydraulic power pack vessels' pressure condition and its pumps start/stop commands
- Diesel fuel system signals like oil level in tank and pump start/stop commands
- Power generation conditions. This may vary from oil level in day tank to generator set start/stop signal to its temperature, oil pressure, etc.
- Instrument air signals including dry air vessel pressure, compressor start/stop signals. Only dry air vessel is used to activate instrument actuators. Wet air is used for housekeeping activities
- Various HVAC signals including air inlet/damper open/close conditions, fans, and motor start/stop signals

During emergency conditions like gas detection, this system can automatically or manually issue ESD command.

C. Typical Signals in FGS

Signals transferred to fire and gas system normally come from process areas. Majority of them refer to detecting toxic, flammable/combustible gases in the vicinity of equipment. To avoid triggering an action by an instrument failure or malfunction, usually they follow a voting system. In this case, if two detectors in an area send the same signal the action will be triggered. Type of triggered action depends on the location and event.

- Platform abandon signal (normally is push–button manual)
- Man overboard signal (push–button manual)
- Combustible gas detection
- H_2S (toxic gas) detection
- Flame detection

- Smoke detection
- Deluge valve signals

During emergency fire detection conditions this system can either automatically or manually trigger ESD command.

If hazardous conditions get out of hand and to rescue operators, abandon platform shutdown (ASD) command may be issued manually. Based on command hierarchy, each higher level command first checks that its lower stage command has already been initiated. That is ASD can start after initiation of ESD, which shall be preceded by PSD. Initiation of PSD is only possible if LSD has been triggered before it. In reality, sequence of events that may lead to ASD has already initiated lower-level shutdown commands.

Old control systems showed main equipment in a schematic manner on a large wall panel. Lights showed working condition of each item. Very little information could be transmitted. General faults alarm/malfunction of equipment were highlighted with blinking or lights. In emergency conditions lights were accompanied with audio alarms. In majority of conditions operators had to attend field and get information from installed gauges. This wall panel was called mimic panel. Fig. 2.8 shows part of an

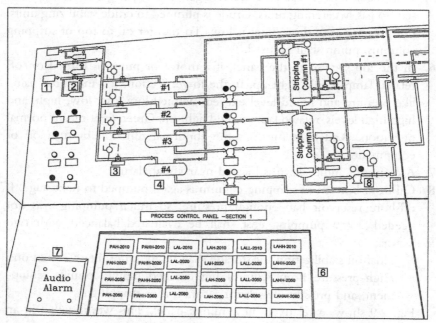

Figure 2.8 Old mimic panel onshore CCR.

old-type mimic panel. This was installed in an onshore plant control system constructed more than 40 years ago. In this mimic panel lights are either on (green or red) or off. Measured values of pressure, temperature, and liquid level could not be shown in online condition.

1. Three oil lines, coming from four offshore complexes with approximately similar compositions, are shown. Crude from two platforms merge in a subsea header.
2. Each line has its dedicated pig receiver. Effluent from pig receivers is directed back to the common manifold. A motorized valve controls flow direction on each line.
3. Products from three incoming lines are combined in the common manifold and then divided between four separators. Flowmeter installed on the line can regulate flow to each separator. For example, when during any platform shutdown volume of incoming flow reduces, only two or three separators will be left in operation.
4. Incoming crude is distributed between four separators (3 phase). Associated gas lines (from top of the separators) are shown by light color (dashed) lines and hydrocarbon with dark–colored (solid) lines. Water line is not shown. It is transferred to produced water treatment system.
5. Four second-stage separators transfer gas to the sour gas scrubbers to be sent to gas sweetening units. Crude is pumped to crude stabilizing units. Oil pressure at onshore inlet is low. To transfer oil to top of stripping columns, pump shall be used.
6. In the top section of the panel, if a motor or pump is turned on/off related lamp would light up. In the lower section, if a measured value like pressure and liquid level exceeds defined low low, low, high, and high high levels, related lamp would light up. Green light means normal operation. Red light means some abnormal function or turn off of equipment.
7. A common audio alarm is located near visual alarms.
8. Oil from bottom of stripping column is again pumped to next stage. If offshore reservoir had enough pressure, so much pumping was not needed. Extra pumping cost shall be evaluated/balanced with two factors:
   • Final oil stabilization for storage shall be in atmospheric conditions.
   • High-pressure flow requires higher–rated piping, pumps, equipment, and pressure vessels for separation or stripping.

Fig. 2.9 shows a relatively new mimic panel. This was installed in an offshore platform less than 15 years ago. In addition to the mimic panel

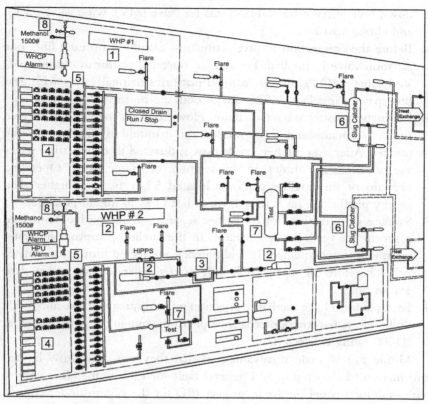

Figure 2.9 Newer control panel offshore CCR.

(installed on the walls), it had control system monitors displaying necessary information on the screen (installed on operator desk).

1. This production platform collects fluid from two wellhead platforms. One is bridge connected to it. Due to very small distance, pigging is not required.

2. Satellite platform is more than 5 km away. Water depth at this region exceeds 65 m. Therefore pig launcher package on the wellhead and pig receiver package on the production platform were installed.

3. For the satellite platform an area is marked to show infield pipeline passing under water. Crude from satellite platform has much lower temperature than crude from nearby platform.

4. Each wellhead platform has seven operating wells. Both have 16 slots. Free slots locations are reserved for future usage. The first box shows Xmas Tree with lights for each valve. These lights show status of

down hole safety valve (DHSV), master valve (MV), wing valve (WV), and choke actuators.

5. Before the connection to production/test manifolds on each line, one isolation valve is installed. For remote operated platforms motor operated valve (MOV) and for manned platforms manual valves with gear box may be installed. In this design both are motorized.

6. Production platform has two trains. Flow from test manifold is directed to test separator and from production manifold is directed to slug catcher. After slug catcher, crude flow is directed to dehydration package, which is not shown in this segment of mimic panel. Operating pressure of the satellite platform located 5 km away is higher than nearby platform. This is to overcome head loss in infield pipeline.

7. Satellite platform has horizontal test separator while bridge–connected platform has vertical test separator. In the satellite platform oily water treatment is not installed. Infield pipeline is selected form exotic material. Therefore all separated liquids in test separator are returned to export line.

8. Each platform has a methanol injection point upstream choke and inside drill well. It has to counterbalance platform shut-in pressure, therefore #1500 rating is selected.

Mimic panels' main disadvantage is that they can only show limited information like start/stop and general fault for each item. New control systems draw a panel similar to project PID for the system under consideration in OS monitor. All information like pressure, temperature, flow, liquid level, etc. of any instrument can be shown on the screen online. Still, for some packages only a few signals may be transmitted.

However, using the mimic panel the operator has a general view of almost all facilities in front of him. With the control console each time a single PID is displayed. Of course operators can have as many consoles as they like. Depending on plant complexity and availability of rapid operator action in the field, several consoles for DCS may be installed.

It has to be highlighted that now even the newer control panel is replaced with large display units showing each PID (see Fig. 2.10). This control system was installed in onshore CCR. Numerical data or status information is either displayed online or by clicking on related icons. Operator may select necessary PID for display. Emergency signals will override selection and related alarm will be generated. On this page flare knockout (KO) drum is shown. Temperature, pressure, and flow data are shown quantitatively on a continuous basis. Buttons are located at bottom

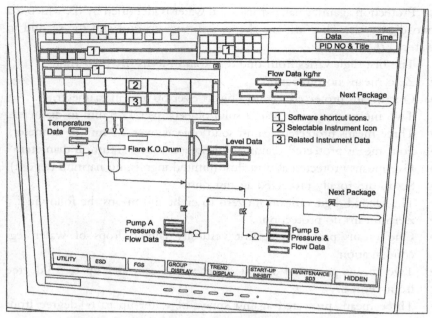

Figure 2.10 New monitor showing PID page, onshore CCR.

of screen to change displayed system. Shortcut icons at top of screen can submit required info immediately.

2.2.1 Control Design Basis

Control design basis defines what conditions/requirements shall be followed in platform control instruments. The same applies to electrical design basis. Both documents shall define the rating for equipment installed in different platform locations. The following main items are listed:

- Ingress protection (IP) degree
- Hazardous zone classification
- Hydrogen sulfide requirement

For unmanned platforms the control system shall operate as an autonomous body but have the capability to transfer control to operators. Control system gets the required information through field instruments. These instruments are exposed to environmental factors like heat, humidity, water, and small particles. At the same time their operation may generate heat or electrical currents that in the presence of ignitable gases may be dangerous. Therefore two ratings are specified. One is for environmental conditions and the second is for hazardous elements.

Protection against small particles and water is identified by IP, which is a two-digit code. The first one shows resistance to dust and the second to water [17].

The first digit varies from one to six and means the following:
- Zero means no protection.
- One means protected against objects over 50 mm in diameter.
- Two means protected against solid objects over 12.5 mm in diameter.
- Three means protected against solid objects over 2.5 mm in diameter.
- Four means protected against solid objects over 1.0 mm in diameter.
- Five means protected against dust (limited ingress, no harmful deposit).
- Six means totally protected against dust.

The second digit varies from zero to eight and means the following:
- Zero means no protection.
- One means protected against vertically falling drops of water, eg, condensation.
- Two means protected against direct sprays of water up to 15 degree from vertical.
- Three means protected against sprays of water up to 60 degree from vertical.
- Four means protected against water sprayed from all directions (limited ingress permitted).
- Five means protected against low-pressure jets of water from all directions (limited ingress permitted).
- Six means protected against strong jets of water.
- Seven means protected against the effects of immersion between 15 cm and 1 m.
- Eight means protected against long periods of immersion under pressure. For this case the expected immersion depth is to be identified.

The hazardous zone definition shows the possibility of presence of combustible or explosive substances. Codes have specific guidelines to specify numbers for different zones. Although the specified digits may differ in standards, there is more or less a general agreement on the definition of zones and requirements to be followed for equipment installation in each zone. For example, American Petroleum Institute (API) — RP 505 [18] gives distances away from the leakage source in any vessel or equipment that shall be classified to a specific zone.

Class I, Zone 0 shows that ignitable concentrations of flammable gases or vapors are present continuously or for long periods of time.

Class I, Zone 1 shows that ignitable concentrations of gases are likely to be present in normal operation or may be present due to repair/maintenance actions or equipment malfunction or operator mistake.

Class I, Zone 2 shows that ignitable concentrations of flammable gases or vapors are not likely to occur in normal operation and, if they do, will exist for a short period of time.

Grade of release is divided into continuous, primary, and secondary. For Zone 0, continuous release is defined as more than 1000 h per year; for Zone 1, primary release is between 10 and 1000 h per year; and for Zone 2, secondary release is less than 10 h per year. This means that for Zone 0 more than 11.4% of the time ignitable gas is present.

Gas presence based on type of gas is classified as Group I, IIA, IIB, and IIC.

Combustible liquids are classified based on their flash point to Class II, IIIA, and IIIB.

Electrical and control devices placed at each of these locations shall have proper specifications and be approved by certifying authorities as type-tested equipment. For more explanation on the definitions of the zones and the required equipment characteristics, please refer to the relevant standards. IEC 60079 [19] defines the requirements for each group.

Group I is used for electrical apparatus for mines susceptible to firedamp. In mines the representative gas is methane. Firedamp is a mixture of gases, composed mostly of methane, that is found underground, usually in mines.

Group II is for applications in other industries. Tables 2.1–2.3 are reproduced from DNV OS – D201, API RP505, and IEC 60079 [17–19].

Table 2.1 Gas group table

Gas group	IIA	IIB	IIC
Representative gas	Methane/propane	Acetaldehyde/ethylene	Acetylene/hydrogen
Reference test pressure (Kpa)	1350	2500	4000

Table 2.2 Combustible liquid class

Combustible liquid class	II	IIIA	IIIB
Flash point °C	$37.8 \leq T < 60$	$60 \leq T < 93$	$93 \leq T$

Table 2.3 Gas or vapor ignition temperature

Temperature class	T1	T2	T3	T4	T5	T6
Ignition temperature of gas or vapor (°C)	>450	>300	>200	>135	>100	>85

The term representative gas does not mean it is only for this gas. The International Electrotechnical Commission (IEC) has an extensive table that classifies different gases in each group. It is understood that equipment classified as IIB is safe for installation in IIA locations. Similarly, IIC equipment is safe for installation in IIB and IIA locations.

Group IIA covers acetone, ammonia, ethyl alcohol, gasoline, methane, and propane.

Table 2.3 shall be interpreted properly. It means for T1 equipment the maximum temperature of the surface of the enclosure may reach up to and be less than 450°C, while the same in T6 equipment will not exceed 85°C. The gas in class T1 has a much higher ignition temperature and is therefore less dangerous than in class T6. In other words, equipment rated as T6 is superior to others rated as, for example, T1 and can be used instead of each other from high to lower number.

Equipment intended for use in explosive atmosphere is marked with Ex. Different means of protection may be used. Flameproof enclosures are marked "d," pressurized enclosures "p," enclosures filled with powder "q," oil immersion by "o," increased safety by "e," and intrinsic safety by "i." As per IEC 60079 − 11, intrinsically safe "Exi" is an apparatus that in normal operation or fault conditions does not produce a spark or thermal effect that is capable of causing ignition of a given explosive gas atmosphere.

2.2.2 Control Philosophy

All platforms shall be automatically controllable. Whenever operators are on the platform they shall be able to control it as they wish. Therefore all control aspects shall be switchable from automatic to manual. However, it is not economical to keep personnel on all platforms. Therefore the majority of platforms are remote controllable. The remote control may be from onshore or a nearby offshore complex facility. But whenever crew visit the platform, for any reason like maintenance or normal operation, they want to be able to take control.

Change from remote to local and automatic to manual control shall be through very well-defined and strict procedures. The status shall be clear at any time. The authority/responsibility of requesting this permit and granting it shall be clearly defined. Using passwords to enable initiating this process minimizes unwanted intrusions.

The control philosophy document describes the architecture of this control, how the signals are transferred, the hierarchy of the control sequence, where and what is displayed, to whom and how the decision–making authority shall be distributed, etc.

In normal operation, platforms are controlled by the process control system (PCS). In abnormal emergency or fire ESD or FGS takes command. Two separate/independent hardwares are used for PCS and ESD/FGS. Always ESD/FGS overrule other signals and have priority. Normally ESD/FGS system is redundant and its cabling is such that if fire or explosion in one side of platform destroys cables, transferring instrument signal cables from the other side of the platform with a separate route transfer the signals to the control system.

This document describes control levels for all process and utility systems on the platform. PID shall be exactly compatible to this document. It defines local, process, and emergency shutdown cases for all equipment.

In addition to the controlling system, this document specifies what measures have been foreseen for the fire and gas system. In some projects two separate documents may be issued, one for control and one for ESD/FGS. In addition, control philosophy may be combined with control design basis.

2.2.3 Control Block Diagram

The control block diagram is a drawing that shows control connections and interfaces. Connection of field instruments to operator station(s) in control room shall be shown. Main and secondary control rooms may be defined with a specific hierarchy. In a complex, in addition to the control room located on the wellhead platform, it is possible to transfer all signals via network over connecting bridges to the control room located in the nearby living quarter. In addition, signals can be transferred to onshore control room via subsea cable. In each case control hierarchy shall be clearly defined. At any time control signal can be issued only via one station. This station shall be master and others shall act as slave. With the progress in instrument technology it is possible to have all signals anywhere in the connecting network. Normally when operators are present on a platform, for safety reasons they shall have command. Therefore even if the master control station may be located somewhere else, after personnel have arrived on the platform they will call Central Control Room (CCR) and ask for control to be transferred to the platform.

2.3 ELECTRICAL

The electrical discipline has the responsibility to calculate required power consumption for all platform usage, provide and distribute power to the users. Provision of electrical power in a platform may be via subsea cable, diesel or gas turbine generators, and even for minimum facility platforms via solar batteries. In addition, they have to provide UPS for specifically defined emergency services.

For power distribution one of two policies may be selected.

The first approach is to provide a safe area in the switchgear or technical room. This way normal switchboards may be utilized. This is considered as a cost advantage, but it has a disadvantage that numerous cables have to pass from technical room to the consumers. This approach is generally followed in many platforms.

The second approach is to locate main switchboards in the safe area but provide some smaller size distribution boards in the process area. In this case the distribution boards shall fulfill specified hazardous zone requirements, but cabling length will be considerably reduced. Of course large-size cables to transfer power between main and local distribution boards will increase.

2.3.1 Hazardous Area Classification Drawings

These drawings divide platform layout to each of the hazardous zone classifications. In Section 2.2.1 the definitions of hazardous zones were given. Now the extent of each zone in the plan has to be determined. Normally this document uses information produced in an area classification table. In some projects both documents may be generated by safety discipline. A typical hazardous area classification drawing is shown in Fig. 2.11.

Fig. 2.11 shows a platform level with technical room and two generator rooms located in the utility area. Some equipment is located in the process area. Process area includes well bay area, manifold area, two free-water knockout drums, test separator, wellhead control panel and power pack unit, flare gas skid, and flare ignition panel. As can be seen, Zone 1 (cross-hatch), Zone 2 (one-line hatch), and safe area (blank) are identified. In addition, since jack-up rig will be used for well maintenance and

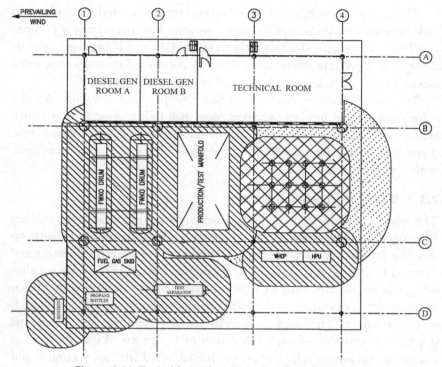

Figure 2.11 Typical hazardous area classification drawing.

wireline operations, an additional area (dashed lines) that is considered as Zone 1 only during this activity is marked.

Zone 1 connected to Xmas Tree extends 7.5 m from its center. Area extension is terminated when they clash with firewall. From the edge of the wall the same extension is still continued.

For all equipment and their connecting valves that may be a potential source of release, some information has to be determined. This may include:

1. Location, whether it is enclosed/ventilated or not
2. Operating conditions (pressure, temperature)
3. Containment conditions (gas, liquid, flash point, grade of release)
4. Distance away from it, which shall be considered as hazardous and classified Zone (0, 1, 2)
5. Class of gas or liquid (IIA, IIB, IIC, IIIA, IIIB, etc.)
6. Temperature class ($T1$ to $T6$)

The extent of a certain zone-circling equipment is determined based on code recommendations. For example, an area 3 m away from a pressure vessel in a nonenclosed/adequately ventilated area will be considered as Zone 2. If a pressure safety valve (PSV) is connected to this vessel, zone extent will be considered to be 3 m after PSV release point.

This information is used to prepare the previous drawing. All the equipment located inside a specified zone shall follow zone requirements. However, as said, to avoid mistakes in installation for small instruments, clients may instruct to procure them as compatible for the most severe zone in the platform layout.

2.3.2 Black Start Philosophy

This may be a stand-alone document or contained in a chapter of another document like "Electrical Design Criteria." The intention is to sum up several scenarios and calculate the maximum required electrical power and duration to start up a platform. Black start case (meaning platform is either being run for the first time or is being started after a complete shutdown and blowdown operation) is the most severe condition. In this case there is no power source. Therefore no detector or safety sensor works. Operators at CCR are unaware of any malfunction of equipment. Procedures set in this document ensure sufficient power is established and major control and safety sensors are operative before starting actual operation. Scenarios for partial train shutdown normally may not be so critical.

In many cases gas produced by the platform may be used as the driving force for the generators. Therefore in black start case (zero production) the duration needed to achieve continuous and stable production shall be determined. Sufficient amount of diesel fuel has to be provided in this duration before platform-produced fuel can be utilized to run generators. In addition, turbogenerators shall either be able to consume dual fuels or auxiliary diesel generator with sufficient capacity shall be provided. This auxiliary generator may come from a standby barge or vessel that is normally present during major start-up activities. Always an electrical shore connection is available nearby the boat landing.

Electrical motors have a very high current demand during start-up. Hence the sequence of starting heavy load motors shall be clearly defined. This helps avoid generator shutdown due to overload. Some of the heavy load motors include:
- Pumps discharging to a high-pressure line
- Pumps discharging high-liquid volume

- Compressors: Normally water/gas injection compressors are very high demand. They have their own turbogenerator and do not use platform power generation capacity.
- HVAC units: It is worth noting that even when a platform is unmanned, to ensure safe working of many control instruments HAVC shall be active. These instruments cannot operate properly at high temperatures.
- Hydraulic pumps operating shutdown valves

On the other hand, since safety equipment depend on platform power, their operation is also needed during start-up activities. Therefore several logical scenarios shall be examined and the critical case defined.

2.3.3 Single-Line Diagram

A single-line diagram (SLD) shows electric flow from generators to the consumers. Based on type of electrical current and usage, it may be divided into several categories. Each board category is shown in a separate diagram. Based on number/diversity of connections, each diagram may have several sheets.

The first diagram will show the main bus and feeding to each individual switchboard. This will only differentiate the board number, power level, intention of the board, and general information.

Subsequent drawings show each switchboard's architecture and the feeding connections. This will give detailed and specific information on the feeding bus rating, switchboard internal architecture for each connection, consumer tag number, and description of the connection and its power. Process and utility switchboards may be separated.

One or several separate boards are normally dedicated to lighting. The total power in this switchboard(s) may not be high but the number of consumer points is. Lighting switchboards contain low voltage equipment and are frequently maintained. Therefore they should have sufficient spares and be easily replaceable.

For normal maintenance operations, welding outlets are provided in different locations of the platform. These outlets may be fed from a specific switchboard.

Emergency systems are also fed from a specific switchboard. In addition to related equipment, it feeds the UPS and charger systems. Based on safety requirements, some emergency systems shall work continuously after total platform shutdown and abandonment for specific time spans. For example, navigational aid systems shall work at least 96 h after platform shutdown.

These consumers are fed from a dedicated switchboard. Their battery pack and the chargers are also connected to the same board.

In fact the single line diagram shall include all consumers and will be the basis to design switchboards and platform cabling.

Fig. 2.12 shows part of a SLD. For clarity, only a small portion of SLD, showing five feeders, is shown. This power distribution system has two 100% diesel generators. One power line is coming from each generator set. Each power line has three live and one null line, each 240 mm^2. They are connected to the same bus.

Feeders in Fig. 2.12 include two 40 kW (230 V AC UPS), two 37 kW (air compressor duty and standby), and one 33 kW (backup monoethylene

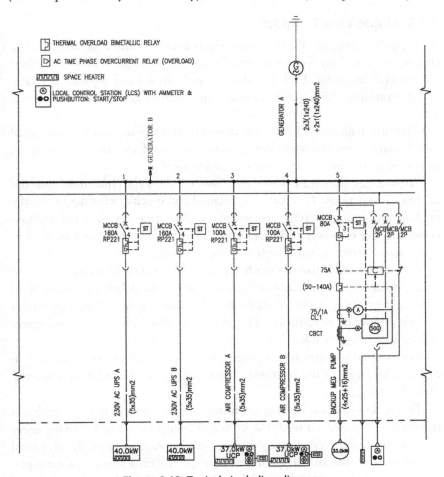

Figure 2.12 Typical single-line diagram.

glycol (MEG) pump). Diesel generator control circuit is also omitted. Each consumer has a molded case circuit breaker (MCCB), thermal overload bimetallic relay, and AC time phase overcurrent relay.

Required total power for lighting in each elevation may be lower than some single consumers. However, it has many feeders. Therefore dedicated switchboards will be used for lighting.

In process area maintenance welding may become necessary for piping and equipment. It is customary to install welding sockets in several points at each level. Again, a dedicated switchboard may be used for all welding outlets.

The shown SLD belongs to a wellhead platform. In a large-production platform these power consumptions may be considered as low.

Switchboards are potentially hazardous equipment. If ignitable gas is present nearby them, electrical currents may induce sparks leading to explosion and fire. Two approaches to install switchboards were explained earlier and are further developed following.

In the first approach, the switchboard room is located in a safe area with proper safeguards. Then switchboards themselves need not be explosion-proof rated. The benefit is that switchboards become smaller, more compact, and therefore cheaper. However, this requires cables for each item to be routed from switchboard room to field. If a large number of cables have to pass long distances, this solution may become expensive.

The second approach is to use intermediate switchboards in several locations. One main cable brings electrical power from the switchboard room in the safe area to the one in process area. This way cable length is considerably reduced but the switchboard located in process area shall be explosion-proof rated and properly air conditioned.

2.3.4 Electrical Load List

Electrical load list determines consumed loads by equipment. This is used to calculate the required generator capacity and prepare a single-line diagram to design switchboards.

Those consumers that are continuously in use during normal operation shall be differentiated from intermittent consumers that may work for a short duration. Intermittent users may also be classified based on their power consumption and frequency of operation. For example, there are feeding pumps (like fuel oil or freshwater) that are used only when a supply boat berths to provide fuel, chemicals, and other consumables. They work only 1 or 2 h at each 2-week interval. Some of the intermittent loads may be high. For example, the compressed air system is turned off after air pressure inside wet/dry receivers reaches to high level. After air receiver pressure drops to a certain level, one of the air compressors may start again. Due to further pressure drop, the standby compressor may also start. In spite of air compressor intermittent operation, it may have a high-electrical demand. Some batch injection pumps may also work frequently and inject at lines with very high pressure. These type of intermittent consumers may have considerable power demand.

Electrical motors driving mechanical equipment are normally the highest consumers. Mechanical power of a pump is calculated by:

$$P = Q \times p/\eta$$

$P =$ Mechanical power (kW)
$Q =$ Discharge (m^3/h)
$p =$ Pressure head (bar)
$\eta =$ Factor $\times$ Efficiency

Alternatively, pump power demand formula can be written as:

$$P = \rho \times g \times Q \times h/\eta$$

$\rho =$ Unit mass
$g =$ Acceleration of gravity (m/s^2)
$h =$ Fluid head (m)

In calculating the previous formula, discharge unit shall be considered as m^3/s and pressure as Pascal. If m^3/h and bar units are used, the result shall be divided by 36 to give kW. If gallons per minute (GPM) and PSI are used, the result shall be divided by 1715 to give power in horsepower. One horsepower is equal to 745.6 W.

This is the consumed mechanical power. The electrical motor needs more power because the efficiency in transferring electrical to mechanical power and from shaft to actual mechanical energy is less than 1. Conservatively a factor from 0.7 to 0.8 may be used.

Pumps discharging to process line should overcome high pressures and may have small discharges. Pumps injecting chemicals to process vessels are of this type. Corrosion inhibitors, demulsifiers, reverse demulsifiers, etc. fall in this category. For example, closed-drain pump discharge may be as low as 5 m^3/h but for pumping to the export line the pressure head may be as high as 130 bar.

Pumps used for normal washing services/housekeeping activities shall have high discharges but need small pressure head. For example, service water pump discharge may be more than 50 m³/h, but the required head may be up to 6 bar. Fire water pump may discharge more than 500 m³/h water with 20 bar pressure. If service water is used for local fire extinguishing, then it shall follow the pressure requirement as well.

Assuming both pumps' mechanical efficiency to be 0.75, the calculated power for service water pump is 11.1 kW and for closed drain pump is 24.1 kW. It is seen that fire water power demand is much higher than others and is equal to 370.4 kW. Fire water pump has a specific safety requirement. In all living quarters a fire water pump is required, which is powered by a dedicated diesel driven motor. This diesel is not used for any other purpose. In addition, there should be another fire water pump as backup with 100% capacity.

The calculated power is not used in the pump data sheet (DS). No motor shall continuously work with full power in design condition. Normally 75—90% of its nominal power may be used. This means calculated 24.1 kW may be procured as high as 30 kW.

For very low-capacity motors, a nominal power from minimum available ranges will be selected, which may be several times more than needed.

When summing up the powers to determine generator capacity, all consumers shall be listed in a table. Then those that will be in continuous operation and intermittent consumers are identified. Based on engineering judgment and operators' experience, a factor is selected to determine the total load considering simultaneous operation. This factor may vary from 30% to 50%.

For example, a compressor providing compressed air will pressurize wet and dry air receivers to assume 10 bar. Normal working pressure of pneumatic instruments is 4—5 bar. Until pressure of air receiver vessel drops below, say 8 bar, compressor will not start. During start-up or maintenance conditions, when a lot of air is consumed in opening and closing valves, compressor starts/stops much more frequently than in normal operations. This means an intermittent electrical consumer may work much longer at different stages of platform operation.

The same may happen for other motors. As another example, a gas platform methanol injection pump is not working during normal operation. Only during well start-up is sufficient volume of methanol injected upstream to the master valve to prevent hydrate formation when gas first enters the wellhead. Although methanol injection volume may not be so high, it shall inject to a pressure equal to wellhead shut-in pressure, which may be as high as 330 bar or more.

To ensure minimum power consumption levels, some packages may run continuously. It is understood that continuous running may reduce those packages' operating life. On the other hand, it will improve generator working condition.

As stated before, in calculating required power of different equipment several efficiencies shall be considered. These efficiencies may include:
1. Pump operating condition compared to maximum/optimum
2. Mechanical efficiency transferred from shaft to the liquid
3. Mechanical efficiency from motor to shaft
4. Electrical efficiency from power source to motor

Electrical motors require very large currents during start-up. This may be several times their nominal capacity. Usually pump manufacturers mention this factor in a related technical catalog. Values up to seven times normal current may be expected. However, it depends on the suction and discharge conditions. A pump that is already facing high pressure at discharge requires larger current compared to a pump that has no load in discharge. Selecting generator capacity will be based on scenarios developed in start-up philosophy.

2.3.5 Generator Sizing

Offshore platforms' power demand is supplied either by subsea cable from onshore power station or via diesel or gas turbine generators installed on the platform. For diesel generators, providing suitable fuel supply like diesel storage/its continuous supply and for turbine generators, gas dehydration/sweetening (to use platform gas) becomes vital in platform operation. If fuel is supplied by supply boats (during rough seasons when access to platform is difficult) adequate supply shall be provided on the platform storage. A redundant power supply for safety systems shall be provided on the platform. This is normally provided by a battery system. But its capacity is limited.

Proper generator sizing is a vital task in platform design. Chemical energy in fuel (fluid or gas) is transferred to mechanical energy by burning in engine cylinders. Up and down movement of cylinders is transferred to shaft rotating movement. Rotation of alternators' magnetic core induces electrical current in surrounding coil, which is transferred to switchboards and consumed as electrical energy. Several factors impact generator efficiency. Complete fuel burning, frictionless transfer of chemical energy to rotating movement of magnetic core, and impact of induction currents all influence total generated electrical power.

Power generation system (diesel or gas turbine) used in an offshore installation shall satisfy several criteria:

1. Shall be robust and be able to supply normal continuous demand of the offshore installation for long-term operation during all weather conditions. Generators are very large equipment. Other than replacement of small items and overhaul repairs, total replacement is very costly. Their design life shall be of the order of reservoir hydrocarbon supply.

2. Shall have suitable contingency. Schemes like 2 × 100%, 3 × 50% and 4 × 33% have been used considering factors like available plot plan area, impact of loss of production due to power shutdown, availability, possibility of maintenance, and total power demand. Loss of production due to power shutdown may become very costly. Selecting a redundancy scheme is further explained in Section 2.8.8. New start-up requires considerable man power and equipment mobilization, which considerably increases the costs.

3. Normal power demand shall cover a sufficient range of generator capacity. Too high or too low power consumption may reduce generator operating life considerably.

4. Generators shall be able to cover temporary high-power demands. This issue shall cover both an impulse surge in electrical current during heavy motor start-up or higher power demand for longer duration. The first one may last only a few seconds and the second one shall not be more than 1 h. Generator control panel may shut it down if load exceeds a certain limit.

5. During start-up of a single consumer, electrical current will have an impulsive peak rise. Either soft starter shall be used to reduce the high peak or generator shall be sized for this condition. This peak may be as high as seven times normal current. Each vendor provides this coefficient for its own electromotor. During calculations a high value is selected. This value shall be confirmed during vendor data review stage.

In selecting generator capacity for the first step the electrical engineer will sum up all loads acting continuously. In the next step, all intermittent loads are calculated. A contribution factor (like 30%) is selected. In some cases different factors may be selected for different types of intermittent consumers. Continuous loads plus selected percentage of intermittent loads constitute generator normal power rating. A contingency factor of 10—15% may also be applied. Project team cannot stop until all loads are accurately known, then procure generators. Therefore when load estimates are based on assumptions, higher contingency can be applied. After getting final vendor data, a reanalysis can be made to verify generator sizing. In this case lower contingency shall be applied.

In the next stage they shall consider applicable start-up scenarios and calculate the high demand of large consumers. The maximum power of different scenarios will be selected as generator power. After that platform sparing philosophy shall be taken into account. For low loads 2 × 100% spare may be selected while for higher demands, which may necessitate turbo generators, 3 × 50% or even 4 × 33.3% sparing may be selected.

2.3.6 Heat Tracing

Water vapor exists in the gaseous material. Gas temperature may reduce either due to expansions, which results in pressure reduction, or cold ambient temperature. Due to temperature reduction, a gas volume may reach vapor dew point. Dew point is the temperature at which the water vapor may form droplets at that pressure and moisture content. This may be dangerous for some equipment like generators, flare tip, etc. The following mitigation measures may be used.

- Reducing the gas line diameter will increase gas velocity. This has two impacts. First, it reduces the time that the gas is exposed to ambient temperature. Gas temperature reduction at this short duration shall be calculated. It may become less than that needed for cooling down to ambient temperature. Second, the increased velocity increases the possibility of moving the droplet and preventing nozzle blockage. However, the increased velocity causes more head loss, which increases pressure reduction. Moving away a droplet solves the problem of water accumulation and blocking the passage. The disadvantage of this solution is that impinging droplets on compressor or pump blade causes erosion. Besides, after reducing pipe diameter, total discharged value may reduce.
- The next solution may be to increase gas temperature by a super heater such that after cooling down it is still above dew point.
- Insulating the gas pipe may reduce its conductivity. Again the time that gas is exposed to the environment may become less than required to cool down its temperature to dew point.
- The last method is to add heat tracing. In onshore plants this is usually by steam produced in the boilers. In offshore platforms normally boilers are not used. Therefore electrical heat tracing is used. This increases the power consumption.

Electrical heat tracing increases the electrical power consumption, and whenever process calculations show that it is needed it shall be included in platform design.

Section 3.2.2 (Compressed Air System) gives a curve showing moisture content in air based on different temperatures and relative humidity.

2.3.7 Cables and Cable Routes

A major scope of work in electrical, control, instrument, and telecomm includes cable material takeoff (MTO) and cable routing. Both power and signal cables shall be safe from hazards.

Cable route equipment consists of cable ladders, cable trays, and their related supports. Normally galvanized carbon steel, glass fiber-reinforced plastic (GRP), or stainless steel (SS) material may be used.

GRP is much easier for handling but is brittle. For material in the open space, UV-resistant GRP shall be used. During the time span when a cable route is installed until the last cable is placed and before the location becomes secure, workers may unintentionally damage it. Damage may happen by placing heavy objects on them, hitting them with hand tools, simply sitting on them during periodic rest, etc.

Stainless steel is heavier but for very hot regions (like near flare) may be the only choice. Distance between supports in stainless steel material may be selected much longer than GRP supports. In general, SS routes have better load-bearing capacity. Manufacturers' standards shall in each case be carefully studied.

Galvanized trays, ladders have lower quality than SS material.

Cable type shall be selected based on zone and package requirements. Armored, flame retardant, or fire resistant cable may be used. Systems that need to be available during emergency conditions shall have higher quality and dual configuration.

Each cable may consist of several strands. The conductor may be solid core or braided from copper. Insulation covers each core. Insulation, armor, and sheath layers may cover all cores. Based on cable location, the covering material may be able to resist temperatures up to 750°C. In addition, covers shall be either flame retardant or fire resistant. Flame retardant material does not allow fire on one point of cable to spread to other locations, while fire resistant material does not catch fire up to a certain temperature.

To avoid emitting poisonous gases, sheathing material shall be halogen free. To prevent producing a large quantity of smoke and impair visibility during escape or repair, cable sheathing shall be of the no-smoking type.

Each wire has two conductors (covered by insulation) to transfer electrical current. In addition there is one conductor to transfer unwanted current away from the electrical system. This is the earthing conductor. Its function is to minimize danger due to faults between current-carrying conductors and other metallic items. Inside the junction box (JB) two current-carrying conductors of a wire are connected to two nearby ports. The third termination point is left for the earthing conductor. All earthing terminals plus JB earthing are connected to the earthing system. Platform steel body is connected to piles and jacket structure. Piles are fixed in seabed. The welded metallic structure of the platform acts as a natural rod. Lightning protection system is also connected to the platform body.

One of the main problems encountered in fabrication yards is a shortage of cables. This may happen due to several reasons either in design or in fabrication stage:

- Cable length may not have been accurately calculated. For example, an item is located at the south of the platform two elevations above the point at which its cable shall terminate. In cable length calculation, a common mistake may be to consider only the length required in the plan and forget the required length to pass two elevations. Another mistake may be to include direct pass and forget actual cable route arrangement. The presence of obstacles like walls or vessels, etc. may necessitate a longer route.
- Some items may have been overlooked. Large and main items are always considered. Only small items or portions of a package may be overlooked. For example, a package may have several sensors or switches that are located away from the package. They may be shipped loose by the manufacturer and have to be installed in the yard. The designer may overlook them.
- Total number of items like lighting that are distributed all over the platform may be estimated less than actual.
- During construction some cable drums may get damaged. This may affect their cables as well.
- Some cables may be mistakenly cut less than required length. This piece may become useless for other cabling.
- During cable pulling or retrieving some cables armor may get damaged.

Designer and EPCIC contractor shall always bear in mind that machinery and personnel idle time due to cable shortage during fabrication or offshore activities is more costly than:
- Procuring additional cable lengths
- Performing design review audits on cable MTO in engineering phase
- Strict check on cable transport from storage area to yard for cutting, cable laying

2.3.8 Contingency/Sparing

In electrical, control, instrumentation, and telecommunication disciplines the meaning of sparing and contingency for bulk items have some similarities but in certain aspects there are differences in the meaning of the same word in other disciplines. For packaged equipment these words somehow may have the same meaning.

In cable list/MTO the length of each cable size is calculated. Assume that for a certain 5×45 mm^2-type cable 140 m length is required. For procurement a certain percentage may be added for inaccuracies in calculating cable length or change in cable route in yard on cable trays or other issues. But the manufacturer has to wrap cables around a drum. A minimum length is determined by manufacturer based on each cable size. This may necessitate procuring maybe even 300 m length. Assume that for 3–core 1.5 mm^2 cable 1900 m is needed. For cuttings during fabrication and extra lengths, 10% is added. Therefore for this size 2090 m is needed. However, the manufacturer minimum drum length is 400 m. For the extra 90 m this means either buying six drums or using less contingency. On the other hand, for the same project maybe some 250 m 5–core 1.5 mm^2 cable is also needed. In this case, the consultant has to combine cables of similar sizes in one group. It means the extra length for 3–core can be used from extra length of 5–core cables. It will be more beneficial if two branches of a 5–core cable are left unconnected. But the full cable length of one drum is utilized. The same approach cannot be used in structure and piping. Their special approach for bulk items will be described in related sections.

For switchboards, marshaling cabinets, multicable transits (MCTs), and control hardware there are always some spare spaces left. Inside a switchboard or marshaling cabinet some racks are installed. Required items are installed inside these racks. Some additional space may be left blank. Therefore if a new development is intended, new switches, relays, etc. can be placed in this empty space. This extra empty capacity enables adding required facilities/connections if in the future some revamping or new facilities are added to the platform. In the future it will be very difficult to cut a hole and weld a new MCT. Instead the rubber inside unused MCT hole will be removed and a cable will be passed. Or a relay or circuit breaker can be added in the free space of a switchboard.

2.4 HVAC

Normally offshore platforms have a harsh environment. In addition to toxic, flammable/explosive gases and hydrocarbon material the cold/hot, humid weather, direct sunlight, and radiation affect both operators and machinery. In the open area the designer depends on natural ventilation, however, in the rooms, sufficient ventilation and air conditioning shall be provided. Environmental factors like storms are evaluated by the structural group.

The platform area is very congested. Several types of equipment, piping, vessels, skids, etc. are installed in a limited area with small access corridors and maintenance area nearby each other. Some clients require performing a computational fluid dynamics (CFD) analysis to prove natural ventilation is sufficient to carry over gases or reduce flammable gas concentration to acceptable levels. In lieu of CFD analysis, some clients accept performing simplified analysis based on available wind data. A sample method is described in Section 2.4.1.

Each working shift is 12 h. After long working hours operators shall have a proper rest. Therefore living quarter HVAC shall be very carefully designed. During platform shutdown proper measures shall be foreseen to provide suitable comfort level for the personnel remaining on the platform.

HVAC in the technical rooms, engine rooms, and battery rooms is also very important. In some platforms electrical switchgears, telecommunication and control equipment are separated and placed in dedicated rooms. The most important room is CCR. It is the brain of the platform. Some of the sensitive control equipment in this room may malfunction in an excessively hot environment. Their best operating conditions are determined by manufacturers and platform design shall provide the required working conditions. In many cases the consequences of control equipment malfunction may be disastrous. Therefore platform may shut down just for HVAC not working properly in the CCR. Considering huge losses due to platform shutdown, it is cost-effective to design a proper HAVC system. Normally sparing for HVAC equipment in CCR is 100%. This gives proper repair time to the operator in case one HVAC set is damaged. The purpose of the HVAC system can be summarized as follows:

1. Maintain acceptable working/living environment (temperature, humidity, fresh air, dust quantity, gas presence, and pressure) for the occupants.

2. Maintain acceptable/nondetrimental indoor conditions for installed equipment
3. Maintain a minimum overpressure to prevent ingress of hazardous gases
4. Prevent the accumulation and buildup of gases to dangerous concentrations.

2.4.1 Ventilation/Natural Air Change

In areas where operators are working or sleeping like CCR, offices, cabins, etc., fresh air is needed. In battery rooms, during boost charging hydrogen is released. Internal air shall be continuously discharged and fresh air replaced to ensure hydrogen concentration is below explosion limit. For battery room design, required fresh air volume may be more than others. For an explanation on fresh air requirements in battery rooms please see Section 3.2.5 "AC/DC UPS Systems."

Fresh air volume to provide breathing oxygen is much less than actual fresh air demand of an area. Air consists of approximately 21% oxygen and 78% nitrogen. The remaining 1% consists of other gases. Only about one-fourth of oxygen in air is consumed during breathing. Forty liter per minute fresh air is sufficient for each operator at normal working conditions. However, additional fresh air is needed for comfort condition. In addition to providing oxygen for breathing, fresh air removes humidity due to individual perspiration and body odors. Smoking is not allowed in offshore platforms, therefore huge air change to remove cigarette smoke is not considered. Reducing oxygen partial pressure in air reduces its absorption rate in blood.

For suitable air quality, in addition to keeping proper oxygen concentration, perspiration by operators shall also be removed to keep humidity in a tolerable level for operators and instruments. This may be achieved by keeping the humidity level between 40% and 70% R.H. [20].

British Standard "Code of Practice for Ventilation Principles..." [21] has tabulated an extensive list of criteria for fresh air volume. The most conservative approach is to provide the highest volume. However, it is not so stringent and an average value can also be selected. Some of the commonly used criteria are given here:

• Five to eight liter per person per second
• One liter per sec per square meter of area
• From half to one-and-half room volume air change per hour

Assume a CCR is 7 m long, 3 m wide, and 4 m high. Three operators are expected to work in this office. The first criterion requires 15—24 L/s, the second requires 21 L/s, and the last one requires from 11.7 to 35 L/s air change. For this CCR, selecting 25 L/s air change (if other criteria are not considered) will be okay.

All rooms are designed to be gas tight. Windows are either not provided or always kept close. Therefore for the rooms only mechanical (forced) ventilation is considered.

In the process area contaminant gases may be emitted. This may be from leaks or drains. In this region only natural ventilation is provided. It depends on wind velocity and temperature difference. Airflow due to wind comprises the major part.

Measured temperatures under sunlight and shade have a few degrees difference. When sunlight hits a steel floor, the air temperature exactly above it will increase. Over water, this may not be the case. Therefore it is not recommended to calculate airflow for temperature differences above 6°C. For enclosed spaces with only one opening, flow due to temperature difference is not anticipated. Two openings located with a center-to-center elevation difference equal to H are needed to enable flow [21].

$$Q_d = C_d A \sqrt{2gH_d \frac{T_I - T_O}{T_O}}$$

Q_d = Airflow based on temperature difference (m^3/s)
C_d = Constant coefficient = 0.62
A = Airflow opening area (m^2) incoming/outgoing assumed to be equal
g = Gravitational acceleration (m/s^2)
H_d = Elevation difference between centerline of incoming and outgoing opening (m)
T_I = Average indoor temperature (°K)
T_O = Average outdoor temperature (°K)

Wind velocity induces a considerable air change rate. Minimum air velocity can be calculated from wind rose. API RP 505 proposes to use 0.5 m/s as the minimum air change velocity. This velocity is almost always present. Wind rose in Section 2.7.1 shows 4.5% calm period. Even with no wind blowing, temperature difference may induce necessary ventilation. The author has not seen any platform or onshore plant with forced ventilation in the process area. This is different from enclosed spaces, which always depend on mechanical ventilation. Wind-induced ventilation is calculated by:

$$Q_w = C_d A_w U_r (C_p)^{1/2}$$

$$\frac{1}{A_w^2} = \frac{1}{(A_1 + A_2)^2} + \frac{1}{(A_3 + A_4)^2}$$

Q_w = Airflow-based wind velocity (m^3/s)
C_d = Constant coefficient = 0.61
A_w = Airflow opening area (m^2)
U_r = Reference wind velocity (m/s)
C_p = Surface pressure coefficient, based on area dimensions (w, l, and h) and sides that have openings can be extracted from BS5925 [21].

2.4.2 HVAC Philosophy/Calculations

Three main HVAC design goals include:
1. Provide fresh air through forced or natural ventilation.
2. Reduce (in hot area) and increase (in cold area) air temperature to comfort levels.
3. Provide either overpressure or under pressure.

The human body is cooled by sweating. In a dry environment, sweat drops vaporize and provide a comfortable condition for working. In offshore areas normally humidity is 100%. This means air is saturated with water vapor. Humidity is the presence of water vapor in air.

At a certain temperature (say 24°C), if humidity is zero the body can cool itself at a higher rate, therefore to the human body it feels like the air is cooler. At the same temperature if the air is saturated with water vapor the human body cannot cool itself and therefore it will feel like the air is warmer. I am not sure of the figure but I have seen a text stating that the difference from 0% to 100% humidity makes about 6°C difference in feeling the temperature. This means that at 24°C with 0% humidity the body can cool itself down faster. Therefore it feels like 21°C while at 24°C with 100% humidity the human body cannot cool down and may feel like it is 27°C.

Normal humans may need about 40 L/min fresh air for respiration. Fresh air, as we have said, has about 21% oxygen. Only about 25% of the oxygen in inhaled air is consumed. Therefore any HVAC package for a confined space with working personnel shall consider to inlet sufficient fresh air. To reduce consumed power for cooling, a major part of internal air can be circulated.

Overpressure is implemented in a space in which the designer wants to prevent potential ignitable gas/smoke ingression from outer space to inside the enclosed area. For example, in a technical room or living quarter, etc. overpressure is maintained by continuously blowing fresh air from a safe area. Underpressure is implemented in a space in which the designer wants to prevent hazardous gas/odors from spreading to nearby spaces or from gaining ignitable concentrations. This is done by continuously blowing out inside air to a safe location. For example, hydrogen generated from operating batteries is discharged to a safe area. Foul odors generated in bathrooms and from toilets are also discharged to open environment. This will prevent its ingression to adjacent bedrooms.

The number of persons in a room, type of activity, and heat generated by equipment like switchboards, panels, lighting, etc. constitute internal heat input.

Some generators may be placed outdoors and some indoors. Outdoor units may be more expensive because they shall meet hazardous zone requirements. Indoor units may be cheaper but they impose certain requirements on enclosure ratings that may compensate or even exceed outdoor units cost. Fresh air supply for a generator unit consists of three parts:

1. Fresh air supply for combustion
2. Fresh air supply for cooling system
3. Fresh air supply for room ventilation

Combustion air and room ventilation does not require a large volume supply. For generator cooling normally a closed-circuit water system is used. This is similar to your car's cooling system. Water is circulated by a pump in the motor. The heated water is cooled in the radiator by fresh air supplied with a fan. A storage tank compensates for the lost water. To enable reaching higher temperatures without boiling, the system is pressurized. Radiators provide a very large contact surface to maximize cooling efficiency. In spite of that, if outside air is already hot (say 40–45°C) air temperature change after passing the radiator does not exceed a few degrees. Required cooling air volume depends on outside air temperature. In most cases design is based on worst conditions. Therefore for hot outside temperatures very large cooling air volumes may be required. As an example, if outside temperature is assumed to be 45°C, for a 330 kW generator, 11.7 m^3/s fresh air is needed. It is understood that with cooler outside air, temperature change will increase and consequently reduce required air volume.

2.4.3 HVAC Drawings

Several HVAC documents and drawings shall be prepared. Three main drawings include flow diagram, duct and instrumentation drawing (DID), and layout. A brief description of each is given.

Flow Diagram: Shows all locations in which airflow to or out of them are expected. Volume of fresh air, extracted air, recycled air, passage from diffusers, etc. are all clearly stated for each area. Fresh air entrance and exhaust air shall be balanced. Some rooms may be designed with either positive or negative pressure compared to atmospheric pressure. This pressure difference is very small, ranging from 30 to 50 pa. It has to be noted that atmospheric pressure is about 100 kp. This means pressure difference is about 0.05%. This may be achieved by pressure loss through louvers and dampers.

Duct and Instrumentation Drawing: DID for HVAC acts as PID for process. It includes all equipment/instrumentation and their alarm/control signals. Some of the major information in DID to be transmitted to CCR may include:

- What is duty system running status?
- When shall standby system start?
- What are temperature, humidity, and pressure status inside an enclosed area?
- What are toxic gas detector (TGD)/combustible gas detector (CGD) findings? Shall a mitigation measure (like closing dampers) start?
- What control and monitoring signals are transferred to CCR and (in response to them) what signals shall be transferred to UCP for action?

Since HVAC is an important package, normally all operation signals are transferred to its UCP. CCR only monitors its operation. As usual for other packages, ESD and FGS signals are transmitted to CCR and their command signals overrun UCP.

HVAC Layout: This drawing shows actual location of all HVAC equipment. Final dimensions as given by manufacturer, wall/floor penetrations, maintenance, or removal access shall all be clear. Support may be defined in other documents. For example, all equipment (mechanical, process, electrical, safety) supports can be designed in structural drawings. It is important to note that typical support drawings are covered in each discipline. Structural drawings cover specific supports based on each equipment general arrangement drawing and weight.

2.5 INSTRUMENT

In order to control the platform, all important data shall be transferred to the central control room through distributed control system. These data are gathered either from field instruments or packages. Instrument data range from pressure, temperature, and flow of a process fluid/system to possible gas/oil leakage to excess heat or flame and to utility system information, etc. Many different types of instruments with specific functions are used to gather all the required information and transfer them to the control system. They are placed either in-field or in-line. Instruments are gauge, switch, or transmitter type.

A gauge only shows the value of the measured parameter in the field. A switch not only shows the value but can also trigger an action based on the preset limits. A transmitter can show the value and transmit a signal to the control system. Then the control system analyzes the received information and starts the necessary action. The action is based on the cause-and-effect diagram and may vary from triggering an audio/visual alarm to stop/start of a pump, vessel or equipment, or a train to the ultimate stage of platform shutdown.

2.5.1 Instrument Items

Instruments can measure temperature/pressure value, flow quantity, corrosion, and liquid level, detect gas/smoke release or flame, analyze existence of a specific chemical, and signal an action. Some of the major instruments and their functions are listed here:

- Sand probe function is to measure presence of sand in the liquid. The status can be continuously transmitted via signals to the control system. Normally gaseous material does not have so much momentum to bring a considerable quantity of sand to the surface, but liquid material can. The probe can be placed inside the flow. It operates based on the change in electrical resistivity due to surface erosion by sand impact.
- Corrosion probe function is to measure corrosion in the piping system. Its action is similar to the sand probe.
- Corrosion coupon is a piece of metal similar to the piping material placed in the flow and left unattended for a specific duration. Then it is taken out to laboratory and measured to find corrosion rate.
- Level measurement instruments may be of different types. They can be installed in storage tanks (atmospheric) with very little or no turbulence. They can be installed in pressurized vessels with considerable flow agitation or they can be installed in caissons that are connected to the jacket

and their level is affected by seawater level. In turbulent flow, stand pipes may be connected to vessel wall to measure the calm level.

- Pressure measurement instruments are used both for piping line pressure and vessels. In-line pressure measurement instruments may use a venturi concept while those used for vessels may be more of a hydrostatic type.
- Temperature measurement instruments are used for process and utility media in addition to room working conditions. Normally piping is designed for the maximum heat expected from the reservoir. Some control packages may malfunction in heated environment. Therefore interestingly one of the platform shutdown conditions may be due to excessively high temperatures inside control room.
- Flow measurement instruments can be in-line or installed outside it. They are installed anywhere that flow reduction or stoppage may cause a hazard to safety or deviation from acceptable environmental limits or process specifications. Wherever measurement of volume for export is required, a metering system will be installed.
- Different types of analyzers may be used. Some of them are moisture and oil content analyzers.
- Signalers show the position of a valve or passage of a pig, etc.
- Gas detectors cover both explosion hazardous gases like H_2 and toxic gases like H_2S. H_2 may be present in battery rooms and H_2S in all sour service products.
- Fire detectors range from heat to smoke to fire detectors. They shall have suitable sensitivity to distinguish natural phenomena like light reflection on the sea surface from the actual fire.

Based on their function these instruments can be placed in or near a flow line, on a vessel/tank, near a vessel, in a closed compartment, etc. Some instruments shall be visited from time to time and the sample examined in the laboratory to understand the result. Others may transfer a signal or trigger an action automatically. Some instruments need a driving power either as electrical energy, pneumatic power, or hydraulic oil. Operating power or alarm signal may be transferred via a cable. Based on location, different cable types may be used. The driving force like compressed air or hydraulic oil may also be transferred via a tubing system.

2.5.2 Loop Diagram

Loop diagram is one of the documents that helps instrument cabling and termination in the construction yard and is later used by operators during maintenance. Process document "Cause and Effect" shows triggered action due to each event (cause). Instrument cable schedule shows cable number connecting source and destination of each equipment. In that, cable type, number of pairs, size, and route are given. Cable route may pass over several trays in different elevations or through walls, etc. Cable schedule is in a list format.

Loop diagram has a schematic format. It starts from the field instrument as source and then shows which pairs of the cable are connected to which ports of the instrument. From there the cable is routed to the next junction box. It shall show which pairs are connected to which termination point of JB. Several instruments may be connected to one JB. This means that several cables enter to it. Incoming and outgoing connections shall be clearly identified. From JB maybe one cable with larger number of pairs is connected to PCS marshaling cabinet.

Control system fabricator cannot connect all signals via cable to its system. In relatively medium–sized platforms perhaps 3000 signals shall be transferred to PCS. Perhaps two or three marshaling cabinets take source cables from different JBs. It shall be exactly clear which connection of marshaling cabinet transfers what signal. This is needed to induce correct response from control system.

Correct termination of cables is crucial during commissioning and operation. As an exaggerated example, if in an offshore platform cables transferring flowmeter information are wrongly connected to the temperature terminal, the control system will always measure too high temperature and will shut down the platform. Lesser consequences may escape detection during commissioning. This may later become dangerous during operation. For example, assume gas detection signal in room A shuts down damper in room B. This effect may not be detected during commissioning. During operation explosive gas may enter room A with serious consequences.

2.5.3 Instrument Cables and Cable Route Equipment

In this book four disciplines (control, electrical, instrument, and telecom) have been separately explained. However, in many engineering teams at least control and instrument may be combined in one department. Their cable types including core material and protection shields are also more or less similar. Some electrical cables have very high current rating. This means that their cross–section area is much larger. On the other hand, signals of several field instruments may be transferred by one multipair cable. But the essential cable specifications are similar.

The same issue is applicable for cable route equipment like cable trays/ladders (SS/GRP/galvanized). FGS instruments' cable route will have full redundancy. For some, tubing vibration effects may be so high that using SS material becomes mandatory.

2.5.4 Instrument Tubing

Tubing is normally small-diameter seamless pipes made of high-strength material like carbon steel, copper, stainless steel, titanium, Hastelloy, Monel, etc. They have higher flexibility compared to piping, therefore require less fitting. They can be bent to form angles and fit in the required space. They may be placed in bundles or supported on trays without need of special supporting systems. Table 2.4 gives guidance on the selection of tubing based on material, media, and temperature range.

To facilitate tubing selection and usage, several tables have been provided that show the required wall thickness per internal pressure. It has to be noted that increasing temperature will reduce allowable stress. Copper and aluminum have the lowest resistance against temperature. At a temperature rise up to 200°C they will lose 50% and 60% of their resistance.

Table 2.4 Tubing selection guidance

Tubing material	Application	Temperature range °C
Stainless steel (SS316)	High-pressure, high-temperature, generally corrosive media	−255 to +605
Carbon steel	High-pressure, high-temperature oil, air, some specialty chemicals	−29 to +425
Copper	Low-temperature, low-pressure water, oil, air	−40 to +205
Aluminum	Low-temperature, low-pressure water, oil, air, some specialty chemicals	−40 to +205
Monel 400	Recommended for sour gas applications well suited for marine and general chemical processing applications	−198 to +425
Hastelloy C − 276	Excellent corrosion resistance to both oxidizing and reducing media and excellent resistance to localized corrosion attack	−198 to +535
Inconel alloy 600	Recommended for high-temperature applications with generally corrosive media	−130 to +650

Downloaded and summarized from Parker—Hannifin website.

Temperatures up to 150°C have no impact on stainless steel material. The same temperature will reduce 10% of carbon steel and about 20% of Monel resistance.

2.6 MECHANICAL

This discipline is responsible for procurement of the main packages on a platform. For each package they shall provide data sheets, specifications, material requisitions, and technical bid evaluation (TBE). Normally procurement stages can be classified as follows:

- Prepare package or parts of the package particular specifications and data sheets, receive client comments and implement.
- Prepare general applicable specifications and data sheets like minimum skid package requirements, minimum nozzle loads, and pressure vessel general requirements and coordinate with other disciplines to issue related documents like electrical equipment in packages specifications, instrument and control specifications for packages, etc.
- Coordinate with piping and structure to fix estimated package dimensions and weight.
- Issue specifications (SPCs)/data sheets (DSs) in a material requisition package with other required documents to EPCIC contractor to get vendor quotations
- Review vendor quotations with help from other disciplines and issue technical clarifications
- Issue TBE and based on EPCIC contractors' instructions issue material requisition for purchase.
- Review received vendor data and comment on related subjects.
- Implement vendor final data in the project engineering documents.

2.6.1 Data Sheets

Data sheets define the basic information required to define each package. Normally it follows standard formats. For static equipment (like vessels), rotating equipment (like pumps) and packages different formats are used.

A typical Vessel data sheet may include:

- Design data like operating and design temperature/pressure, fluid type and main specifications, major nondestructive testing (NDT) requirements, environmental loads, etc.

- Material type for shell, head, nozzles, flanges, gaskets, internals, skirts, saddles, external attachments, stiffener rings, and anodes. Some of them may not be applicable to all vessels.
- General construction items like type of head forming, support, whether manway is required or not, etc.
- Operating conditions may only be applicable for process vessels. It may include fluid composition, temperature, pressure, molecular weight, expected GOR, water cut, etc.
- Vessel general arrangement in consultant data sheet is preliminary. However, as per PID it will show location, order, number, and calculated size of nozzles, vessel preliminary dimensions, etc. For storage tanks, vendor is only responsible for mechanical design and size is determined by consultant. For process vessels, design is very much dependent on the internal design, which is normally vendor proprietary design. Therefore vendor shall design everything including accurate determination of nozzles' location.
- All specific project requirements on material selection, fabrication procedures, NDT requirements, postweld heat treatment (PWHT), painting, corrosion protection, handling, hydrostatic testing, etc. shall be given as "Notes."

A typical pump data sheet may include:
- Operating data like operating/design temperature and pressure, maximum and minimum fluid discharge rate, net positive suction head, etc.
- Liquid specifications like specific gravity, viscosity, etc.
- Environmental data, which may be very brief and refer to another document.
- Construction data like type of pump, nozzle size, and rating, etc.
- Material type for different parts of pump.
- Power consumption and available feeders, maximum overrating capacity.

A typical package data sheet may include:
- All of the previous items for vessels and motors inside package battery limit.
- Specific instructions for procurement and installation of electrical, instrument, and control items inside package.
- Specific instructions for interface with platform electrical, instrument, control, and piping.
- Specific instructions for future maintenance and repair requirements.

2.6.2 Specifications

Specifications are documents that describe technical and some contractual items applicable in design and purchase of a package. For example:

- Applicable standards that are both technically and contractually binding
- Design data will be normally specified in data sheet but main design criteria and guidelines will be given in specifications
- Material selection guidelines for different parts of a package, for example, does it have to satisfy The National Association of Corrosion Engineers (NACE) requirements? Are there any additional criteria due to H_2S presence, etc.?
- Applicable NDT requirements like type/frequency of NDT, repair actions, etc.
- Project inspection requirements like certifying authority, third-party authority, purchase supervisor, etc.
- Project factory acceptance test and handling/transport requirements

Package-related specification and data sheet are the two main documents showing vendor SOW. For example, in pressure vessels specific internal lining (cladding) to protect against corrosion may be instructed. Specific welding, fabrication recommendations, or restrictions may be given. One common issue for clad pressure vessels is the maximum allowable FE content at the intersection of two clad strips. This limitation will impact electrode selection, welding procedure, number of cladding layers, etc.

For process vessels, in addition to mechanical design specific guarantee for process design may be required. This shall only be done by a competent process house designer.

2.6.3 Material Requisitions

Material requisition in itself is a contractual and procurement document. It may be divided into two separate commercial and technical documents or a single document. Normally contractors or vendors have their own fixed format for sales or purchase terms and conditions. This document is prepared by their procurement and financial department independent of the technical department. Therefore they may issue it separately from technical material requisition (MR) in a request for quotation (RFQ) or purchase order (PO) letter. MR content in RFQ and PO is slightly different.

- Reference to vendor original quotation and clarifications/meetings. Items confirmed by vendor will constitute their obligations (only in PO).
- SOW definition for the number and pieces of packages/items in vendor responsibility

- Particular documents like SPC, DS directly applicable to the package
- General documents like environmental data. Project specifications for electrical, instrument and control items, project painting specs, project layout to show package boundary limits
- Sample documents like ITP, inspections, etc.
- List of suppliers' documents
- Specific information to be provided by vendor like center of gravity (COG), handling procedures, etc.
- Priority of documents in case of conflict
- Payment terms

2.6.4 Interface With Other Disciplines

Mechanical discipline cannot perform the intended activities without technical support from other disciplines. Each package has process, instrument, electrical, and piping information that need to be reviewed by specialists in each discipline. This is done via interdisciplinary reviews. Of course mechanical discipline is responsible to summarize/combine them on the issued document by vendor.

Interdisciplinary activities may cover a wide range. It may include data from:

- Process, like inlet and outlet crude composition and data
- Electrical, like power source (AC, DC, 400 V, 230 V, cable or gland size, etc.)
- Instrument/Control, like signals from detectors (TGD, CGD), gauges/switches/transmitters (temperature, pressure, flow), alarms on package malfunction to controlling (start/stop) commands, etc.
- Piping connections, like drains, hydrocarbon fluid/gas, vents, etc.
- Structure, like access and supporting system

This information can be input in project specifications and data sheet or checked in vendor document or explained in the meetings.

Majority of packages and all vessels are procured by mechanical discipline. However, they take the required specifications and data from other disciplines. In many projects mechanical discipline procures the most expensive items of a platform. Costwise procurement of a platform's equipment may be more than 50% its total value.

Each package needs specification and component data sheets. After upgrading, they are included in material requisition to be issued to get vendor quotation. This is the first step in purchasing packages.

Two different approaches are followed in preparing data sheets. In one of them mechanical, process, and electrical disciplines each prepares separate data sheets. Instrument may add their requirements/input in electrical data sheets. In this approach, managing to get data sheets issued on time by different disciplines is very difficult. Then client shall comment on them and the response be issued to upgrade data sheets before using it in the material requisition. To the author's understanding it is very difficult (if not impossible) to get approval of data sheets for all packages in a timely manner. For example, for a certain package, mechanical data sheet will be approved while process data sheet is not ready. For another package, electrical data sheet will be ready but others not ready. This improper coordination prevents issuing material requisitions.

The other approach is to combine all disciplines data sheets in one document. The discipline responsible for the package will circulate the relevant data sheet between other disciplines. They will input their information and only one data sheet having all the required information will be issued to get client approval. This way material requisitions may be issued properly.

Package specifications and the general specifications shall also be prepared by consultant and approved by client. A material requisition shall have all documents whether directly or indirectly related to the package. Vendor shall have full information to provide a complete package.

After receiving vendor quotations, mechanical discipline shall coordinate to transfer required data to relevant disciplines and get their technical clarifications. Normally consultant is not allowed to directly contact vendors. Some EPCIC contractors have strict financial regulations, which do not even allow their own engineering team to know vendor proposed cost.

After clarifications are complete, TBE shall be issued. Commercial department of EPCIC contractor also issues their commercial bid evaluation (CBE). In some cases the end user may reserve the right to reject some of the vendors even though their proposal may be technically acceptable. Normally each operator has its own approved vendor list (AVL). Comparing TBE/CBE and client comments (if any), EPCIC contractor shall select the winner. After this stage consultant is authorized to issue material requisition for purchase.

After issuing purchase order vendors start to design their package. At this stage close cooperation between consultant and vendors to check documents and approve their conformity/interfaces with each other and as per platform layout is necessary. After approval, vendors can start fabrication.

2.7 PIPING

Piping discipline is responsible for the 3D model. Platform model consists of many different items. For example, it may show main, secondary, and part of tertiary structural members, vessels, equipment/packages' location and layout, control valves, piping path, electrical and instrument cable tray or ladders, lighting, rooms etc. The model helps several purposes like:

- Enables clash check in order to avoid unwanted change in piping route, which may result in rework and/or material loss/shortage.
- Ensures equipment/instrument elevations are as per plan.
- Ensures suitable maintenance and handling spaces have been foreseen.
- Ensures safety equipment is located near the place it should be.
- Facilitates extracting piping MTO.
- Facilitates extracting piping isometric drawings.
- Facilitates preparing piping supports drawings.

With advent of 3D modeling software like Plant Design Management System (PDMS) and Plant Design System (PDS), all of the purposes listed here can be achieved in virtual space. This is a great advantage, which facilitates material ordering, accelerates construction site work orders, and eliminates clashes and rework in yard. Every single item located on the platform can be modeled exactly. Of course, the accuracy of the model depends on the accuracy of the input information.

The first step is selecting complex or platform layout and the second step is equipment layout.

2.7.1 Complex/Platform Orientation

As explained in chapter "Introduction to Offshore Platforms," a group of offshore platforms each serving a purpose (production, processing, providing utility services, LQ, etc.) that are connected to each other and controlled by one operator team is called an offshore complex. Selecting different platforms' location and orientation is based on their intended purpose and environmental considerations. Several disciplines are involved in decision-making.

Well/s' geographic coordinates are determined by reservoir studies. Other disciplines are not involved at this stage. After fixing the location of the production platform, its basic layout criteria will be determined. Some of them have to be decided by the operating company. Decisions are governed by evaluating financial aspects. For example, any production well requires some well maintenance work. This is not a daily activity. Some wells may need maintenance only once per several years. In shallow and medium–depth water (up to ∼120 m) the operator may decide to do maintenance by jack-up. In deep water, either a permanent or temporary rig may be used. They may have their flare system or install a burner boom on the topside. Jack-up day rate is high and a mobilization fee is needed. In addition, it needs a supporting fleet. Capital expenditure (CAPEX) needed for installing a permanent drilling rig is very high. Also, it needs a supporting fleet for operation. In some platforms while provision is made for future jack-up operation, the structural system is designed for temporary drilling rig installation. Based on these requirements criteria like well bay area location, jack-up approach, or rig/or burner boom installation point may be decided. Then the process and utility area may be identified. Chakrabarti [5, Chapter 10] summarizes API RP 14J guidelines on this issue.

At this stage, process, piping, and structure become involved. Based on environmental conditions like wind, waves, and currents, location of process platform and living quarter may be decided. Process platform shall be located in a direction that winds blow away the gases and hazardous materials from others. Living quarter shall be at windward side. This means fresh wind shall blow first to LQ and from there to wellhead platform (WHP) or production platform (PP) or flare.

Graphical representation of wind distribution in different directions for each velocity group is called wind rose. A sample wind rose is shown in Fig. 2.13 and Table 2.5. This was originally derived for an offshore location in the Persian Gulf. Some figures are slightly modified while keeping the relative ratio in each direction more or less constant.

In old books [22,23] wind rose was shown as a continuous curve for each speed group. This concept better matches actual measurements. Now they are shown as boxes only in the reported direction. Check wind rose in Fig. 2.13. This representation does not mean 32% wind is exactly blowing only from NW direction (45 degree to left of north) and at say 50 degree to left of north there is no wind. In fact all measured winds blowing from 22.5 to 67.5 degree left of north have been included in NW region. Assuming uniform distribution in the 45 degree area is useful and simplifies calculations. In reality if percentage of wind blowing from W is compared with N it is seen that N has a higher rate of wind. This means wind-blowing concentration toward N is higher than toward W. This does not happen abruptly. It is a gradual shift in the measured sectors.

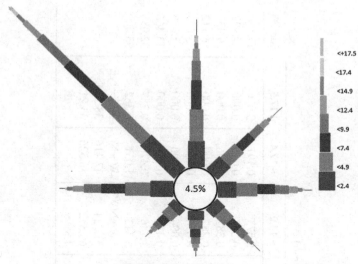

Figure 2.13 Wind rose.

Wave rose more or less follows a similar trend. It is to be understood that due to the effect of swell waves (which are generated long distances away from the location), wind and wave rose may not be exactly similar.

Wind rose governs toxic gas, smoke, heat, and radiation dispersion. Wave rose governs structural orientation and boat landing location.

Current is divided into two main portions. One is current due to wind and the second is due to tidal variations. Diurnal or semidiurnal tides may occur in different areas. Offshore platforms far away from shorelines normally experience one low/hightide cycle per day. Current direction reverses at these two conditions. Only in one project the author has seen tidal current charts predicted for 1-, 10-, and 100-year return period for a new offshore location [24]. In a majority of the reports only highest astronomical tide (HAT) and lowest astronomical tide (LAT) are reported for structural calculations.

In chapter "Introduction to Offshore Platforms," two approaches were explained. The first approach ensures smoke, heat, and radiation are in prevailing wind direction and blown away from the complex. In brief this means, "maximum time away." It has to be noted that "prevailing wind" does not mean maximum wind velocity. It means direction that the maximum percentage of wind is blowing from it. Although in a majority of cases maximum wind velocity may also blow from the same direction, in general this may not be true.

Table 2.5 Wind rose table

Geographical direction	Wind speed groups (m/s)								Sum %
	0–2.4	~4.9	~7.4	~9.9	~12.4	~14.9	~17.4	+17.5	
N	3.725	4.393	4.107	2.674	1.433	0.573	0.191	0.096	17.190
NE	2.292	2.579	2.292	1.624	0.764	0.287	0.096	0.000	9.932
E	2.101	2.388	2.197	1.528	0.764	0.382	0.096	0.000	9.455
SE	1.337	1.433	1.433	1.051	0.573	0.287	0.000	0.000	6.112
S	0.955	0.860	0.860	0.573	0.287	0.096	0.000	0.000	3.629
SW	0.860	0.955	0.860	0.573	0.287	0.096	0.000	0.000	3.629
W	2.865	3.343	3.152	2.197	1.242	0.478	0.191	0.000	13.466
NW	6.685	7.927	7.545	5.253	2.770	1.146	0.478	0.287	32.088
Sum %	20.819	23.875	22.443	15.471	8.118	3.343	1.051	0.382	95.500
Sector	N–E	NE–SE	E–S	SE–SW	S–W	SW–NW	W–N	NW–NE	
Sum %	23.254	17.477	12.654	8.500	12.176	31.324	47.416	38.200	
Centroid	39.387	41.311	37.231	40.070	58.632	60.332	46.326	35.213	
Centroid is taken from the first direction			Calm Weather %		4.5				

A review of wind roses generated for two points in the Persian Gulf (approximately 250 km away from each other) shows the difference. In the first location ($28°8'N$ and $50°4'E$) wind is blowing 52.59% from Northwest direction and 16.34% from Southeast. However, maximum measured velocity in Northwest direction is 23.35 m/s while from Southeast is 25.5 m/s. In this area maximum wind velocity is blowing from a direction exactly opposite prevailing wind.

In the second location ($26°46'N$ and $52°5'E$) wind is blowing 33.6% from Northwest direction and 6.4% from Southeast. Maximum estimated 100-year return (1 min) wind velocity in Northwest direction is 36.7 m/s while from Southeast it is 35.2 m/s. In the second area, maximum wind velocity and maximum blowing duration are in the same direction.

Normal wind blowing measurement and statistical extrapolation (for higher return periods) is performed in 45 degree sectors. For example, all winds blowing from $-22.5°N$ to $+22.5°N$ are reported to be from north ($0°$). It is clear that wind distribution in this 45 degree sector is not uniformly distributed. Some of the recent environmental measurements have been reported in 22.5 degree sectors. Instead of eight direction measurements/calculations, the report is generated in 16 directions.

The first proposal of the author is to use "centroid" concept instead of "center." In centroid calculation it is assumed that half of the wind percentage blowing in each direction belongs to the right of the arrow and half to the left. For example, in Table 2.5 prevailing wind is blowing 32.088% from NW sector. It is assumed 16.044% is blowing from WWN to NW and the remaining 16.044% is blowing from NNW to NW. By this assumption, 45 degree measurements are further divided to 22.5 degree sectors. In this smaller sector, wind blowing is again assumed to be uniformly distributed. Reduction of sector size may increase accuracy.

In Table 2.5 the centroid of a sector is defined by an angle from its first geographical direction. For example, NE sector is $90°$ wide. Its center is $45°$ from north. Calculations show that centroid is located $39.4°$ from north. This direction has about $5.6°$ difference with normal assumptions, which shows centroid is more inclined toward north. The reason lies in the fact that wind blowing from the north is 17.19% while blowing percentage from the west is 13.47%. In this table maximum difference is about $15.3°$ (toward NW direction) for SW—NW sector. The reason is wind from SW is only blowing 3.63% of the time while wind from NW is blowing 32.09% of the time. Table 2.5 shows $90°$ sectors with $45°$ overlaps. For each specific wind rose, values of centroids shall be calculated.

Selecting 90 degree or narrower sectors can be studied using CFD analysis. If release source is near the platform with low agitation, the dispersion sector will be smaller. For longer distances and turbulent conditions, gases will be distributed in a considerably wide sector. At the same time gas concentration reduces by inverse of second power of distance. Therefore very large sectors may be useless. Flares are placed at relatively long distances from production platform or offshore complex. The angle connecting flare to platform or a complex may not be so large. For example, if a platform has 35-m width and is placed 160 m away from flare, the connecting angle will be approximately 12°. This angle will increase considerably if two PPs or one PP and one LQ are bridge-connected to each other.

To determine different platforms' locations and global orientation of an offshore complex, several main criteria shall be investigated. They include at least the following:

1. Flare/vent location shall be such that flame, heat, radiation, smoke, and gases are blown away from the platform.

   • First method to select orientation may be to place flare platform downward of prevailing wind or downward centroid of the prevailing wind sector. Radiation and dispersion studies shall be performed to find a suitable distance from platform. This is to ensure that whenever wind is blown toward platform, heat radiation ratio or concentration of gases reaching platform are reduced in level or diluted enough not to cause any harm to human safety or operational condition.

   • Second method (which may be used when bridge connection or submarine line to flare platform places it sufficiently away from complex) is to place it at right angle to prevailing wind directions (centroid of prevailing wind sector). This may be used in a condition when wind in perpendicular direction toward offshore complex is not very high.

     Both approaches of placing flare downward of prevailing wind direction or perpendicular to it have been used in South Pars Phases in Persian Gulf and have been working for many years.

   • The author's second proposal (when CFD analysis is not done) is to use a third method and place flare in the centroid of the sector with "minimum time toward" platform wind. It has to be noted that sectors having "maximum wind away from" or "minimum wind toward" platform may not necessarily happen in opposite directions of one line. In Table 2.5 prevailing wind sector is WN and wind blowing percentage in this 90 degree sector is 47.42%. Centroid of this sector is located 1.3 degree toward north compared to its center direction. SE—SW sector blows only 8.5% toward platform. Centroid of this sector is located about 4.93 degree toward SE. These two directions form an acute angle equal to 38.77 degree. This difference was calculated considering centroid directions. If they were placed at center directions, the difference would have become 45 degree. Since selected eight directions have 45 degree overlap with others, total sum of blowing wind plus twice calm weather value sums up to 200%.

2. Well bay location shall be easily accessible by jack-up and away from boat landing. This means that it cannot be on the side with bridge connection to flare or boat landing. In a complex a boat landing is mainly needed on LQ. This does not mean boat landing is not installed on wellhead platforms. All three platforms shown in Fig. 1.1 have a boat landing. Some of the platforms in Fig. 1.2 do not have a boat landing. The main reason is difficulty in maneuvering to berth on internal platforms.

3. Process and utility areas shall be separated from each other to enable creation of a safe area. This means that with respect to wind, process area shall be located downward of the utility area or in perpendicular line to it. This ensures that in normal conditions hazardous gases are blown away from utility. In the process area sources of ignition (like fired vessels, diesel engines, etc.) shall be located away from fuel sources like storage tanks.

4. Pipeline and riser approach shall be away from jack-up approach and free from dropped objects. Jack-up heavy spud cans may damage pipeline. Operators exert utmost care not to drop anything from pedestal crane on the pipeline. Similarly, supply boat captains normally do not drop anchor during berthing. In spite of all these efforts, consequences of a damaged pipeline are so high that it is always better to keep pipeline approach away from boat landing side.

5. Helideck location shall have 210 degree obstacle-free zone. Helicopter approach direction shall be such that wind is not pushing it toward platform. This means choppers may land only with a working engine. The same philosophy is applicable to supply boats. It means currents, which are more or less in the wind direction (other than tidal currents), and winds shall push them away from boat landing. This is to ensure any

tugboat in platform vicinity with a shutdown or damaged motor will be drifted away from platform. It is to be noted that pedestal crane shall be directly installed on top of the boat landing and its elevation violates helicopters' 210 degree obstacle-free zone. Therefore these two conditions cannot be satisfied on the same platform. However, since supply boat arrival is more frequent than helicopter and it brings more cargo, normally the boat landing satisfies this criteria and helipad is designed in an angle that 210 degree obstacle-free zone permits.

6. Location of emergency evacuation vessels (lifeboat/life raft) shall be such that after dropping to sea, wind/waves and current help to move them away from platform. This will ensure if lifeboat motor is damaged escaping personnel will not be trapped under platform legs. At the same time, personnel who are using life raft will be able to paddle away from platform without excessive effort.

7. Main escape route arrangement shall avoid hazardous area. At least two independent escape stairs shall be predicted. They shall be at the furthest points away from each other. This ensures operators at any location will find their nearest route to safety. Besides, if there is any accident like fire or explosion that block access to one of the escape routes, the other one is still accessible.

It is very difficult (if not impossible) to fulfill all these requirements simultaneously. In most cases a compromise between pros and cons of a proposed layout is accepted. Criteria with highest priority will satisfy required conditions and remaining items selected as much as possible near to project demands. From Table 2.5 it can be seen that in this location maximum wind is blowing in a sector from north to west. Centroid of the maximum blowing sector toward platform is located 46.3 degree from the west. This direction is fairly aligned in prevailing wind direction, which is northwest.

One possible solution for this area may be to orient the platform north and jack-up approach toward geographical northwest and boat landing in northeast. Another solution may be to keep platform north and boat landing in the same location but for the flare to be in south direction. A third solution may place flare in the direction of minimum wind toward platform. This direction is shown to be SE—SW sector. Centroid of this sector is about 4.93 degree from south to SE. After approving platform or complex layout, engineering team may proceed to prepare equipment layout.

2.7.2 Equipment Layout

Preparation of equipment layout is a major task that involves several disciplines and requires a proper interface management system. Although piping discipline will be responsible for the final issue of the document, almost all disciplines are involved. Although layout design will not start from scratch, it needs several rounds of back-and-forth information between disciplines, client, and finally vendors. General dimensions and layout are fixed in conceptual and basic design stages. The best goby is similar platform layouts. Architectural, safety, mechanical, structural, process, electrical, and instrument are all involved [2,5,18,25−28].

Layout design may start from a block design showing general dimensions and location of packages. At this stage, platform process flow diagram (PFD) is followed. Considering location of wells, piping discipline obtains general dimension of Xmas Trees and packages from drilling contractor, mechanical, safety, and instrument disciplines and tries to follow crude flow path as much as possible similar to the PFD. At this stage input from architectural, mechanical, and safety is required to distinguish and separate escape routes, maintenance spaces, accommodations, firewalls, etc.

In selecting a package location, their role in PFD shall be investigated. For example, if flow from vessel 1 to vessel 2 is by gravity then vessel 2 can be located in a lower level. This may be applicable for flow from diesel day tank to generator or freshwater tank to consumers like kitchen, accommodation area, etc. If a crude temperature has to be increased or decreased via heat exchanger, then they have to be located near each other to reduce piping length.

Platform modeling follows crude route. Crude is transferred to surface either with reservoir pressure or via artificial lifting. It passes drill pipe and enters Xmas Tree. One safety valve is below seabed and two safety valves are located within Xmas Tree block. It then passes choke valve and connects to either production or test manifold. Before connecting to manifold, a PSV is located as safeguard against well overpressure due to pressure fluctuation. Based on platform PFD, if processing is not done in the platform, crude may be directly routed to export line. If processing is done on the platform, then crude may pass through first- or second-stage separators. Gas may pass dehydration stages before being compressed for injection or export. Crude may pass coalescing vessels to separate water. Extracted oil will be directed for export. Oily water will be treated before dumping overboard. Several utility systems will be used to enable proper process handling. Main utility systems include power generation, wellhead control panel, compressed air system, fuel gas, etc. Flare is considered to be a process system. Without flare becoming fully functional, wells are not started up.

In the second stage, components of each package like horizontal/vertical vessels, pumps, compressors, etc. are defined and their dimensions are given. Those components that shall normally be fabricated by vendor on a single skid are identified. This arrangement will be informed to vendor. It is vendor's duty to confirm if this is okay or if they need to be separated. However, before starting any fabrication by vendor, piping discipline shall confirm/approve suitability of the proposed layout by vendor.

1. For each package size, number and dimensions of major interconnections and tie-in coordinates to other platform facilities, packages, and utilities are identified. Inside package vendor has freedom to route piping but shall combine similar piping inside package and provide connecting tie-in point at predefined flanges at package edge. Assume a package has several drains of different sizes. It is impossible for platform piping to provide a dedicated connection for each drain. Vendor shall combine all drains with similar pressure and material type in a manifold and provide a single flange at edge of package. At this stage some relocation/rearrangement may become necessary.

2. Xmas Tree size affects well bay area dimensions. Sufficient space to remove valves (or their actuators), flanges, connecting lines and free operator movement shall be considered. One important issue is up and down movement of Xmas Tree due to fluid temperature. When crude flow enters platform piping system, drill pipe reaches to operating temperature. Operating temperature depends on reservoir conditions. In majority of cases it is much higher than ambient temperature. When crude flow stops, conductor and later drill pipe temperature approaches seawater. This temperature variation may cause up to 10 cm movement. Within this range there shall be no clash between Xmas Tree branches and platform structure, piping, trays, etc. Well intervention becomes necessary after a few years of operation or even earlier. For this purpose, space above Xmas Tree to hatch covers shall be left unobstructed.

3. Cables (if any) and tubing connected to Xmas Tree valves shall also be able to tolerate this movement. Leaving a free cable length is very easy. The main point is in hydraulic oil tubing connected to actuators, if installed in a cantilever configuration they may tolerate displacements very easily.

4. Sufficient area shall be foreseen around packages/vessels for issues like:
- Space to move transportation trolley and store goods. Assume trolley dimensions to be 900 mm W × 1900 mm L. One may transfer chemicals in 20-L bottles with this trolley. Before discharging bottle contents to the package, first trolley shall be removed from escape route, then operator shall be able to stand alongside it and move bottles one by one.
- To move trolley from a point to the other point of the same deck in platform and maneuvering in end of corridors, care shall be exercised to provide some clearance. For the example trolley at least length equal to $1.9 + 2 \times 0.05$ and width equal to $0.9 + 2 \times 0.05$ shall be considered as shown in Fig. 2.14. For this condition, calculations show if both corridors have 1.2 m width, maximum length of a trolley with 1.0 m width (which is able to pass) is equal to ∼1.4 m. The optimum passage angle is obviously 45 degree. If width of one corridor is increased to 1.3 m, the same trolley may have a length of ∼1.53 m and the optimum passage angle will be 41.87 degree.

Optimum angle (α) for passage is the one giving zero value for derivative of length (L') in the following formula. Maximum trolley length that is able to pass this corridor is found by inserting related values in formula for L. Parameter definitions are as shown in Fig. 2.14.

$$L' = \frac{-a\cos^3\alpha + b\sin^3\alpha + c(\cos^2\alpha - \sin^2\alpha)}{\sin^2\alpha\cos^2\alpha}$$

$$L = \frac{a\cos\alpha + b\sin\alpha - C}{\sin\alpha\cos\alpha}$$

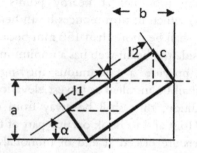

Figure 2.14 Maneuvering trolley in corridors.

These equations can be solved by trial and error and are very useful in calculating minimum width of a door to allow equipment entrance to any area.

It is worth mentioning that operators do not follow these calculations in routine daily activities. By tilting trolley or transfer from one to another, these limitations can be overcome to some extent. But it is a useful tool in defining door and corridor widths.

- Space to remove internals from manholes of each vessel is needed either at ends or alongside each vessel. In all vessel specifications (which need PWHT), a condition is set for design of internal parts so that they are installable with bolts and removable from manhole opening. For heat exchangers, tube bundles may be removed as a single piece from top or bottom of the vessel. Electromotor part of pumps shall also be removable.

5. For pigs, receiver/launcher space for pig handling shall be considered. Intelligent pigs need much larger space and lifting devices. A special trolley may be needed to place intelligent pig on top of it while pig launcher door is opened to enter the pig.

6. Generators and compressors need space for removing and reinstalling cylinders. Moving parts normally need more frequent maintenance. First, engine cap shall be removed and then cylinders can be lifted using hoist. Since several cylinders are placed in a row and they shall be moved to a handling space after the engine, installing a monorail on top of it will be much easier.

Fig. 2.15 shows a hoist and trolley with 2-ton safe working load (SWL) installed on a monorail. Chain collector box, hook, and flapper lock are clear.

Fig. 2.16 shows side view of a hoist moving on monorail above a typical diesel generator. Monorail beam height depends on its structural design. Supporting monorail at nearby points can reduce design bending moment, which in turn reduces beam height. For this figure, minimum height shall be more than 150 mm because part of the hoist that is moving inside the beam web has a minimum height of 115 mm. Monorails shall provide a continuous moving path for trolley. Therefore they shall be installed at lowest elevation of roof structural members. In addition, they shall be away from other obstacles like pipes, trays, etc. Hoist and its hook occupy part of free elevation below monorail. Stoppers are placed at end of monorails to limit its movement and prevent hoist falling. In this sketch from 4500-mm room

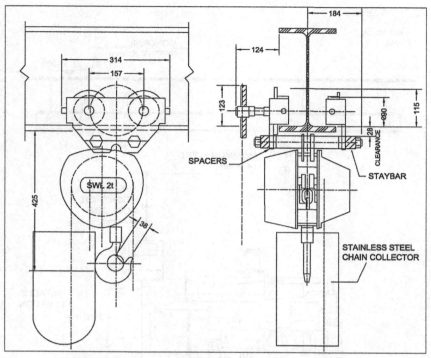

Figure 2.15 Hoist with trolley.

height only ~1270 mm is available for free transportation of parts. For each specific condition actual dimensions shall be checked.

7. Normally electromotor of submerged pumps is installed above deck while its driving parts are much deeper inside caisson. To transfer rotating action, a long shaft is needed. Based on available space between deck where electromotor is installed and deck above it, shaft shall be divided into several pieces flanged together. Sometimes it may be necessary to install a hatch cover and connect a lifting device to the next deck.

8. Firefighting hoses' movement in the access corridors is another necessity. When empty, 2″-size hoses can be easily moved. But when valves are opened and water with perhaps 10–16 bar pressure enter the hose, it will become very difficult for handling and may hit instruments.

9. Free space may be needed on top of chemical storage tanks. Some of them are filled with pumping from a tank stationed on the supply boat.

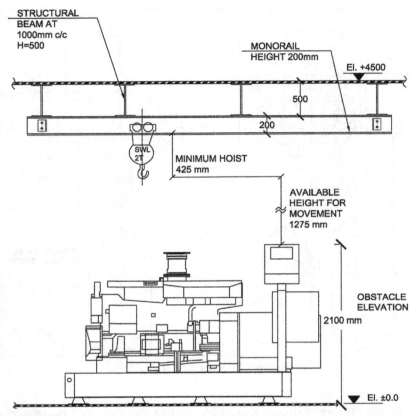

Figure 2.16 Cross-section of a typical hoist-and-trolley movement on monorail.

For others tote tanks shall be transferred to a higher elevation and gravitational flow or hand pumps may be used. In each case hazardous properties of chemicals shall be carefully checked. For each chemical a material safety data sheet is issued. This document contains information on chemical components, properties, and their effects on humans if inhaled or swallowed or in contact with skin, eyes, and sensitive parts of the body. Before any action, all safety precautions for chemical handling shall be ready.

10. High-pressure gas cylinders are used for different purposes like CO_2 for firefighting and N_2 for blanketing or activating solid material flow. They are stored in 200 bar cylinders. Although N_2 can be generated from air, in many cases cylinders are filled onshore and transported via supply boat. Cylinder removal from supply boat, transfer to storage

area and from there to package location, and removing/replacing empty bottles shall follow operation procedures. Sufficient space for required maneuvers in safe conditions shall be provided.

11. Well bay area layout shall be carefully planned to enable access to different gauges, valves, piping, etc. Xmas Tree design may need one or two elevations for access. Operators may have to do different actions like monitoring a gauge, manual opening/closing of valves by turning wheel, opening flange bolts to remove a valve or connection, etc. Two operators can help each other to place items lighter than 300 kn on transfer trolley. For heavier objects, winches and hoists shall be used. Future well maintenance operations need access from top. Normally hatch covers are placed above each well at highest elevation. Access to inside well is provided from top. A cylindrical area with a diameter larger than Xmas Tree is needed to be free from any obstacle and clash starting from top of Xmas Tree to main deck.

12. Choke valves may weigh more than 1700 kg, BDV may be up to 800 kg, Normally in a multi-well platform these valves are placed in a row. A monorail with suitable access platform can provide a good handling route.

13. A vessel has a considerable number of nozzles. They may consist of inlet/outlet of gas, water, and hydrocarbon to temperature/pressure and level measurement gauges, manholes, etc. Each nozzle may have connecting flanges, piping and valve ranging from small 2″ drains and vents to large 24″-access manholes with their heavy lifting anchors. Valves like PSV, SDV, and BDV may be placed around vessels. Their function may be pressure relief, isolation, blowdown, drain, sampling, etc. Required access for them may be different. But all have one thing in common, which is enough space for bringing repairing tools and performing various maintenance activities. In some cases an area equal to vessel diameter shall be free.

14. Offshore platform equipment may either contain flammable material like storage tanks or have ignition sources. These two shall be separated as much as possible or a barrier placed between them. Barriers may include steel wall, firefighting equipment, etc. For a specific volume, rectangular tanks occupy less space. This shape is normally used for atmospheric tanks. For high-pressure vessels, a cylindrical shape is suitable to reduce wall thickness.

15. Duty generator exhaust is very hot. MONEL material may be used, which shows about 25% change in yield stress even above 500°C.

Their high temperature may be used in heat exchangers, otherwise they may be cooled down by natural ventilation. If hydrocarbon or any flammable material contacts this hot surface, it may reach ignition point and start a flame. Therefore in addition to protection for human contact, the piping route above it shall be carefully selected to prevent any leakage and hydrocarbon drop above it.

16. Some equipment like ESDVs shall be operative during fire condition. They may need a fire-protection coating. Since valves may undergo periodic maintenance visits, fire-resistant coating shall be of removable type. Therefore spaces both in front and sides shall be reserved for removing and reinstalling this jacket.

17. Valves (especially those on main export line) may be very heavy. A 32″ size ESDV (rating 900) may weigh more than 20 ton. Transferring it by trolley from top of secondary beams may not be possible. Transport trolley capacity and secondary beam strength may both be the limiting criteria. Therefore parts of it may be removed. Remaining heavy parts may be lifted using winches or hoists connected to padeyes preinstalled on structural points. To distribute load up to four padeyes may be used. Load share and chain direction may vary during movement. Worst conditions to sustain the load shall be used in design. In addition, sufficient headroom shall be allocated. Changes in chain direction from start to stop shall be studied. This will govern padeye direction to ensure minimum transverse load is applied on it.

18. Packages' skid is normally welded to deck plate. Two limiting conditions in this design is an empty package during offshore transportation and an operating full package affected by seismic loads. In both cases design criteria identifies acceleration at base elevation. To apply lateral loads in skid design, specific acceleration at actual location shall be selected.

19. Any package may need repair, which is done during temporary (pre-planned) shutdown. Isolation valves shall be installed at inlet/outlet of all tie-in points to platform. In addition to packages, removable spools shall be placed at inlet/discharge of all pumps to enable their removal.

In the third stage, as-built vendor data will be reviewed and checked with the model. Normally no relocation is acceptable at this stage. However, there may be some small changes in piping routes/fittings/sizes, etc. These changes shall be incorporated in the model and, if needed, related documents revised.

2.7.3 3D-Model Preparation/Review

The intention of this review is to make sure proper access is provided for maintenance and operation and clashes are removed or minimized for fabrication. This model provides piping isometrics and MTO. After detail design stage and later in construction engineering stage, the same model will be further refined and used to generate spool piece designs and work orders.

Normally two stages for 60% and 90% model review are foreseen. In some cases (based on the contract) client may require additional review stages. The distinction between the two review stages is not so clear. Some take it as per project timing intervals. Some take it as project progress. Based on the author's experience it is better to define some minimum requirements for engaged disciplines. Piping is the main discipline responsible to arrange the review, prepare clash report, and implement the comments. Structural shall prepare their model in some 3D software like BOCAD and then transfer to piping 3D software like PDS or PDMS. In addition, electrical/instrument shall model their cable trays/ladders and their supports. Main supports are modeled with their accurate size and dimensions. Small supports are only modeled by a box to prevent clashes, then typical drawings are prepared to enable yard fabricate them. Mechanical discipline is also involved in review. Mechanical team will provide equipment data to be used in modeling by piping. But electrical and instrument themselves shall prepare cable trays and assign cable identification tags.

Exact/universal definition of model review stages is very difficult if not impossible. Final model shall be based on as-built information. This may be prepared long after detail design is finished by construction engineering team. As-built information prepared by fabricator shall include both vendor data and actual fabricated platform with all structural, piping, and other disciplines' input. Construction stage models shall be accurate enough to prevent major clashes. In some projects vendor data may be late. They may have clashes that by coordination of disciplines shall be resolved. Although review stage definition depends on engineering contract and is not a worldwide accepted norm, the following definitions may be agreed upon. Instead of an engineering contract, these definitions may be placed in an execution procedure. In this case, based on actual conditions both sides may agree on certain definitions.

A. In a 60% model review all packages (based on data available from received quotations), all pipes, tubes, trays (above a certain size, other than field-run items) and all structural items have been modeled. Using quotation information and the term "ALL" may cause some problems. In bidding stage normally several quotations are received. Which one shall be implemented? What should be done if all packages other than one have been received/implemented? A solution may be to use information from the vendor giving the largest-size package. In addition, a deadline may be placed to limit maximum delay for 60% stage.

B. In a 90% model reviewed vendor data from selected vendors shall be implemented. In this stage all structural items like tertiary members, stiffeners, ring/star plates, etc. shall be implemented. Actual clashes in yard happen on small-sized elements that are considered to be unimportant and therefore not modeled.

Consultant, EPCIC contractor, and client shall all understand these criteria are not set to stop the project. If data from a majority of the main packages are implemented but a few packages are not available yet, the project team shall perform a model review. This enables issuing isometrics and starting piping fabrication. It can be said that clashes are an inevitable part of platform construction. Consultant shall try to minimize them as much as possible. For the remaining clashes, a solution can be found later in the fabrication stage.

Almost always the engineering model (used in detail design office) is different from the construction model (used in fabrication yard). Main items included in a model may include the following:

1. Tagged equipment (packages). Their general arrangement and some major items will be shown. All interfacing items like connecting piping, footprint, and supports connecting to structure will be accurately modeled. The largest item that may be removed during maintenance shall be specified by vendor. Consultant shall foresee suitable area and handling equipment for largest removable item.

2. Tagged vessels. Normally they have several nozzles that shall either be connected to piping or an instrument. In addition, manholes are required for access to internal items. For large vessels, two manholes away from each other are mandatory. Access platforms to reach top nozzles may be constructed by yard. Since most of these vessels are PWHT, any supporting clip shall be welded in manufacturer's factory. After PWHT nothing can be welded to vessels. Separators that operate at high pressures will have thick walls. Their empty weight may exceed 100 ton. Therefore supporting saddles shall be designed carefully for all transportation and operating conditions. Since they may experience

high-thermal stresses they shall be allowed to expand. This means that one side of the saddle shall be fixed and the other side shall be a roller type.

3. Tagged valves. All MOV, ESDV, control valves, etc. shall be modeled. This includes face-to-face dimensions, arrangement and dimensions of actuators, and access for their maintenance and removal. Some of these valves with large diameter and higher class rating will be very heavy up to 30 ton. Only parts of them may be removed for maintenance. Large-bore manual valves require gear boxes to enable easy opening. It must also be shown that it does not clash with other items.

4. Escape, evacuation, and rescue equipment like lifeboats and life rafts shall also be modeled. Not only their dimensions to fit in platform layout but also access to lifeboat and suitable muster area nearby it shall be checked.

5. Main escape routes shall be modeled. Normally they shall have 1.2-m width and 2.2-m height. No obstacle shall be present in this corridor. Other escape routes shall also be shown but may be smaller and can have dual usage like maintenance area.

6. Switchgears inside electrical room and boards in technical room do not necessarily need to be modeled. Their plan dimensions are checked in 2D-layout drawing. In 3D model they may be shown as boxes.

7. Location of lighting shall be shown. This is required to check probable clashes. Checking sufficient brightness (lux level) will be done by special electrical software. One of the common clashes happens between lighting pole support stiffeners and structural members.

8. Piping headers like production, flare, and export headers will be modeled. Each header with connecting valves and pipes requires a considerable area for access and maintenance. Normally headers are large-diameter pipes with several connecting branches. Each branch will have its own isolation valves. They may require access platforms.

9. Platform piping is the major part of a model. As per each contract requirement, all pipes above a certain size will be included in a model. Industry practice is to model all pipes equal to and greater than 2″. Smaller sizes are considered field run and may be modeled as per designer requirement to get more accurate MTO or contractor requirement to get refined 3D model, but normally they are left to be field installed. This gives better flexibility to fabricator.

10. Piping supports are divided into two categories. The first one is special pipe supports, which transfer considerable loading to structural

members. These supports may sometimes be exactly incorporated in the structural model or at least their loads are accurately applied. Other supports that have small loads may be modeled as uniformly distributed loads. For the 3D model, both types are accurately modeled.

11. Electrical cable routes will be different from instrument and telecommunication cable routes. If they are placed nearer than a specific distance they will induce noise on control signals. Based on power ratings this distance may be different, but 200—300 mm is industry practice. Cable routes equal to and above 200 mm are normally modeled. In some cases smaller sizes may also be included in the model.

12. Electrical and instrument cable route supports normally exert small structural loads. Small-size supports are not actually modeled and only a box-shaped presentation is given.

13. Safety items like fire hydrants, hose reels, fireboxes, extinguishers, etc. will be modeled. This is only to ensure that they have proper access. Their coverage can be checked in 2D drawings.

14. Although not imperative but to ensure proper coverage, instrument detectors may also be shown. This may include gas, heat, flame, and smoke detectors.

15. Sprinklers are normally used in accommodation areas. In process area water nozzles shall be shown. Their location and water-spray angle shall be checked. Required water-discharge volume is calculated as per standard regulations. Manufacturer specifications show water-spraying angle and installed elevation shows coverage area.

16. Main junction boxes are also shown. This is to ensure cabling from them to the user is possible and that there is proper space for their supporting.

17. Instrument tubing is normally not shown. They are 8-mm tubes that operate different instruments. In some cases large-size tubes up to 1.5″ have also been used. They are normally field run. In cases in which a large number of tubes pass from an area, a tube tray shall be modeled.

18. Online instruments are shown. This may vary from pressure, temperature gauges to switches and transmitters. Transmitters may have a probe installed in-line and then the signal taken away by a cable to the transmitter body, which may be located in a more accessible location.

19. Main and secondary structures are modeled to check clashes and possibility of supporting. While piping, electrical, and instrument items in
3D model are used for fabrication, the structural model is not used for
fabrication. Therefore although the structural model is very accurate
for the location and size of members, it does not accurately model internal stiffeners and the end connection.
20. Tertiary structures like handrails, stairs, ladders, and access platforms
may also be shown in 3D model. To prevent unwanted clashes,
ring/stiffener plates or star plates at joints shall be at least modeled as
a box. One common clash may happen between piping, electrical/
instrument cable routes with structural padeyes, stiffeners, etc.

Recently clients have tended to also include field detectors in the
model. The intention is to see if they have proper coverage area. The
author does not find any advantage in this. Two-dimensional drawings may
serve the purpose better. However, always new technology provides better
facilities.

2.7.4 Material Selection Study

Material selection study uses the results of heat and material balance (HMB).
The presence of corrosive/lethal material like CO_2 and H_2S necessitates use
of special alloy material. It is normal to follow NACE standard requirements
for carbon steel material in offshore platforms. In addition, since it is very
difficult to distinguish between NACE qualified and non-NACE material, in
order to avoid mistaken use of unsuitable material in the wrong place even
for locations in which NACE standards are not required, client may insist to
use NACE-qualified material. Use of other alloy material like Inconel with
higher Cr content to have better corrosion-resistant property is necessary in
some locations. Different grades of stainless steel and copper nickel material
with GRP are also used. Selection of the material type is based on the
composition of the crude oil or gas.

This document will be used in filling data sheets for piping, fittings,
equipment, vessels, valves, etc. both for process and utility items.

Specifications for procurement of piping bulk items and piping material
specifications are two documents that require "material selection study" as
prerequisites. Both documents (or extracts from them) may be included as
an appendix for piping material procurement.

Piping materials have considerable diversity. The copper nickel manufacturer may not produce Inconel or stainless steel pipes. Therefore to
procure all parts of a single platform piping, the procurement department

may have to communicate with several totally different manufacturers. While piping discipline combines all material specifications in one document, each manufacturer may need only portions of it.

2.7.5 Piping Stress Analysis

Platform piping is an important part of this analysis. High pressures, toxic or in some cases lethal fluid, high temperature variation from very hot design temperature to below zero blowdown condition, etc. require special care. Two different analyses are required. Line size calculation is normally done by process to calculate diameter. Piping stress analysis is done by piping discipline to calculate piping internal stresses.

Piping route modeling is quite complicated. Many factors shall be considered:

- Fabricator yard shall be able to fabricate spools, connect them to close each loop circuit, perform NDT and hydrostatic tests, etc.
- From operation point of view, each line shall be individually accessible for checking and at the same time shall have suitable safeguards against high temperature, etc.
- From maintenance point of view, any point in a specific piping route shall be replaceable with suitable isolation from other parts of the plant while other trains follow their normal operation.

Large-size piping exerts heavy loads on the topside structure. Design of a proper supporting system to transfer vertical loads to the structure and at the same time allow proper expansion against temperature is very important. In addition to the supporting system the piping route shall also provide suitable flexibility to allow expansion against increased temperature. Temperature loads may be very heavy. Assume a carbon steel pipe $\Phi16'' \times 1''$ ($\Phi406.4 \times 25.4$) 5 m long (fixed against expansion at both ends) undergoes $50°C$ temperature increase.

Steel expansion ratio $= 11 \times 10^{-6}$

Steel pipe cross-sectional area $= 304 \text{ cm}^2$

$\Delta L = 11 \times 10^{-6} \times 50 \times 5000 = 2.8 \text{ mm}$

$\sigma = E \times \varepsilon = 2 \times 10^6 \times 50 \times 11 \times 10^{-6} = 1100 \text{ kg/cm}^2$

$F = A \times \sigma = 304 \times 1100/1000 = 334 \text{ Ton}$

Inconel pipes have approximately similar modulus of elasticity but about 20% higher thermal expansion. Therefore more force is required to keep them in place.

As is evident, preventing piping against temperature changes during normal operation will exert heavy loads to structure. A suitable piping design may require some trial and error to find proper support and flexibility. In some cases, due to space limitations providing expansion loops becomes very critical.

Fig. 2.17 shows an expansion loop over a bridge to flare. In fact, this picture shows small loops above flare knockout drum in addition to the large expansion loop above supporting bridge connected to flare tower. To allow proper action of expansion loops, piping supports shall also allow relative movement of the pipe to its support. This may be achieved by connecting low-friction material (like Teflon) on the supporting pads.

Since during emergency flaring temperature may rise heavily, low friction material shall have suitable resistance to temperature. Although this line is more than 180 m long in plan and is 62 m high, temperature at the expansion loop may reach above 100°C. Teflon has very good thermal resistance. Even some cooking dishes are fabricated with Teflon coating to prevent food from sticking to the dish due to burning.

In addition to operating loads and temperatures, pipe stress analysis shall take other environmental factors like wind, seismic effects, etc. into consideration. Different load combinations shall be generated and for each one induced stress shall be less than allowable values.

2.7.6 Isometric Drawings

Isometric drawings are prepared based on piping route in 3D model. Consider a portion of a piping route from Xmas Tree to production manifold and from there to process vessels. Line from wing valve mating flange connects to choke valve. Pressure safety valve is immediately connected to choke. PSV release shall be in a safe place. Production and blowdown headers are installed alongside each other. Blowdown valves are connected to flare manifold and from there to flare knockout drum. If a

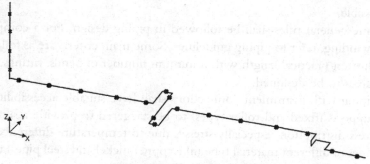

Figure 2.17 Expansion loop.

platform has several slots, for each one a line shall be modeled to ensure allocated space is sufficient. This is regardless of the fact that some lines may be reserved for future usage.

If two trains are designed on a platform, interconnections shall allow directing crude from any well to the other train. Installation of two trains may be due to large production rate or different composition between wells. In the second case, connection to both headers is not needed. To enable blowdown and immediate depressurization, all wells shall be connected to the flare header. To remove water or carried-over hydrocarbon droplets, flare header is connected to flare knockout drum. Wells are high pressure while flare knockout drum is low pressure. Considerable pressure reduction causes large temperature change. Therefore an expansion loop shall be provided exactly before entrance to flare KO drum.

Each modeled line may be divided into several isometrics. Each isometric drawing starts from a piping point, showing its horizontal and vertical branches and ends to another point. This end point will be the starting point of another isometric line. All X, Y, Z coordinates of turning points, supports, and in-line equipment are given. MTO for all items shown in isometric drawing is also given. This includes pipes, fittings (reducer, elbow), gaskets, flanges, bolts/nuts, and valves.

In the fabrication yard each isometric line may be divided into several spool pieces. Spool piece selection will consider several factors. Dimensions of the selected piece shall be such that:

1. Its fabrication in workshop is possible.
2. Spool pieces that have to be galvanized may be immersed in galvanizing pool.
3. Hydrotesting of spool piece is possible.
4. Nondestructive testing of the fabricated spool is possible in workshop.
5. Transferring and installation of spool piece in the final destination is possible.

Some general rules shall be followed in piping design. For a complete understanding, refer to piping guidelines. Some main criteria are as follows:

1. Shortest practical length with minimum number of bends, fittings, and valves to be designed.
2. Piping with instrument connections shall have suitable accessibility.
3. Supports (fixed and roller types) to be staggered to provide a suitable stress distribution, especially stresses due to temperature differences.
4. Pipes of different material type (like copper nickel and steel pipes) to be separated to avoid galvanic corrosion.

5. Large-diameter pipes to be easily accessible for repair work.
6. Drains from a pipeline to be from the lowest point and drains connected to a pipeline (for example, a header) to be from above it. This will prevent return of drain material in addition to placing non-return valves.
7. All flare/vent lines to have positive slope. This will prevent any fluid transport to flare tip.
8. Crests between two vent points and troughs between two drain points to be avoided. This will prevent gas or liquid traps.
9. If any fire-rated bulkhead or wall is penetrated, the penetration shall have similar rating.
10. When pipes of different diameter (D, d — both in inches) are connected, if $d > \frac{D}{2} - 1$ use manufactured branch tee. In fact, if the American Society of Mechanical Engineers (ASME) recommendation to have 30 degree staggered longitudinal weld lines is considered, the two recommendations follow the same philosophy. The angle connected from large pipe center to smaller pipe edge makes an angle $\beta = \sin^{-1}\left(\frac{d}{D}\right)$, which with increasing D (inches) approaches to 30 degree.

2.7.7 Piping MTO and Contingency

Piping is the second discipline that starts in platform fabrication. Piping model cannot be completed before receiving as-built vendor data. But if any fabricator waits to receive as-built vendor data and then start piping bulk material purchase, they will lose considerable time. Software like PDMS generates MTO. This will be based on the modeled piping route, and necessary isometric drawings can be generated from it. However, this is not final. Some isometrics will change after receiving final as-built data. To address this change, piping bulk items will be purchased with some contingency. Selecting suitable contingency factor for piping material depends on several items. Therefore in a single project different contingency factors may be used.

• Material cost: Some materials may be very expensive. For example, Inconel or authentic stainless steel material with high-pressure rating (class 1500 and higher) are very expensive.
• Modeling accuracy: Platform piping is accurately detailed in 3D model. All pipe sizes below 2″ are left to be field run. In spite of that, some last minute changes even in larger-size pipes due to unforeseen clashes may require rerouting and therefore may change pipe MTO.

- Standard size: Pipes are manufactured in standard sizes. After defining the spools, some of them may leave pieces of unused pipe lengths.
- Manufacturing tolerance: Some pipes will not be cut to the accurate standard size. They may be longer or smaller. Manufacturer is selling them based on total weight.
- Damage during packing/delivery, transportation, storage, and handling in yard: Final delivered bundles may not be accurate as per requested MTO. In addition, some damage may occur during transportation by vendor, storage and handling in yard for relocation, or transfer to cutting/welding site may damage some portions of steel material.
- Damage/mistakes in cutting/fabrication: This may occur when a spool is rejected due to any possible cause such as mistakes in branches' length, misalignment, weld rejection, etc.

It is obvious that for large sizes in each class fewer contingencies are selected. They are handled with care and due to their higher strength are less prone to damage. For different material 5–10% contingency may be selected. For some items no contingency or only one piece may be selected.

2.8 PROCESS

To some extent process is the main discipline defining platform operating conditions. The first step in designing a platform is to clearly identify stages that the crude will undergo. Inside the reservoir and under tremendous pressure and considerable heat the product is a mixture of light/heavy hydrocarbon and gases containing some portion of water and salt. After coming to the surface, with each change in temperature, pressure, and velocity it may undergo a change in composition or a separation between different materials. The first step may be water and gas separation. Pressure is needed to enable crude transmission through subsea pipeline. Although the flowing pressure of the product is very much less than reservoir shut-in pressure, in some cases still the designer may need to reduce platform operating pressure to avoid procuring costly high-pressure equipment. Presence of water with gases like CO_2 increases corrosion rate tremendously.

Major design data like line sizes, vessel dimensions, design and operating pressures, temperature, etc. are calculated by process discipline. First, process design basis and philosophy, platform process flow diagram, and then utility flow diagram are agreed on with the client. Based on them, heat and material balance calculation is performed. Based on these results, other documents like line/vessel sizing, PID, blowdown and radiation calculations, etc. will be prepared.

Reservoirs are either classified as gas or oil producers. Normally an oil reservoir also has water and gas. The gas coming up with oil is called associated gas. Its volume is determined by gas—oil ratio (GOR). Volume of water inside crude oil is determined by water cut. Gas reservoirs also have water in the form of moisture, which condenses when coming to the surface in addition to free water and hydrocarbon condensate.

Fluids in each field may be collected from two reservoirs or different elevations of the same reservoir. Content and composition of these products may be quite different from each other. For example, one reservoir may have very low H_2S and the other has high H_2S; even sometimes in one field one reservoir may produce only gas while the other produces oil. For example, South Pars is a gas region, however, it has an oil layer as well. Even in the same well, after several years of production the composition may change. For example, water cut or gas ratio may increase. One of the main reasons for this phenomenon is pressure reduction. After extracting a major portion of reservoir crude its pressure reduces. This will cause:

- Gas dissolved in crude may escape the solution due to less pressure.
- Due to pressure reduction, fluid in the nearby pores migrates toward lower pressure zones. Water migrates more easily than oil due to lower viscosity.

Crude composition is defined in mole percent. The main portion of gas is methane and after that ethane. In oil fields heavy hydrocarbons constitute a higher portion of crude. The main elements affecting material selection and safety are H_2S and CO_2. H_2S is toxic and its release shall be carefully monitored and proper safeguards implemented. CO_2 is corrosive and proper corrosion monitoring and prevention methods are required. Section 2.9.4 gives a table showing impacts of exposure to different H_2S levels. Table 2.6 shows three reservoirs in the northwest and middle of the Persian Gulf that have considerable differences in H_2S and CO_2 levels. Columns 2 and 3 are from different layers of the same oil reservoir. Columns 4 and 5 show two gas reservoirs about 250 km away. It is seen that even in a single field, H_2S and CO_2 ratio of different layers may vary.

In addition, these reservoirs have different pressures. Several factors shall be considered.

- Temperature and pressure are given inside reservoir.
- Oil layers are at 2150 and 2440 m depth. Sampling depth for column 4 is at 1183 m and well bore depth for column 5 is at 2805 m.

Table 2.6 Crude composition in three reservoirs

Component (mole fraction)	Northwest of Persian Gulf (oil)		Middle of Persian Gulf (gas)	
	Filed 1	Field 2	Field 3	Field 4
H_2S	Nil	3.08	4.16	0.67
CO_2	0.77	3.18	7.58	2.0
Nitrogen	0.08	0.00	6.35	3.37
Methane	44.96	42.80	81.60	81.80
Ethane	7.06	2.52	0.23	5.28
Propane	4.64	5.96	0.06	2.16
Temperature °C	94	99	134	107
Pressure Bara	255	274	674	360

- Methane portion in oil reservoirs is almost half that of gas reservoirs.
- Sum of the mole percentages in oil reservoirs given in this table is about 57% and for gas reservoirs is above 95%. The remaining portion is composed of heavy hydrocarbons. In oil reservoirs they include more than 42%.
- Compositions are given for dry crude.

From a corrosion point of view, normally the most stringent condition is used for material selection. However, piping hydraulic design may impose different solutions.

- Two separate trains may be used to handle each crude from well to export.
- If pressure loss to storage/stabilization tanks allows, the higher pressure may be reduced to use lower-grade piping. This is possible only in short distances. Assuming 15 bar head loss in process lines and 20 bar in export line, this scheme is possible.
- Gas fields have much higher pressure than oil fields. Choke shall drop pressure to platform operating value.
- Oil field has much higher specific gravity and viscosity.

Pressure reduction is normally preferred. However, it also has its own limitations. Total pressure at offshore platforms shall be able to cover below head losses:

- Offshore platform-piping head losses
- First-phase separation pressure reduction requirements (if done offshore)
- Export line head loss from offshore platform to onshore plant
- Onshore plant head losses for separations/stabilizations and other process activities
- Onshore plant to storage area head losses

This list is mostly applicable to crude oil. Gas storage is normally in small quantities and in liquid state.

Crude pressure is a main design criterion. Its meaning shall be clearly understood. Reservoir may be some thousand meters below seabed. Pressure inside the reservoir is needed for drill pipe and enhanced oil recovery (if required) equipment design. This pressure drives crude to the surface. In addition to friction loss due to flow, the static head of crude column also reduces pressure at the surface. Therefore at the down hole safety valve (installed at 80–100 m below seabed) or SSV (first safety valve in Xmas Tree) two different pressures may be experienced. When valves are closed, measured pressure is equal to reservoir pressure minus static weight of crude column. This is normally referred to as shut-in pressure. The second is this pressure minus head loss due to flow. This is flowing pressure. Pressure drop in choke valve further reduces it to platform operating pressure. As said elsewhere, a PSV protects downstream choke from overpressure surges.

As an example, in one of the South Pars gas field development phases the following levels were used:
- Reservoir pressure at 2805 m depth below surface = 361 barg
- Reservoir shut-in pressure at DHSV level = 301 barg
- Reservoir flowing pressure at MV = 240 barg
- Platform operating pressure after choke = 125 barg

Onshore receiving facilities were designed for much lower pressure levels. About 120 bar, pressure drop in choke is very detrimental for platform/export pipeline CAPEX.

2.8.1 Process Design Basis

Consultants may follow different approaches here. In some projects basic design data are given in one document and engineering design criteria in another document. In others both are combined in one document.

Process design basis summarizes platform basic design data. Among them are the following:
- Crude composition at different stages of platform life including molecular weight and percentage of light/heavy hydrocarbons, CO_2, H_2S, N_2, toxic elements, etc.
- Water cut, condensate ratio, associated gas ratio
- Crude specific gravity
- Production rate at different stages of reservoir
- Total and each well's normal production rate, maximum flow, turn-down rate, etc.

- Reservoir and platform design parameters like temperature, viscosity, and pressure (shut-in, flowing, and operating) at different stages
- Environmental data like wave, wind, temperature, radiation level, rainwater, etc.
- Seawater and produced water salt content
- Process description of platform operation
- Description of process equipment including each component size, sparing, and design parameters
- Description of utility equipment including each component size, sparing, and design parameters

Some parts of this summary list are taken from other disciplines' specifications. In fact, this document shall be carefully reviewed by other disciplines before issuing to client.

In addition to this, process design criteria may be issued. This document summarizes the main criteria/guidelines that are used in process calculations. Some of them may include the following:

- Maximum crude speed to avoid noise/vibration generation
- Maximum crude speed to avoid erosion
- Minimum crude speed to avoid settling
  (These guidelines may be submitted for gas and liquid or in two-phase flow conditions.)
- Chemical injection rate to prevent hydrate formation or corrosion in the line or foaming in vessels
- Droplet settling size (retention time) and demulsifier injection rate to achieve settling time
- Maximum oil content discharged to sea
- Isolation guidelines
- Sparing of all process and utility equipment
  These criteria will be used in sizing vessels, process lines, etc.

2.8.2 Operation Philosophy

Operation philosophy is different from "operation manual." In philosophy, platform operation is explained in brief. Reservoir master development plan is used in preparing this document. It describes previous investigations and developments (if any) in this field. Intended development including production target, number, and type of platforms (satellites location, central complex, LQ), offshore processing extent, means of providing utility services (like electrical power, etc.) are all explained in this document.

Number of trains, type of main process equipment and their operation including set pressures, HH, H, L, and LL settings for pressure, temperature, and liquid level are defined. In general, the following items may be covered:

- Gas/liquid and liquid/liquid separation means and extent (quantity and ratio)
- Sweetening and/or dehydration performed offshore
- Water treatment before disposal to sea
- Flaring or venting in emergency and normal conditions
- Electrical power generation type (diesel, gas) and extent/type of fuel (gas/liquid) treatment on the platform or supply from onshore, sparing configuration, and required utilities for power generation
- Drains to sea or closed drain
- Firefighting and mitigation measures
- Pneumatic systems for instrument operations
- Hydraulic systems for main valves (DHSV, SSV, WV, choke, and ESDV) operation
- Purging and blanketing means/conditions

In the operation manual for all packages on the platform an explanation of their operation, routine maintenance, and repair procedures are given. Step-by-step vendor recommendations for operation/checking and repair of each package are given. A package may consist of tanks, vessels, pumps, motors, valves, instruments, etc. A complete reference to manufacturer spare part numbering is also given. Set points of all valves and the expected function of each one are also provided. In addition, reference is given to each package repair manual.

2.8.3 Heat and Material Balance

The process engineer starts from the composition defined for the crude oil and gas. Extracted crude shall pass different stages on the platform as specified in PFD. At each critical node crude composition, pressure, and temperature are calculated. To cover different conditions, up to 16 cases may be defined. They may include design at start and end of the reservoir production, winter and summer cases, rich and lean conditions, and maximum and turndown flow rate. Heat and material balance shows all results in tabulated format at specific points marked in PFD.

Based on each project's design, data client/consultant may agree to fewer case studies or include intermediate cases and increase the number of conditions to be examined. The outcome of this analysis at each node will be the basis for other disciplines like piping, mechanical, and instrument to define package's material properties, pressure rating, design temperature, etc.

Differences between platforms' design start at this stage. The main factors are crude flow rate, pressure, temperature, crude composition, water cut, GOR, presence of sand, and light hydrocarbons.

Some wells may produce large volumes of oil/gas. Normal profitable values may start from 1000 BOPD or less. It is understood that economic evaluations depend on oil price. In the last 10 years (2005—2015) oil price has undergone considerable changes from 25 to 140 and back to 45 US$/B.

Some wells may have large water cut. It may be desirable to separate water and dispose it to sea. Environmental regulations do not allow discharge to sea unless it is treated to a certain degree. This necessitates oily water treatment facilities.

Some wells may produce sand. Desanding will reduce erosion rate downstream. Product composition may change during platform life. The author has encountered a case in which the original crude had only 2% water cut without sand trace. After 40 years of production, the same reservoir had approximately 50% water cut with considerable sand. In fact the water portion volume in the first-stage separators was filled up to 60% with sand. Approximately every 3 months operators had to shut down a train and clean up sands. This subject will be further explained in chapter "Systems and Equipment for Offshore Platform Design."

2.8.4 Process Flow Diagram

Process flow diagram is the drawing showing general flow of hydrocarbon material (gas/liquid) in the platform. Typical PFDs for wellhead and production platforms may be different.

In small satellite platforms with a short distance to the main platform normally no processing is done. Gas, water, and crude are transported to the production platform. In other wellheads water may be separated. The controlling factor is that the combined cost of installing additional equipment on the platform for water removal (which inevitably leads to larger topside and jacket) plus its increased operating expenditure (OPEX) compared to the cost of using corrosion-resistant pipeline (with exotic material) or injecting corrosion-prevention chemicals and suitable three-phase pumps enabling crude transport without frequent pigging shall be minimized. In some platforms that are long distances away from shore, crude may be cooled down to seawater ambient temperature during transfer. This may cause hydrate formation. Or the water in the product may tend to separate. Since water is heavier than crude, it may settle in the lower portion of the pipeline and while corroding the pipe's internal surface block the gas passage. In this case pigging is required to remove water/slug. Hydrate formation may be avoided by injecting hydrate-prevention material, but water separation inevitably occurs and necessitates pigging.

As a rule of thumb, for short distances it is more economical to transfer three-phase material to shore or from satellite to production platform. For longer distances, water removal is recommended.

Production in a gas wellhead platform starts from wellhead and ends at the export line. This is from Xmas Tree to export riser. Xmas Tree valves, casings, and related indicators may also be shown. A normal PFD may start from Xmas Tree to production and test headers/separators. The product is then directed to free-water knockout drums. Depending on platform capacity, two trains may be provided. This is required to prevent total platform shutdown due to one train's maintenance or shutdown. Gas separated in FWKO drum is directed to export line. Water and condensate are directed to other vessels like condensate coalescer for further treatment.

Condensate is a highly volatile liquid consisting of heavier hydrocarbon portions. When wet it will condense and when dry it will be in gaseous stage. The gas from condensate coalescer will be again directed to the export line and the water will be further treated before discharging to sea. Water treatment does not affect crude composition and is normally not shown in the PFD. A typical PFD for a gas wellhead platform is shown in Fig. 2.18 and for an oil platform in Fig. 2.19.

In Fig. 2.18 fluid from production wells is directed to two production headers and one test header. This means platform has two trains. For clarity, one of the trains is deleted. Test separator is intended for measurement only. Any well can be cut from production manifold and directed to test manifold. This is done by isolation valves installed in each flow line. Production headers direct the flow to two free-water knockout drums (first phase separator). In this PFD three phase separators (gas, condensate, and water) are shown. Separated gas is directed to gas export header. Water is directed to oily water treatment package and condensate is directed to condensate coalescer vessel. Treated water will be disposed to sea. This will save the export pipeline from corrosion.

It has to be noted that condensate and gas both contain water as moisture. Higher gas temperature may contain water vapor to the saturation point. Along the export pipeline, due to pressure and temperature

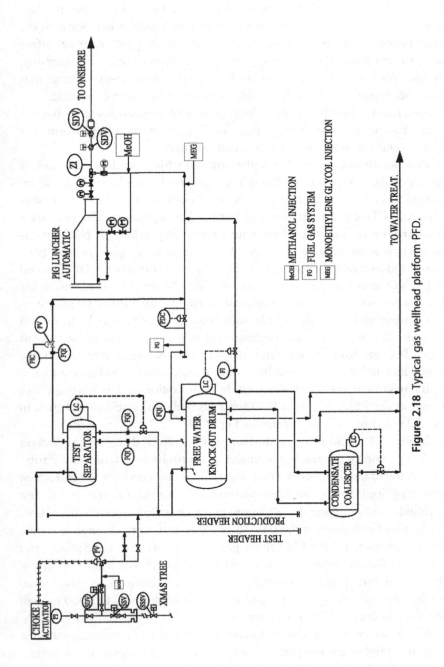

Figure 2.18 Typical gas wellhead platform PFD.

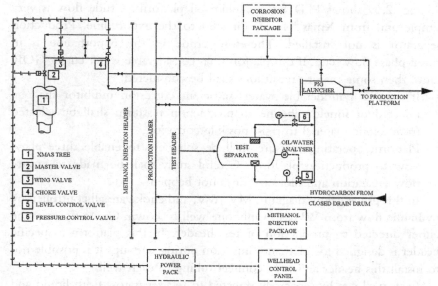

Figure 2.19 Typical oil production platform PFD.

reduction water vapor may condense to liquid. Condensate also contains some water as emulsion. This water may separate. Therefore even if a major portion of water is separated from crude, injecting corrosion inhibitor to the carbon steel export line is necessary. Although short-length infield pipelines may be constructed in Inconel, for long export pipelines from platform to onshore facilities it is uneconomical. In this PFD an automatic pig launcher with its motorized valves and two ESDVs are shown at the start of the export line before connecting to the export riser.

Water and condensate ratio depends on crude composition. Normally it is uneconomical to have a separate condensate export line. Therefore after final water removal condensate is again directed to the gas export line. This stage requires careful examination of pressures to enable gas/condensate proper mixture and export. Normally a pressure control valve at gas discharge of free-water knockout drum controls the pressure in condensate discharge line of condensate coalescer. It will induce sufficient head loss to enable gas and condensate mixture in the export line.

In oil platforms a large volume of associated gas may also be encountered. In some cases it may be used for power generation. In a project the associated gas was not only used for power generation but also the remaining part was injected to satellite wells to boost their oil production.

Fig. 2.19 shows PFD for a satellite oil platform. Crude flow is very simple and from Xmas Tree is directed to the export line. Production separator is not installed. Therefore crude in the export line is in three-phase flow state. If this platform has considerable water cut or GOR ratio, then some major precautions shall be considered:

- If there is considerable water, sufficient corrosion inhibitor shall be injected or infield pipeline to production platform shall be selected from exotic material to resist possible corrosion.
- Platform operating pressure shall be sufficient to enable three-phase flow to production platform. Careful study shall be made to ensure flow separation and blockage does not happen.

In this process flow diagram, MV, WV, and choke are all controlled by hydraulic flow from WHCP. Only one well is shown. Crude flow can be either directed to production or test header. In this platform a specific header is designed for methanol injection during start-up. It is possible not to install this header and only connect tubing to each well.

Each well can be separately directed to test separator. Both liquid and gas phases from test separator are returned to production header. In addition, oil flow from closed-drain drum is also directed to this header. If this satellite platform had only one well, the pipeline to the production platform could have been used as a test line and installation of test separator would not have been necessary. Large CAPEX for test line installation shall be compared to cost of test separator and OPEX for sending operators during well-testing periods.

In production platforms, PFD may be very complicated. There may be several trains like crude export, water injection and gas injection, or gas lift. Crude may undergo gas/water separation to increase transport pumps' and pipeline efficiency. Water injection may require several stages of oxygen and living organisms' removal before injecting to reservoir. Gas injection also requires several stages of dehydration, pressurizing, scrubbing, etc. before transferring to injection lines. These are all required to avoid well contamination. Gas lift may also undergo the same process.

In each PFD nodes at which the crude composition is calculated are shown. These are critical points that impact design of vessels, equipment, and material selection. Change in pressure, temperature, and composition impact material selection, size of vessels, and selection of equipment.

2.8.5 Piping and Instrumentation Diagram

One of the most important documents is the piping and instrumentation diagram. PID summarizes information obtained from several documents in a single drawing. Other disciplines will use the presented data in their documents. For each system on the platform a PID shall be prepared. Considering complexity of a system, the required information may need several sheets. PID as a minimum shall give the following information:

A. Flow of crude from start to the end of each system
B. Sectionalization and isolation of the connecting piping for future maintenance and operation
C. Size, class rating, and material of construction of the piping, valves, and vessels
D. Power of motors, pumps, compressors, generators, etc.
E. Instruments needed for control and monitoring of the package
F. Generated signals and required action for each one (local or to ICSS)
G. Blowdown facilities for each package/vessel

In summarizing these data, information is taken from other documents. A brief list may be:

A. Process flow diagram
B. Isolation philosophy
C. Various process sizing calculations
D. Material selection study
E. Platform control philosophy
F. Cause-and-effect diagram
G. Platform operation philosophy
H. Blowdown calculation

The data provided in PID will be used in many documents and shall match the information given in data sheets for packages/equipment/instruments. These will be the basis for procurement of packages. If any information is repeated in two documents, care shall be exercised to ensure they are equivalent. Always priority of documents (in case of conflict between submitted data) shall be clear. Safe and proper operation of the packages is a prerequisite for proper functioning and handover of a platform. Therefore PIDs always shall be updated to include latest information.

PID is intended for a minimum A3 printing size; when printed in smaller size it would be very difficult to recognize information. For presentation purposes in this book and to show flow entrance and exit lines' drawing in book-printing size, the sample PID is divided into two sections.

Fig. 2.20 shows the PID for free-water knockout drum vessel part of a gas/ water separation system. This is the first vessel after production header.

For clarity the majority of signals, instrument/valve numbers, and piping material specification break lines are removed. From line numbers, characters identifying package number, material specification, rating, and line

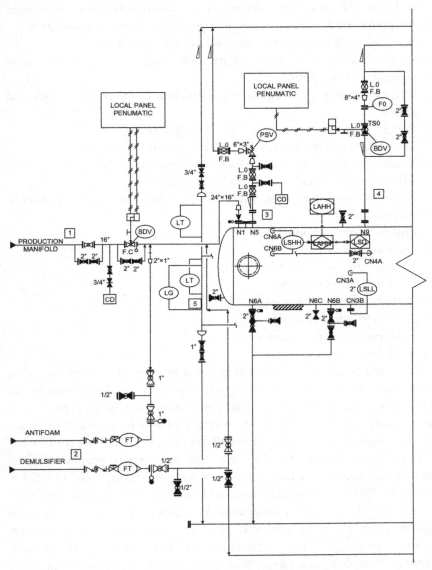

Figure 2.20 Typical free-water KO drum PID.

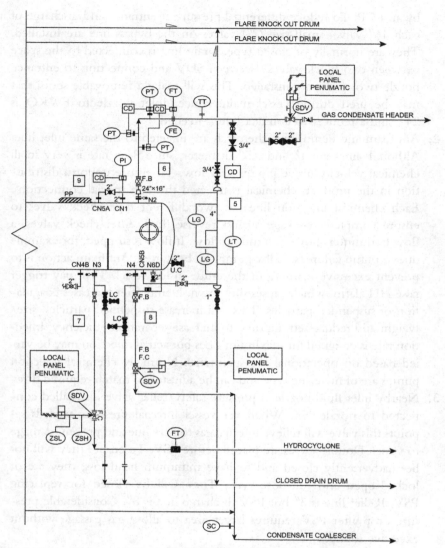

Figure 2.20 Cont'd.

sequential number are deleted. Only line size is shown. Major inlet/outlet connections to this vessel and their function are explained.

1. Gas, water, and condensate enter from production header to the vessel through nozzle N1. This line is equipped with a fail-close shutdown valve (16″). Due to high H_2S ratio, double block configuration is used for bypass or drain. Shutdown valve is pneumatic and operated

by an LCP. To reduce differential pressure at entrance and discharge of each 16" valve, small-size (2") valves on the bypass line are installed. They are normally of closed type. Drain line is connected to the space between two block valves. Between SDV and connection to entrance nozzle two spades are installed. This will make a removable spool that may be used during vessel maintenance. Inlet nozzle to FWKO is 24", and a reducer to connect line is needed.

2. Antifoam and demulsifier chemicals are injected to the same inlet line. Although lines are 1" and 0.5" diameter, since flow rate is very small chemical velocity inside pipe is very low. To ensure uniform distribution in the product, chemical is injected through special connections. Each chemical injection line has two different type check valves to ensure complete blockage against reverse flow. After check valves, a flow transmitter shows continuous flow. If flow is stopped, for example due to pump failure, standby pump can be started. Antifoam action is to prevent excessive foaming of the crude inside vessel. This may trigger false HH alarm, which causes shutdown. Demulsifier increases coagulation of suspended particles. This will increase suspended particles' size/weight and reduce settling time to increase separation efficiency. Injection rate is designed for maximum cases but actual injection may be varied based on operational measurements. Normally chemical injection pumps are of metering type and can be adjusted to match required flow.

3. Nearby inlet fluid nozzle, a pressure safety/relief valve is installed connected to nozzle N5. Whenever vessel internal pressure exceeds set points this valve will relieve internal gas to flare line and prevent damage to vessel. Double valves are placed before PSV. To ensure they will not be inadvertently closed and to have minimum head loss, they are of locked-open and full-bore type. They will be closed for replacing PSV. Relief line is 3" but PSV discharge line is 8". Considerable pressure drop after PSV requires larger area to allow gas passage without excessive sound.

4. A blowdown valve is installed (through nozzle N9) to discharge vessel inventory for maintenance purposes. This BDV is pneumatic and operated by LCP. Operating signal to these LCPs can be manually initiated at site or issued through ESD. Blowdown line is 4" but discharge line to flare knockout drum is 8".

5. A level transmitter controls hydrocarbon level (nozzle CN5A, 5B) and another one controls water level (nozzle CN4A, 4B). At HHL hydrocarbon contents in the vessel are released to flare KO drum line. Release

to flare prevents train shutdown. Some parts of PID are not shown for clarity. Each FWKO drum is connected to a dedicated hydrocyclone in deoiling package of oily water treatment system. Interface signals control water level in the two systems and allow discharge or limit it to prevent overflow. All level transmitters are connected with 2″ line to vessel.

6. Discharge line to gas header is connected to nozzle N2. Discharge line is also equipped with another SDV. This SDV is pneumatic and operated by LCP. After this SDV a pressure control valve (PCV) will be located. PCV is not shown in this PID. It is needed to match gas pressure from FWKO with condensate pressure from condensate coalescer. Pressure, temperature, and flow transmitters are installed on this line. Similar to inlet line, spade is installed to enable vessel isolation during maintenance. Discharge nozzle is 24″ and reduced to 16″ before SDV.

7. Water from each FWKO drum is discharged to dedicated hydrocyclone through nozzle N3 via 2″ line. Another pneumatic SDV operated by LCP is installed in this line. A flow transmitter controls flow. If water level is less than LL, then flow can be closed. FWKO drum is one of the most important vessels in this platform. A complicated logic governs water and hydrocarbon level in the vessel and packages located downstream of it.

8. Hydrocarbon from each FWKO drum is directed to the dedicated condensate coalescer via nozzle N4 through a 6″ line. A pneumatic SDV operated by LCP is also installed in this line. Oil level shall be controlled to prevent possible gas passage to oil system.

9. Contents of FWKO drum may become very hot. Operating temperature in this special case may become as high as 90°C. Operators have to work around this vessel. Chances of contact are high. Therefore personnel protection shall be installed.

Free-water knockout drum (three-phase separator) is the first vessel in production line of a gas platform. Pressure drop inside this vessel is negligible compared to other downstream vessels. Gas separated from liquid in other vessels shall also be connected to this line. PCV function is to check pressure downstream itself and regulate the pressure to prevent gas flow in other vessels. Separated water is routed to oily water treatment system and condensate is routed to condensate coalescer for further water removal.

FWKO is a big vessel. Due to high pressure and large diameter this vessel has thick walls. A great deal of piping and many instruments are connected to this vessel. Their routine maintenance and check requires sufficient space for personnel and equipment movement. Some gauges may be installed at higher elevations. Suitable access platforms shall be installed. Due to congested areas, lifting lugs are needed to enable moving heavy parts.

Fig. 2.21 shows PID for flare knockout drum. In spite of FWKO drum here at inlet and gas outlet lines there is no SDV. At water outlet to closed-drain drum only one SDV is installed. This is to ensure that at all conditions passage to flare is unobstructed. Some main connections include:

1. Inlet line collects flare discharges from different packages and vessels via main 24″ flare header connected to nozzle N1. Inlet nozzle is 36″ and will be connected to header with a reducer. Flare headers may start from small sizes and increase diameter with connection of new lines. Flare and blowdown lines that transfer contents of high-pressure vessels to atmospheric low pressure may experience temperatures below freezing point. Special carbon steel material shall be selected to withstand this low temperature.

2. In this vessel three level transmitters (001A, B, and C) are installed. A voting logic is defined for the operating system to prevent frequent platform shutdown due to false alarms. These transmitters check H to HH level. If two of these three transmitters detect HH level, PSD command will be initiated.

3. Gas outlet is connected via nozzle N2 to flare tip. In the line connecting flare KO drum to flare tip no valve can be installed. In addition it shall have a positive slope without any low point. This ensures water or condensed humidity will not be transferred to flare tip. Outlet nozzle is 36″ and is connected with a reducer to 30″ flare line.

4. Any possible water will be transferred via nozzle N3 to closed drain drum. A level transmitter is connected to this line, which controls the flow. Another level switch checks LL level and via LCP controls the pneumatic SDV. This configuration ensures minimum water level is kept over nozzle N3. This way it prevents gas passage to closed-drain system. All connections to closed drain have a slope, which ensures water moves only in one direction.

5. Fuel gas line for blanketing is connected to this vessel to ensure that positive pressure is always maintained inside vessel. This prevents air intrusion that may increase oxygen level to explosive gas mixture. For maintenance a nozzle is installed, which may be connected to a nitrogen source that will be required for purging. It will push all toxic/explosive gas content out of the vessel and allow operator to open manhole door. One flow and one pressure indicator are installed in this line.

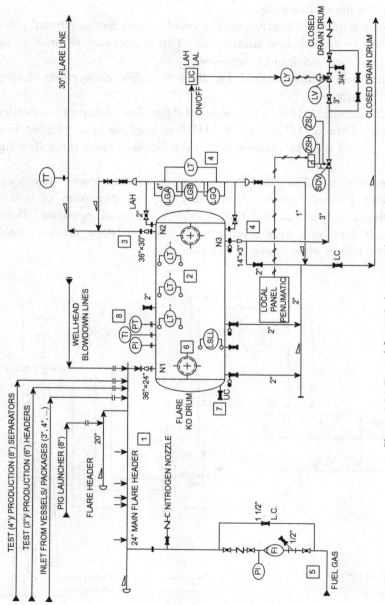

Figure 2.21 Typical flare KO drum vessel instrumentation.

6. Other than a water outlet this vessel has no internal, but it is very long with a large diameter to ensure suitable pressure drop. It has two man-holes for internal cleaning.
7. Pressure transmitter installed on the vessel ensures that its internal pressure never drops below atmospheric. This is necessary to prevent air intrusion that may lead to explosion.
8. Several nozzles are installed for different utility connections during maintenance.

Fig. 2.22 shows PID for a typical HP/LP flare line with propane bottles for ignition. One 12″ LP and one 30″ HP flare line have been installed. For brevity parts of original drawing have been deleted. Note the following points:

1. Flare lines start from WHP and continue to flare tower. This long distance ensures excessive heat, radiation, and dispersion of toxic/combustible gases do not impact platform normal operation. Both HP/LP flare lines have positive slope. Positive slope prevents fluid (water, hydrocarbon) carryover to flare tip.

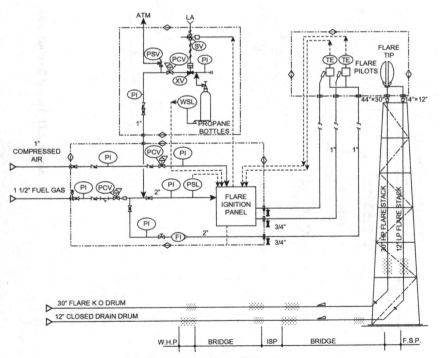

Figure 2.22 Typical flare line.

2. In flare lines no connection, branch, or valve is installed. This ensures leakages due to impaired gaskets, inadvertently left-open valves, etc. will not allow air intrusion (leading to explosive levels) or gas dispersion toward platform.

3. To ensure flare tip will burn gases in subsonic condition a reducer is installed, which increases flare tip size to 44″ (ID 1117.6 mm). Vendor may provide smaller size but shall guarantee final outcome will be subsonic.

4. Several 1″ lines may be connected from ignition package to flare tip. This enables turning on of any flare pilot in case it is extinguished. One 1″ line connects fuel gas as purge gas and pilot to flare tip.

5. Ignition package consists of two separate skids. One contains propane bottles and the second contains local control panel. A weight sensor checks propane bottles and sends alarm when the first skid becomes empty.

2.8.6 Utility Flow Diagram

To enable main process equipment perform their intended functions some utility equipment is needed. For a better explanation on the systems that are classified in this category, the reader is referred to Section 3.2. Utility flow diagram (UFD) is a drawing giving information similar to PFD but about utility equipment. Here again equipment capacity, line sizes, pressure rating, control/monitoring instruments, etc. are indicated in the related drawing. Fig. 2.23 shows a typical UFD for compressed air and inert gas generation packages, while Fig. 2.24 shows portions of the related PID only for compressed air system of this UFD.

UFD starts from intake air to compressor with its after cooler and oil-absorbing filter. Air from here is directed to utility (wet) air receiver. Liquid is drained and air is divided between air dryer package and plant air distribution system. Dryer transfers air to instrument (dry) air receiver. Although this air is relatively dry, a drain is again considered for this vessel. It divides air to instrument air consumers and inert gas-generation package.

Since Nitrogen constitutes about three-quarters of air, its offshore generation for large volumes is relatively cheap. In some plants in which inert gas consumption is not very high, it is provided via high-pressure cylinders transferred by supply boat. Even with a N_2 generation package, some N_2 cylinders are kept as contingency. These bottles can be filled by extra generated N_2.

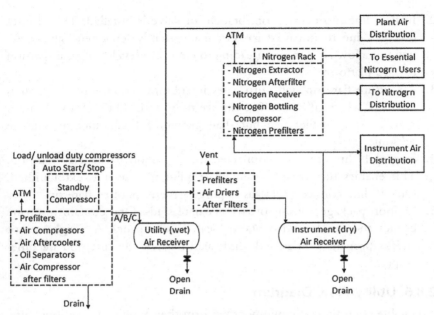

Figure 2.23 Typical compressed air UFD.

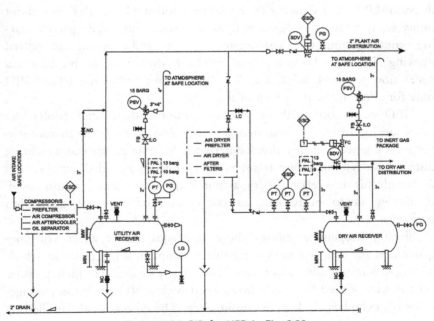

Figure 2.24 PID for UFD in Fig. 2.23.

Manufacturers may have different practices but a majority of vendors that fabricate compressor packages do not fabricate vessels. Instruments to be installed on vessels may be loose supplied by package vendor to ensure their compatibility with UCP.

In the PID 3 × 50% configuration for compressors and 2 × 100% for dryers has been used. Air intake is coming from a safe area. Normally CGD is installed at air intake duct. Since compressor increases air pressure the presence of combustible gas may cause explosion even without an ignition source. Based on signals from CG detectors, ESD closes air intake duct and stops running compressor if gas is detected.

Before air enters a compressor a prefilter removes moisture, particles, and dust. Refer to chapter "Systems and Equipment for Offshore Platform Design." for a brief description of the Charles and Gay-Lussac law on ideal gases. Discharge air is hot. Increased temperature increases air capacity to hold water moisture. Each compressor has dedicated after cooler and oil separator. If oil-free compressors are used, an oil separator may not be needed. In some projects consultant may include a condition in the MR that vendor can propose either system and then EPCIC contractor has the choice to select based on commercial advantages. In any case, instrument (dry) air shall be free from oil particles. Otherwise the instruments may plug.

In this partial PID 3 × 50% compressors were used. This package generates compressed air for a lot of consumers. Three main branches include wet/dry air receiver and inert gas generation. Specific logic is incorporated in UCP based on required air volume and pressure measurement in each of the three systems.

Pressure transmitters installed on wet air receiver control compressors' running/stop command. In this system for wet air receiver, low-pressure alarm is set at 10 barg and high alarm is set at 13 barg. For this vessel, PSV to safely vent is set at 15 barg. For dry air receiver, low-pressure alarm is set at 9 barg and high alarm is set at 13 barg. Wet air receiver is kept at a relatively higher pressure.

For air dryers 2 × 100% configuration is selected. For vessels, providing a spare unit is not customary. Both wet and dry receivers can be constructed as vertical vessels. Due to small volume they have only one manhole.

Vent pipes' ends are arranged in a form to prevent water and moisture in addition to bug entrance. Compressor/dryer packages and wet/dry air receivers may be installed on different skids. Drains of all equipment in one skid are connected to form a common drain header. All moisture drains are collected in a header to nonhazardous open-drain caisson.

For clarity, alarms including start/stop, duty/standby, pressure transmitters low/high, temperature high, etc. are removed. Since this is a relatively large and complicated package, it can have a UCP to control all signals and issue operating commands. In addition, manufacturer has to guarantee package operation. Therefore it is better to allow them to run their own logic software. To connect with the platform control system, specific signals can be transferred.

As much as possible different skids of this package shall be located near each other. This will facilitate power and control cabling and piping connections. UCP can be moved to a safe area, otherwise its enclosure shall have a suitable outdoor rating. If very hot/low temperatures are expected, which may affect UCP operation, it can be placed in an air-conditioned and properly ventilated area.

2.8.7 Flare Radiation Study

Flaring has a very detrimental impact on offshore platform safety and environment protection. In several cases the inventory inside platform piping and vessels shall be emptied as a mitigation procedure to prevent hazard escalation. In this case the best method is to direct the ignitable gaseous material toward a burning point sufficiently away from the platform. The liquid ignitable material will be directed toward the incinerator.

Two documents (which may be combined in one volume) cover flare radiation study and heat, smoke dispersion study. The outcome of these studies ensures that normal platform operation is not affected by this phenomenon.

Flare will be designed for two cases. One is the design case flaring. This covers the maximum platform design capacity, which is called emergency flaring. This may happen in ESD cases. In addition there is operating case flaring, which is called continuous flaring. This may happen in LSD cases and normal operation. The flared volume/rate and duration are quite different in these two cases.

Flare shall be sufficiently away from the platform for several reasons:
- The dispersed heat does not impact working conditions of the platform packages
- The radiation is not so high to prevent personnel from safe working
- The generated smoke reduces to very low concentrations if the wind blows it toward the platform

Two separate layouts may be used. The first one is that the flare platform is put suitably away from other platforms and connected via bridges. In some cases, if the distance is very long an intermediate platform will be installed to support the connecting bridges. The bridges will support main flare line, pilot line, and access. The second layout is connecting the two pipes to flare tip via submarine lines. In this case, flare platform can be installed further away from other platforms of the complex.

Full flaring releases tremendous heat. The heat reduces steel load-bearing capacity. Increasing steel temperature to 600°C may reduce up to 75% of yield stress.

In order to enable structural discipline to perform platforms' permanent analysis, process department performs heat radiation analysis. The environmental condition for flare and its tower may be different from the condition for bridges. Flare stack is exactly below flare tip. If it is not inclined, then calm weather air (no wind) condition may have the worst impact. But bridges are away from flare tip, therefore for them the worst condition may be when a wind is blowing toward the platform.

Radiation study affects working conditions. Personnel may face health problems if they are exposed for a long time to high radiation levels. The hazard may range from bruises to severe burns. This study is mainly important for platform working area. Therefore again the environmental condition shall be taken into account. Here an interface between process and structure is required. Radiation study is done by process engineers. But structural engineers are aware of the wind rose and wind speed. Wind rose shows the percentage occurrence of wind in each speed category from each direction per year. The most critical case may be when wind is blowing toward the platform. The higher the wind speed, the higher the temperature and radiation on the platform. However, it is very improbable that a long return period phenomenon (eg, 100 year) happens at the same time as emergency flaring. Therefore it will be much more costly and incorrect to design for such a case. One proposal is to use two different scenarios. In the first one, design capacity flaring is combined with a low-velocity wind considering directions. Normally platform—flare direction is in the direction of prevailing winds. Therefore wind blowing from flare toward platform is a relatively rare phenomenon. For design case flaring, in many platforms it is not needed to check wind blowing toward the platform. However, in operating case it may be needed. The area wind rose shall be reviewed.

Isothermal contours are needed for several reasons. They determine:
1. Up to which point in structural analysis yield stress reduction shall be considered and how much.
2. Piping stress analysis basic data.
3. Type of paint necessary on steel structure, piping, etc.
4. Type of electrical, instrument cables.
5. Type of electrical/instrument cable route equipment.

2.8.8 Sparing Philosophy

In many projects sparing philosophy is not an individual document. It is included as a chapter of process design basis. However, I would like to give more importance to it and discuss it under an independent topic.

In its broadest sense a financial study is required to decide on best sparing scheme. It is understood that if a package shutdown like a generator leads to platform shutdown, then several days' loss of production is much more costly than installing a spare unit. In addition, for small units using $2 \times 100\%$, sparing is less costly than $3 \times 50\%$. For small capacity units, $3 \times 50\%$ configuration may have some shortcomings.

- Cost of a unit with 50% capacity is much more than half the cost for the same unit with 100% capacity. Some items like supporting skid, cabling, control instruments, quality assurance (QA)/quality control (QC) requirements, etc. are similar. Therefore three packages with 50% capacity if they do not become more expensive than two units with 100% capacity will not be less. This means more CAPEX for this configuration.
- Operating costs for three packages is higher than two packages. This means more OPEX.
- Three 50% units will occupy more space than two 100% units. This will increase platform dimensions and therefore indirectly adds to total cost.

All of the previous arguments may not be applicable for very large-capacity packages.

Unfortunately, the author has not yet encountered a cost-evaluation study on this topic. Instead, the general practice in the region and previous projects' experiences have been followed. Preparing quantitative design criteria for this study is also very difficult. That is why qualitative explanations and general practice in each region is more readily followed. After all it has to be kept in mind that project engineering is not intended for academic research. They have to prepare necessary documents for platform fabrication within a certain budget and time.

In platform design single-jeopardy philosophy is always followed. This means that two similar-type mishaps are not considered to occur simultaneously. For example, if we have three generators on the platform (two operating and one of them considered as spare) the possibility of two simultaneous failures in operating generators is not considered. Or, if gas leakage occurs due to valve gasket failure, the possibility of nearby CGD/TGD instruments' simultaneous failure is not considered.

Single jeopardy consideration is different from cascading effect evaluation. The possibility of a gas release and fire in one location being followed by fire in the neighborhood is another issue that has to be evaluated in hazard identification (HAZID)/hazard and operability (HAZOP) studies. Proper mitigation measures shall be implemented as safeguards against this phenomenon.

It is uneconomical to increase the number of spare units on the assumption that two units may fail simultaneously. This increases platform dimensions, which means additional weight and fabrication costs. This approach will increase CAPEX as well as OPEX.

Sparing is normally related to the number of equipment intended for a single function. In its broadest sense, dividing flow lines between selected trains is itself some kind of sparing. In this case, platform capacity is divided between 2 × 50% trains. The intention is to ensure production is continued even if a malfunction or periodic repair or any other cause requires shutting down a train. It is obvious that platforms with small yield may not require division to separate trains.

Dividing to two trains will increase:
- Platform plan area and dimensions (additional distance for maintenance and repair)
- Platform weight (additional steel, piping, and equipment)
- Platform additional piping and cabling
- Additional field or in-line instruments, valves of different types
- Additional equipment

All in the previous list mean additional cost compared to single-train platform. The gain is continuous production (at lower rate) during maintenance and local shutdowns.

Other than process trains, inside each system there will also be sparing. Some of the main items are listed herein:
- In control systems the first sparing is in transferring FGS/ESD signals. In many projects FGS/ESD signals are transferred via redundant routes from different parts of the platform. Therefore even if fire or explosion

in one part cuts instrument signals, they are transferred to the control system from another route and cause necessary action.

- In sensing gas release, smoke, heat, and flame detection, to avoid triggering functions based on false signals, normally a voting system like 2 out of N (2OON) is utilized. This is in addition to field instrument manufacturers' efforts to increase sensitivity and prevent false alarms generated by natural phenomenon like solar radiation, light reflection, etc.
- Power generation sparing may vary from $2 \times 100\%$ for low powers (less than 2 to 1 MW) to $3 \times 50\%$ and $4 \times 33\%$ for high powers (several MWs).
- Compressors, either low capacity for providing dry pressurized air for instruments or wet air for housekeeping activities, or high-capacity compressors for gas lift/injection, also have similar sparing as power generators. Low-capacity compressors are mostly electrically driven and high-capacity compressors are turbine gas-powered.
- Vessels are not considered susceptible to breakdown. Therefore normally spare is not considered for them. However, in some cases in which a considerable overflow or off-spec fluid (for example, in oily water treatment) is expected and flow may be directed from discharge to intake of the system, a spare vessel may be considered.
- Continuously operating rotating equipment like pumps are suspected to break down and are therefore provided with a spare. Pumps inside packages are 100% spared while high pressure/discharge export pumps/water injection pumps are provided with $3 \times 50\%$ spare.
- Only block valves that are located in high H_2S content lines are doubled. A bleed will be provided between them to direct the flow between them to drain line. Double block and bleed valves of small size may be fabricated in a single block.

2.9 SAFETY

Each platform shall provide a safe working condition for operators. In addition it shall not harm environment by emitting gaseous elements and oil spillages. After many disasters that have taken hundreds of human lives, destroyed the environment and costly complexes, now safety regulations are becoming a major factor in designing an offshore platform. Normally this is considered as health, safety, and environment. In brief this is studied as HSE.

2.9.1 Hazard Identification

HAZID stands for hazard identification. Normally HAZID is performed in a workshop with end user participation. This study covers different platform elevations and areas. To be most effective it has to be performed in early stages of the job to identify possible impacts on the layout and general platform design. Otherwise each party including client and contractor will be reluctant to accept changes.

Normally the main intention is the operation hazards but it can cover construction, installation, and drilling hazards as well. Preliminary construction and installation hazards can be covered in basic design stage to assess selected structural forms. However, the detailed construction and installation hazards are better to be covered in construction/installation engineering phase.

The HAZID meeting concentrates only to identify hazards. Solutions may be found by detailed investigation of the related discipline after the meeting. Similar to environmental impact assessment (EIA), conclusions in HAZID may impact decisions on project development. EIA impact is more general, covers higher-level factors, and may impact project development, but HAZID impacts are limited and may impact methods and layouts.

HAZID is also based on a checklist of potential safety issues. First, three main definitions shall be clarified:

- THREAT is any cause that has the potential to release the hazard and produce an incident.
- HAZARD is any phenomenon that has the potential to cause health problems or injury to personnel or damage to property, environment or reduce/stop production.
- INCIDENT is any event or sequential acts/events that may cause injury, illness, or loss to personnel, environment, and assets.

HAZID meeting is more effective if a series of guide words are used. Some of the guide words to be considered for HAZID are listed following. In each case some of them may not be applicable: Not-ignited hydrocarbon release, ignited hydrocarbon release causing fire, ignited hydrocarbon release causing explosion, toxic/asphyxiating gas release, high pressure, high/low temperature, dropped object, maintenance, confined space, access/egress/ escape and evacuation, extreme weather, radiation, explosives, structural integrity, helicopter operations, construction activity, and ship collision.

One major HAZID application is in risk assessment. By definition risk is considered as event occurrence probability multiplied by event consequence.

Based on this definition a low-probability event with severe consequences may be ranked higher than a high-probability event with low or negligible consequences. In many cases adequate data do not exist. Therefore decisions are subjective and based on individual experience. Tables 2.7–2.9 are useful. See Det Norske Veritas [29] for further explanation.

Table 2.7 Risk matrix table

Frequency	Severity				
	1	2	3	4	5
A	M	H	H	H	H
B	M	M	H	H	H
C	L	M	M	H	H
D	L	L	M	M	H
E	L	L	L	M	M

Table 2.8 Measure of severity

1	Insignificant	No injuries, low financial loss
2	Minor	First aid treatment, on-site release immediately contained, medium financial loss
3	Moderate	Serious injuries, on-site release contained with outside assistance, high financial loss
4	Major	Extensive injuries, single fatalities, loss of production capability, off-site release with no detrimental effects, major financial loss
5	Catastrophic	Multiple fatalities, toxic release off site with detrimental effect, huge financial loss

Table 2.9 Frequency measure

A	Almost certain	10 times per year	Is expected to occur in most circumstances.
B	Likely	Once per year	Will probably occur in most circumstances.
C	Possible	Once every 10 years	Might occur at some time.
D	Unlikely	Once every 100 years	Could occur at some time.
E	Rare	Once every 1000 years	May only occur in exceptional circumstances.

Some other forms of these tables may also be used. Nomenclature of data presented in Tables 2.7—2.9 are different from Det Norske Veritas [29], but the general approach in ranking each event by its probability of occurrence times consequence is the same.

After identifying risks, some measures to reduce their impacts are proposed. These actions may range from early detection, eliminating/ preventing or controlling incidents, and mitigation measures. The mitigation measures may vary from preventing event escalation to completely extinguishing it.

2.9.2 Hazard and Operability

Although HAZOP is categorized as a safety document, it is a multidisciplinary task.

It can be said that HAZOP is a brain-storming meeting performed on selected PIDs. In some cases all process and the main utility system PIDs are reviewed. The intention is to find possible hazards and identify operability deviations/requirements to be included in the system design. In summary, HAZOP intentions can be categorized as:

- Identify any shortfalls in the design that may lead to deviation in operation.
- Identify the ultimate consequences of any deviation.
- Ensure that the processes are designed safely.
- In case necessary, will propose actions to enhance design.

The meeting is guided by a chairman assisted by a scribe. They should not be on the design team to be free from any bias. Process, instrument, and safety engineers from the designer side and in addition to them operation engineers from the client side shall be present. Other disciplines shall be on standby basis. Whenever a clarification is needed, they will be requested for explanation. This explanation may cover selected material, layout, etc.

HAZOP chair defines the extension of the nodes and marks the lines on the PIDs to be reviewed. This will show the path under review.

For each PID a set of standard keywords is reviewed. The main keywords include high pressure, low pressure, no pressure/vacuum, more flow, less flow, no flow, reverse flow, high temperature, low temperature, start-up, shutdown, maintenance, composition, operability, corrosion, and toxicity. All keywords may not be applicable for all systems. For example, terms related to level when there is no vessel are not applicable.

In some cases where the project schedule allows, two HAZOPs may be performed. The first is a preliminary HAZOP to implement all comments on PIDs. The second is for main systems in which a new HAZOP including that particular package vendor is performed. The second HAZOP concentrates on package internal PID.

For each guide word, possible deviations from normal operation are identified with their potential causes. Then each HAZARD consequence is evaluated by the team and recorded by scribe. In the next step the team will decide whether the consequences are so severe that they need an action, whether the mitigation measures already foreseen in the design are sufficient, and finally whether a specific action is recommended by the team. All these items will be recorded to provide an auditable action list to be checked later. These actions may vary including further study of an issue, adding certain instruments, valves, changing/adding connecting pipeline, changes in the operation logic, etc.

The PIDs and all related documents shall be updated based on HAZOP meeting findings. Some basic assumptions/premises are accepted by all sides in performing HAZOP:

- The simultaneous occurrence of two or more causes of a deviation within single node (double jeopardy) will not be considered in HAZOP.
- The simultaneous failure of more than one independent protection device within a single node will not be considered.
- Failure of critical safety items (PSV, BDV, and SDV) will not be considered.

Scribe records all items in a specific table. The recommendations are recorded to enable later review of the actions. If HAZOP is performed at basic design stage, necessary actions can be checked in detail design stage. If HAZOP is done at detail design, after first issue of the report, all disciplines shall perform their assignments as recommended in the report and prepare a response. Second issue of HAZOP report combines all these responses.

2.9.3 Passive Protection

Passive and active protections are used in offshore platforms. Active protection is achieved by means of fire water pump, sprinklers, hydrants, hose reels, chemical extinguishers, foams, deluge valves, etc. Passive protection is achieved by providing specifically rated barriers, walls with the required fire rating, special coatings, etc. These walls shall be capable of tolerating temperature increase as per code requirement. They are classified as A-, H-, and J-rated walls. B-rated walls are only comfort insulation. Standard fire temperature and duration as per ISO 834 (EN1364—1) is shown in

Table 2.10 Standard fire temperature variation

Duration (Min)		3	5	10	15	30	60
Standard fire	A	502	576	679	738	842	945
Temperature (°C)	H	887	948	1034	1071	1098	1100

Table 2.10. With temperature in °C and time in minutes, the following formulas represent standard and hydrocarbon fire test curves. ISO standard fire curve is somehow different from ASTM E−119 standard curve.

Standard Fire: $T = 20 + 345\log(8t + 1)$

Hydrocarbon   Fire: $T = 20 + 1080[1 - 0.325\exp(-0.167t) - 0.675\exp(-2.5t)]$

A-rated walls are for general fire case. The selected compartment shall be able to provide structural resistance against normal operational loads and limit heat transfer for a specified duration. For example, A60 means durability for 60 min against class A fire. H-rated walls have the same condition against hydrocarbon fires. J-rated walls shall resist jet fires. This is caused by sudden leakage and ignition of gas. In many cases this may be accompanied by blasts. Therefore for J-rated conditions a blast overpressure is also identified. Blast pressure reduces with square root of distance. Blast load for a localized impact is much more than a large area. Detailed safety studies are needed to calculate blast load and specify wall/deck ratings.

Structural steel with special coating systems is needed to achieve the required fire rating. Coating may be applied on structural wall, columns, braces, beams, joints, decks, vessels, piping manifolds, and valves. For valves that require periodic inspection, coating shall be of a removable jacket type. Both type of fixed-shape and flexible jackets are available. Required thickness shall be calculated for each case. The important feature of these coatings is that heat transfer from the face adjacent to the fire to the face adjacent to the metal surface is very much limited.

2.9.4 Fire, Smoke, and Gas Dispersion Study

Fire, smoke, and gas dispersion study investigates the potential sources of toxic gas release, its concentration, and impact on human health. In addition, for the emergency case it determines radiation levels hazardous to operators. Major release source may be identified based on the volume of inventory, pressure, or both. Normally, wellheads, separator vessels, etc. shall be studied. Their proximity to escape and evacuation means like lifeboat, life raft, escape route, and stairs are shown. SO_2 and H_2S are toxic gases that shall be considered. NORSOK Standard [30] has given exposure

Table 2.11 Effect of different H_2S concentrations [30]

H_2S concentration (PPM)	Effect
20—30	Inflammation of the conjunctiva of the eye
50	Objection to light after 4 h exposure, excessive tear shedding
150—200	Objection to light, irritation of mucous membranes, headache
200—400	Slight symptoms of poisoning after several hours
250—600	Pulmonary edema and bronchial pneumonia after prolonged exposure
500—1000	Painful eye irritation, vomiting
1000	Immediate acute poisoning
1000—2000	Lethal after 30—60 min
>2000	Actual lethal poisoning

limits for SO_2 and H_2S and effects on people exposed to H_2S. Other technical sources may provide a different list. There is a general agreement on immediately dangerous to life and health (IDLH) limit and short-time exposure limit (STEL).

Occupational exposure limit for SO_2:

- 4.6 ppm—10 min
- 2.0 ppm—8 h
- 1.3 ppm—12 h

Occupational exposure limit for H_2S:

- 100.0 ppm — IDLH
- 15.0 ppm—15 min (STEL)
- 10.0 ppm—8 h

Lower-flammability limit for gas is around 3% by volume. Table 2.11 gives an explanation of H_2S effects on the human body.

For fire cases, radiation levels shall be limited. Its effect may vary from irritation to severe burning. Recommended radiation levels for lightly clothed and clothed personnel are given in Table 2.12 as 100% fatality limit.

Smoke reduces visibility and delays personnel access to escape route. Visibility ranges based on smoke concentration are given here:

- Smoke concentration of 2.3% reduces visibility to 10 m
- Smoke concentration of 15% makes escape routes inaccessible due to toxicity and very low visibility

Table 2.12 Radiation fatal exposure [30]

Exposure time	Radiation level kW/m²	
	Lightly clothed	Clothed
Less than 0.5 min	16	25
From 0.5 min to 1.0 min	10	13
From 1.0 min to 2.0 min	4	8
From 2.0 min to 10.0 min	2	4

All electrical, instrument, and telecommunication cables shall be low smoke to prevent impact on visibility and halogen free to prevent emitting toxic gases.

2.9.5 Explosion Overpressure Study

Explosion overpressure study investigates credible blast scenarios. Based on leak dimensions and the inventory inside vessel, a blast pressure will be calculated on the nearby distances. Other equipment and blast/firewalls separating process from utility area are the main concerns.

One investigation of industry records of accidental releases show a majority of them (61%) are pipe work system. Releases from small-bore pipes include only 18% of this 61%. Other main sources of release include flanges, seals, and valves.

Causes of the release include incorrect installation, incorrect design, degradation of material, and vibration/fatigue. Release source size is selected with actual design pressure of that specific item for overpressure analysis. Blast may affect personnel, assets, and structures. Table 2.13 shows typical overpressure ranges' effect on each.

2.10 STRUCTURE

Offshore environment is very harsh. Huge waves, very fast gusts, earthquakes, seawater's corrosive effect, operation loads, and ship impact affects offshore platform structural integrity. One of the major standards used in design and analysis of offshore platforms is API RP 2A [31]. Normally a platform shall be operative for 25–30 years. The author has encountered platforms being studied for another 25 years after finishing their original intended life cycle.

Topside loads are transferred to seabed through the jacket structure and piling system. There are several types of offshore platforms. Here only fixed-jacket-type platforms are discussed.

Table 2.13 Overpressure impact

Overpressure (bar)	Effect
Personnel injury	
0.03	Will cause injuries from flying debris
0.21	20% chance of fatality to a person in a building
0.35	Threshold for eardrum damage, 50% chance of fatality for a person in a building, and 15% chance of fatality for a person in open area
0.7	Will cause 100% fatality for a person in a building or in the open area
Asset damage	
0.02	50% window shattering
0.05	Collapse of tank roof
0.07–0.14	Connection failure of corrugated paneling
0.08–0.1	Minor damage to steel framework
Structural damage	
0.15–0.2	Concrete blocks shattered
0.2–0.3	Collapse of steel framework
0.2–0.3	Small deformations on pipe bridge
0.3	Paneling torn off
0.35–0.4	Displacement of pipe bridge, failure of piping

Guidelines for Evaluating Process Plant Buildings for External Explosions and Fires by CCPS, 1996.

Based on actual fabrication and installation sequence, platform analysis and design is divided into two different but interconnected parts. They include jacket and topside. Jacket is mainly concerned with structural discipline while in topside design impact of other disciplines shall also be considered.

Structural analyses are divided into two permanent and temporary groups. In some references another terminology (service and preservice) has been used. This does not include local analysis during fabrication. The first group deals with operational and environmental loads like wind, waves, earthquakes, temperature, etc. The second group deals with loads applied during fabrication, transportation, and installation phases.

2.10.1 Jacket Analysis

Fixed jackets are tubular structures extending from the seabed to above water surface. They transfer topside loads to seabed via piling system. Jackets may be in different forms like monopod, tripod, four, and more legged.

Several elements may be distinguished in a jacket. For a description of jacket components, see Dawson [1] and Patel [6]. From seabed above they may be listed as:

- Mud mat: Lowest elevation of jacket in contact with seabed is covered with wooden planks or steel plates to distribute load more uniformly and support jacket temporarily before piling.
- Elevations: From seabed to sea surface horizontal members are welded to jacket legs to provide lateral support for legs and reduce their free height.
- Braces: Inclined members at jacket sides welded in K or X configuration to resist lateral loads (environmental and ship impact).
- Piles: Long, tubular members from connecting point to topside (other than skirt piles) driven from inside legs or skirt to transfer vertical and axial loads to seabed.
- Launch trusses: Welded tubular members in the form of two parallel trusses (embedded inside jacket structure and protruding from one side) to distribute load out, transportation, and launching loads between jacket members and fabrication yard or transport barge.
- Legs: Large-diameter members (thickened at contact points to elevations) to support piles and provide platform stability.
- CAN: Large thick pieces of leg at elevations connecting all horizontal and vertical members.

Joint CANs may face tensile stresses perpendicular to their main rolling direction. Special tests are needed to ensure weakness plane does not develop to initiate lamellar tearing. Therefore material used for joint CAN fabrication shall have Z property (through thickness property; TTP).

There are auxiliary items as well that may not act directly in load-bearing process but are used for corrosion prevention and installation activities. They may include:

- Anodes: Aluminum- or zinc-based alloys welded to jacket members to prevent corrosion in saline offshore environment.
- Rigging platform: A platform fabricated at top of jacket to hold slings and shackles used temporarily during jacket installation.
- Buoyancy tanks: Large cylindrical members bolted to supports welded to jacket legs used to provide necessary buoyancy in addition to legs and braces to ensure floating before upending.
- Boat landing: A structure installed at sea level used for berthing of supply boats to transfer personnel and cargo. It has two levels adjusted to low- and high-tide elevations. Fenders are installed between boat landing structure and jacket legs to absorb ship impact energy.

- Barge bumpers: Are protruding elements that keep barges away from the platform. In case of accidental impact it deforms and thus absorbs maximum portion of the energy.
- Drain deck: This is the first deck above sea level connecting boat landing to topside stair.
- Caissons: Are open-ended large tubular members supported by clamping to jacket members and connected to topside utilities like open drain and seawater service.
- Riser: Is a tubular element supported by clamps to jacket legs or elevations connecting export header to the seabed pipeline.
- J tube: Consists of tubular element allowing pulling of subsea cable to the topside.
- Flooding system: Consists of 2″-diameter piping at each leg with connecting valves to allow leg flooding for jacket installation. Opening of valves shall be possible from sea surface.
- Grouting system: 3″-diameter piping with connection valves to allow inserting grout in the annulus between leg and pile. The system is supported at close intervals to the legs to withstand transportation and installation loads. It has full redundancy so that if a line is clogged or damaged the other one can be used. Each line is connected to leg at certain elevations to enable pumping grout at necessary locations.
- Enclosure diaphragm: Is a rubber diaphragm bolted at end of jacket leg to prevent flooding during upending. It shall be stiff enough to withstand applied loads when jacket is upended on seabed. At the same time it shall be soft enough to tear off with weight of first add-on after welding to first segment pile.
- Launch lug: Consists of two identical pulling segments welded to each launch truss to exert equal first pull to initiate jacket movement at start of launching operation. They have a flapper at end to prevent wire or sling slippage before movement. The same mechanism allows wire disengagement after jacket has passed a certain location on the barge and has gained enough momentum to be launched.
- Lifting padeye: Consists of one main and two cheek plates with lateral stiffeners. Shackle pin is inserted in the main plate eye and lifting slings are connected to it.
- Trunnions: Used in lifting or upending operations to connect slings. After removing lifting load, when wires become slack they can be easily dismantled.

Fig. 2.25 is a four-legged jacket structure in 3D configuration. The following items are shown:

1. This jacket has four elevations below water level and one above it. Lowest astronomical tide is located at 0.0 and MSL at +1.0. First elevation above water is called drain deck. Since it may experience wave impact in heavy storms, major or vulnerable equipment are not placed on it.

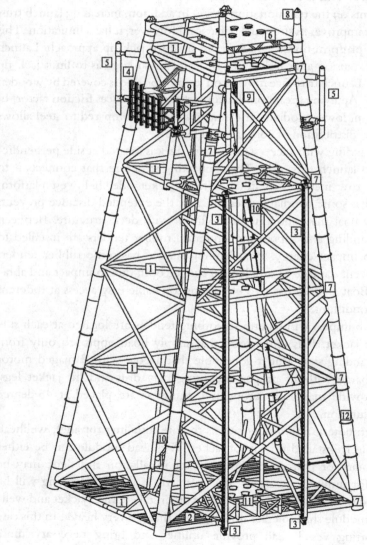

Figure 2.25 Sample jacket in 3D.

2. Seabed in this location is soft. Therefore mud mat covers almost full area of the lowest elevation. It is covered by steel plate. To provide structural strength, intermediate H beams are constructed at about 1 m intervals. At sides they are supported by horizontal elements of jacket's lowest elevation. To reduce their free span, a tubular element connects each panel center point to jacket leg.

3. Launch trusses are shown on vertical face. They extend about 1.5 m from side face. This is to ensure no part of the jacket may collide with steel elements on the transportation barge. In addition, increasing launch truss height improves its structural behavior. However, it has a limitation. This side is platform north, which is also used for jack–up approach. Launch truss extension after jacket face shall not be so large as to limit jack–up access. Launch truss face moving on supporting skid is covered by wooden planks. Applying grease on the contact surface reduces friction factor. In addition, lower modulus of elasticity of wood compared to steel allows some redistribution of concentrated loads.

4. Boat landing with access stair to drain deck is located at side perpendicular to launch truss side. The supporting structure that connects it to jacket row increases its distance from jacket side. If lowest platform deck has some overhang after this row the extended distance prevents supply boat mast and antenna collision with deck structure. Between boat landing and supporting structure, rubber fenders are installed to absorb impact energy. Boat landing surface has some rubber fenders to prevent damage to supply boat surface both due to impact and abrasion. Boat landing has two elevations to enable easy access at different tide conditions.

5. Barge bumpers with energy-absorbing fenders are located at each side of the jacket. It is understood that supply boats approach only from boat landing side. But it is possible that a boat with damaged motor or a barge without proper anchoring may drift and hit jacket legs. To provide maximum protection, bumpers are placed at 45 degree configuration.

6. For temporary drilling operations (before installing topside), wellhead module is installed on top of jacket. Wellhead module can be either temporary or permanent. After jacket installation this deck may be used for drilling operations. If jack–up rig is used, its loading will be limited to live load. If rigging towers are installed, then jacket and wellhead module shall tolerate drilling loads that are very heavy. In this case supporting vessel shall provide utilities and bring necessary mud,

cement, etc. Only pulling load on drill pipes may exceed 1000 ton. For future drilling or well maintenance operations, it is better to design jacket and topside to be able to support this heavy load. This has both structural and arrangement requirements. The advantage is that later on operators can select any of the two options based on availability and financial evaluations.

7. Joint CANs have large thickness. Other parts of leg are thinner. Lateral braces of two vertical faces in addition to horizontal members of that elevation connect to joint CAN. Preferably weld overlap shall be avoided as much as possible. However, if member size and jacket configuration do not allow it, allowable stresses for fatigue resistance shall be reduced. Jacket-supporting piles are driven from inside legs. To facilitate pile driving, inside surface of joint CAN shall be preferably flashed with other segments of leg.

8. After drilling, Xmas Trees are installed to control crude flow. Although they have a robust structure, lifted topside may exert heavy impact loads when swinging. Horizontal impact load up to 5% total topside weight is expected. Stabbing guides are installed at jacket legs to limit topside movement during installation. They extend higher than Xmas Tree highest point. This will enable safe topside installation while Xmas Trees are installed. Since internal dimension of topside legs may vary in elevation, external dimension of stabbing guide shall match it. First segment of stabbing guide shall have ample tolerance to facilitate topside installation. As topside is lowered, its movement tolerance is further reduced. The end point tolerance shall be equal to allowable construction tolerances.

9. Open drain and service water caissons are clamped to horizontal members. They normally end at drain deck level. If jacket is launched their clamps may experience heavy loads in both vertical and transverse directions. One solution is to install them after jacket installation. In this case jacket verticality tolerance in installed condition may impose some difficulties in caisson placement. To avoid diver operation, lower clamp may only be a guide with small tolerance. Large gap means caisson may deform considerably against wave loads. This will bring its structural response close to cantilever condition. Even if caisson and internal piping are installed by jacket itself, pump vertical shafts shall be installed after jacket is fixed in position.

10. This jacket had one internal riser and two J tubes. Only part of riser and J tube are visible in this sketch. A 4″ MEG injection line is piggybacked

to 32″ export riser. Export riser is clamped at jacket elevations to horizontal members and one intermediate clamp to side brace. MEG piggyback line is clamped to export riser at 3-m intervals. This is to protect it against wave force and vortex-shedding effects.

11. During launch or when jacket is in upright position (before flooding or grouting) parts of leg located at lower elevations are subject to very high water loads. This is especially true when leg buoyancy is used to float it during installation. External stiffeners are used to prevent leg wall collapse due to hydrostatic water pressure.

Temporary analysis covers jacket during fabrication, transport, and installation stages. These include load out, transportation, lifting, launch, upending, on-bottom stability, and pile derivability.

Permanent analysis covers the complete platform in its final location with topside installed and all equipment in the operating condition. These analyses include in-place, seismic, and fatigue.

Miscellaneous analysis like vortex shedding, cathodic protection, boat landing, barge bumper, and mud mat design are also part of the job.

The general criterion is that jacket members shall be designed for permanent cases. Then they are checked for temporary cases and if possible either the methods and in the last stage the members are modified.

Fabrication

One jacket fabrication sequence (practiced by many yards) for four-legged structures is to fabricate all legs on temporary supports on the ground. They are placed with distances exactly equal to their distance at each jacket elevation. Braces and horizontal members for the outermost legs are welded. Then using several heavy-lift cranes, each row is rolled up on turning saddles to vertical position. To reduce friction, Teflon pads may be installed on turning saddles. After a few rollings, Teflon pads may wear out. To reduce costs, grease may be applied between jacket leg and inner surface of rolling saddle. Several small nozzles may be installed at bottom of rolling saddle. This facilitates grease injection.

Normally crawler cranes are used for roll-up operation. Crawlers are arranged in the movement direction (perpendicular to row). In addition to exerting smaller and uniform pressure on yard surface, crawler cranes are less sensitive (compared to wheeled cranes) to abrupt vertical movements like sudden soil settlement. Yard surface is more or less level, which minimizes vertical movements.

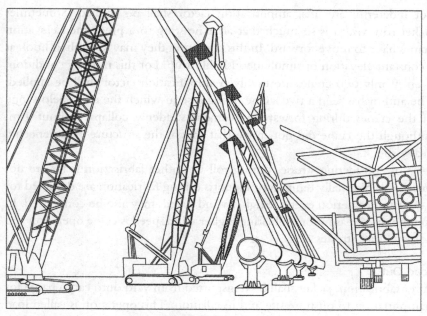

Figure 2.26 Jacket row roll-up.

Horizontal elevations that had already been prefabricated on the ground are then welded above and below to ensure the structure becomes stable. Roll-up operation is shown in Fig. 2.26. Conductor guide slots are visible at front row. This jacket houses 16 slots. Three to four slots are kept for future usage. Extra slots are reserved for future hydrocarbon or injection wells.

This picture shows three cranes lifting one row. In the background of the figure three jacket elevations with their conductor guides have already been erected in upright position. Rows will be welded to them when in final position. Connecting members of horizontal elevation are not constructed. After jacket row is rolled up, actual lengths will be measured for cutting and welding.

Roll-up saddles are also visible in this figure. They are installed on large concrete blocks. At joint locations wooden planks are installed to prepare a safe working bench for welders at elevation.

Simultaneous operation of three cranes requires very good coordination. Therefore code introduces amplification factor for the case when several cranes are working together. Any one of them hoisting faster than others will absorb more load. For onshore lifting no load factor is required.

For indeterminate lifts, amplification factor shall be applied. Sometimes jacket row width is so much that after hoisting to a predefined elevation cranes have to move forward. In this condition they may keep their hook at a constant elevation or simultaneously lift it up. For this moving condition, even if only two cranes are involved amplification factor shall be applied. The author has seen a two-crane lifting case in which the earth below one of the cranes during forward movement suddenly collapsed about 1 m. Although the crane did not move that much, the structure experienced a heavy shock.

Not considering stresses during roll-up, other fabrication loads are not so critical. Normally temporary supports during fabrication are designed for 1-year return period events. Higher wind speeds may also be considered. In fact the main reason for evaluating higher wind speed is crane operation and not structure stresses.

Load Out

After fabrication, jacket shall be transported from yard onto cargo barge for transportation to offshore site and installation. This operation is called load out [1,5,14]. Based on jacket weight and yard facilities, several methods may be used:

- Lifted load out: in which the structure is lifted by a crane barge or a combination of several land-based cranes and placed on cargo barge.
- Trolley load out: in which the structure is placed on several trolleys. They are pulled via winch installed on the barge or crawler crane. In using trolleys it is better to pull them. Therefore even if crawler crane is used, they are connected via slings to the pulling eyes installed on the barge. Crawler crane moves toward yard (away from quay side) and trolleys are pulled toward barge. This has the advantage that it pushes the cargo barge to the skidway, thus eliminating use of heavy mooring lines. This method needs very long slings.
- Buggy load out: in which jacket is placed on one or two buggies. The main advantage of a buggy is that it has a large number of wheels, hence the load on each one is very small. Besides they can turn very sharp angles and are ideal when yard has space limitations. In addition, wheels are hydraulically connected. Therefore whenever one of them settles (which may happen due to yard unevenness) load is transferred to other wheels and buggy elevation is kept at horizontal level. In spite of this, they are very expensive and have low capacities, therefore cannot be used for very large topsides.

Figure 2.27 Lift load out.

- Skidded load out: in which jacket is placed on skid beams either with skid shoe or on launch truss. Skid beam is a steel plate girder, sometimes with double walls, which is laid on skidway surface. Its function is to distribute applied loads to yard tolerable values.

Fig. 2.27 shows a lifted load out. This is an eight-legged jacket with ~35 m height. Crane barge maneuvers perpendicular to quay side and uses anchor winches for stabilizing and next-stage movement. At first stage it has to be near quay side as much as possible. This will enable efficient usage of crane working radius. In second stage, crane barge moves away from quay edge to open space for transportation barge. This movement is done while crane barge is holding topside. This is possible only if waterway has sufficient width. Based on waterway traffic, blocking passage of other vessels is not allowed for a long time.

Main hook is connected to padeyes at internal side of second elevation. This is to compensate crane barge elevation limitation. After offshore installation, diver shall open hooks.

As can be seen this jacket has only four small mud mats. This means either seabed is stiff or jacket is light or a combination of both. Boat landing at side and external riser with protection guards are visible. As is clear, jacket internal span is completely free to a considerable depth. This shows related topside will be installed by floatover method.

After jacket is lifted, crane barge uses its anchor line to move away from quay side. Then transporting barge is moved in the space between quay side and crane barge. After cargo barge is fixed in position, jacket will be lowered on the intended supports.

Fig. 2.28 shows a jacket loaded out using buggy. Some important features are explained:

1. Four horizontal elevations plus mud mat are shown. Mud mat is covered with steel plates instead of wooden planks.
2. Each elevation contains 16 conductor slots. Four slots will be left for future usage.
3. This jacket does not have any launch truss. It will be lift installed. At first stage, crane barge will lift it from transport barge and float it over water. Then hook connections will be changed and crane barge will gradually upend jacket by flooding its legs or if needed buoyancy tanks. Lifting wires are secured on the rigging platforms. Two separate rigging platforms are installed.

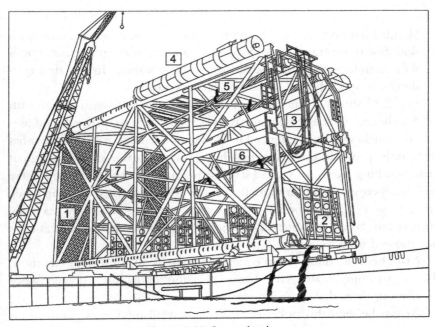

Figure 2.28 Buggy load out.

4. Temporary installation buoyancy tanks at top and slack wires at front elevation are visible in the sketch. Weld lines between plates constructing vessel wall are visible. This tank had a diameter equal to 2.3 m. One perimeter length is equal to 7.2 m. Plates with 1.2 or 2.4 m width were rolled in longitudinal direction to construct this vessel. Rolling tolerances in diameter size or total perimeter curvature are rectified with pulling fixtures from inside each segment. After matching tolerances, fabricated segments are placed near each other such that consecutive longitudinal weld lines have 180 degree difference. Each buoyancy tank has an inlet and outlet valve. To prevent valve damage during installation, it is located inside a protecting segment. Lifting padeyes are installed both on flat ends and on the perimeter. Internal stiffening is installed below padeyes to prevent plate buckling.

5. Two caissons are connected with clamps to first and second elevation. Considering end flanges it seems caissons' internal piping is installed. Submersible pumps will be installed with topside. At horizontal position pump shaft may experience large stresses and get damaged. Caisson structural strength shall be checked for this position. Dead weight increased by dynamic amplification factor (DAF) shall be used for checking. At installed condition caisson is in vertical direction. Therefore applicable load includes water particles' horizontal velocity due to current and wave. Normally caisson diameter is governed by pump/connected piping dimensions. Its thickness is governed by minimum wall thickness and corrosion allowance. Therefore caissons are normally safe against environmental forces. To protect them against corrosion, internal plate anodes may be installed. In addition, at splash zone level corrosion allowance is considered.

6. Riser at first elevation starts from one side but turns and almost diagonally leaves from opposite side at the bottom of the jacket. At each elevation it is clamped to horizontal members.

7. Anodes connected to jacket members and to mud mat are clear. As can be seen they are only installed below splash zone. In this area seawater, which acts as electrolyte, is not always available. For splash zone, special painting and corrosion allowance extra thickness are used.

As jacket moves forward, load is transferred to transportation barge. To keep barge at the same elevation as quay side, pumps deballast water in the tanks. Most probably simultaneous with deballasting at entrance, end tanks located at barge opposite end are also being ballasted. Using buggy is for medium-size jackets and has several advantages:

Figure 2.29 Buggy entering cargo barge.

- Main advantage of buggy is that jacket load is distributed uniformly on the supporting structure over many wheels. This reduces unit load.
- In addition, if an axle undergoes settlement buggy hydraulic system automatically compensates load change.
- The third subject is that buggy has very good maneuverability.

Fig. 2.29 shows a close-up view of Fig. 2.28 jacket transfer via buggy at entrance to cargo barge.

1. Wires connecting cargo barge bollards to quay side bollards are visible at front.
2. Very thick steel plate to enable buggy wheel movement is shown. It can be replaced with stiffened link beams between yard and barge. Its top flange shall have enough width for buggy wheels. Link beam provides rotational flexibility. Buggy can overcome small level differences. In theory barge deck and quay side shall be level.
3. Buggy hydraulic system on each wheel is clear. This is the main advantage of load transfer with buggy. It automatically balances level differences and redistributes load.

4. Installed anode below level −4.0 shows minimum splash zone elevation.
5. Shown piping (connected to leg) is used for grouting. After pile driving, the space between jacket leg internal surface and pile external surface is filled with grout. It starts from the lowest elevation. If a grout seal with sufficient pressure rating is not found, a short duration can be allowed for grout above seal to harden. At this duration no grout shall remain inside injection pipe. After grout passes a certain height, a new injection point or pipe can be used to inject at higher elevations. This will reduce injection pipe and pump-design pressure.

In load out with buggy or trolley the rolling friction factor (less than 1%) shall be overcome while in skidding the sliding friction factor (up to 20%) shall be overcome. Weight of the platform multiplied by friction factor plus safety margin determines pull/push capacity. Applying grease or welding Teflon pads on the skidway surface reduces friction factor considerably. Wooden timbers are placed below skid shoe. Hard wood can tolerate considerable loads. This can be used under launch truss as well. At start of movement, friction is the largest (called static friction). After movement is initiated, friction factor reduces (called kinematic friction).

Codes have given some recommended friction factors for different conditions. Friction factors equal to 8% for kinematic and 25% for static have been used.

The pulling/pushing equipment capacity shall be much more than the calculated required force. In addition, a retrieving system always shall be employed to enable pulling the jacket back to the yard. The retrieving system shall be a fast system. Therefore winches, crawler cranes, or bulldozers may be used. Jacket loadout operation may be divided in two phases: before COG leaves the yard and after it. In the first portion retrieving operation is possible. In the second portion loadout shall proceed under any circumstances.

One hydraulic power pack is used for both jacks. This is to ensure both jacks operate simultaneously. Its capacity shall be sufficient to feed both jacks. Since the internal surface area of jacks is equal, this means equal force is exerted on both sides. In some cases an obstacle may exist in the path. This obstacle may be more friction on one side (uneven grease application) or a remaining welded piece, etc. This causes uneven movement of the jacks. A regulatory valve can be installed in hydraulic oil path, which permits equal oil volume injected in each line. Equal oil volume means equal opening or closing of jack cylinder. This prevents uneven movement.

Crane barge is connected to the skidway. Before jacket exits yard, transportation barge is ballasted down to be well below jacket bottom elevation. After jacket exits a certain distance (normally to the first hard point), barge is deballasted. In this condition barge surface touches bottom of jacket skid and pushes it up. At this stage it does not lift jacket, only pushes it upward to get some positive force on the barge. This is essential so that due to water surface variation (due to small waves) jacket does not hit barge or skidway surface. As long as jacket center of gravity is inside the yard, the same condition is kept by gradually deballasting barge. After COG leaves yard area, both forward movement and deballasting can proceed faster.

During load out jacket may experience several loads:

- The first load condition is the uneven load on the rows. If hydraulic push/pull or winches or crawler machinery is used, there is always a possibility that one jacket leg/row is pulled/pushed more than the other. This induces a skew load on one row causing a couple. In many analyses 10% total pull/push force has been used. Using synchronized hydraulic push/pull jacks lower skew forces, and using two crawler machinery larger skew forces, shall be applied.
- After several years of fabrication in a yard, the soil under skidways becomes compact. However, always during load out the possibility of uneven settlement is considered. Several scenarios of one or two support settlements are considered. Normally 1″ settlement is enough.
- Loss of support may also happen. If either the skid beam or jacket itself is stiff enough, loss of support deflections may be limited to 2″ settlement. In fact, first the support is removed in the structural analytical model. Then the settlement is reviewed and it will be decided to remove it completely or limit it for example to 2″. In one case the author has encountered 4″ settlement criteria.

Transportation

Offshore voyage route and timing shall be carefully selected. Environmental conditions of the route shall be properly studied. For short distances code-specified accelerations shall be used. For longer distances motion analysis shall be performed. Before starting offshore transportation, seafastening members shall be completely welded. They are members that connect jacket or its skid shoe to cargo barge. Seafastening configuration shall be selected in a manner to allow rapid removal during offshore installation.

Correct understanding of the coordinate system moving with a barge (ship) is important. Front side is called bow (fore). Back side is called stern (aft). An observer standing in the stern and looking in the bow direction has the same position as the ship pilot. In this case his right-hand side is starboard and left is portside. For a definition of the origin of this naming system, please refer to naval architectural books. It dates back to the Vikings who used a boat styribord (steering board) with their right hand, which normally was the stronger hand.

The origin of coordinates is normally at the lowest end point. For some ships that do not have a smooth keel, a theoretical smooth plane is taken as reference. Positive X-axis starts from stern and points toward bow. Positive Z-axis is pointing upward. Positive Y-axis follows the right-hand rule. This means positive Y is directing to portside.

Any vessel has six degrees of freedom. Three of them are translational and three rotational. Translational degrees of freedom in X, Y, and Z directions are called surge, sway, and heave. Rotational degrees are called roll, pitch, and yaw. Among them, three are most important on stresses induced in offshore structures:

• Roll is barge rotation around its longitudinal axis. As per code for large cargo barges, 20 degree roll within 10 s shall be analyzed. Static displacement in the same direction is called heel.

• Pitch is barge rotation around its mid-length transverse axis. As per code for large cargo barges, 12.5° pitch within 10 s shall be analyzed.

• Heave is barge movement in vertical direction. As per code for large barges, 0.2 g shall be analyzed.

It has to be emphasized that ship motions are defined based on its midsection.

Different combinations of these movements shall be analyzed [1,5,14].

Fig. 2.30 shows a jacket with marine fleet including cargo barge and two tugboats during offshore transportation. A second reserve tugboat is directed in the opposite direction.

Small waves were visible on the original photo. Unfortunately, they are not clear in the redrawn sketch. Sea State was between Beaufort 3 and 4. Wind velocity at this condition may be less than 10 knots. Significant wave height may be between 0.5 and 1.0 m. At high sea states, transportation fleet has to find a safe haven. High wind velocity and large waves considerably increase the necessary pulling load and endanger transportation barge stability.

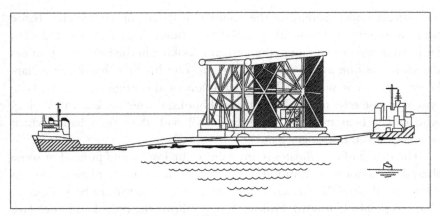

Figure 2.30 Offshore transportation.

Installation

Jacket installation consists of several steps. First, the cargo shall be lifted or moved from barge deck to be placed on the water surface. Then, it has to become upright and rest on the seabed. In the last step, piles shall be driven to stabilize it. In some cases (like vertical lift) two steps can be performed in one stage [1,5,14].

Placing from the cargo barge on the water surface is done either with lifting or launching. Offshore lifting is similar to nearshore lifting. Code only increases the amplification factor. This is to cater both for the increased consequences of any failure in offshore lifting and for crane barge movements. For short jackets offshore lift can be in vertical position. This allows direct transfer from cargo barge to seabed.

Fig. 2.31 shows a vertical lift. In this figure a sheer leg crane barge is used. These barges have a good capacity in their longitudinal direction but are not able to rotate. The transportation barge is brought in front of it. Then the jacket is lifted and the barge is pulled away. In the next step the jacket is lowered on the seabed.

Barge hook had sufficient elevation. If main hook elevation was not enough (similar to Fig. 2.27) padeyes could be connected to lower hard points. To use this configuration, location of internal members shall be checked to ensure slack and tightened slings do not clash with jacket members. Lifting wires are placed at ~60 degree configuration. Since a single-piece wire with sufficient length was not available, two wires were connected via center shackle. To facilitate wire release, connection to jacket can be with trunnion instead of padeye.

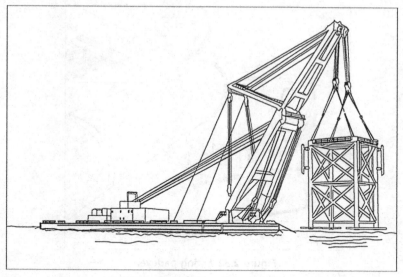

Figure 2.31 Offshore vertical lift.

Jacket barge bumpers and internal/external installed risers are visible. The number of risers show this was a manifold platform.

This jacket was installed at shallow depth. Total jacket height from mud mat to leg pile connection point was ~27 m.

Normally long jackets are placed sideways on the barge. After lifting they are floated on the sea surface. At this stage if leg and braces' buoyancy is not sufficient, two buoyancy tanks may be attached. After jacket is placed on the water surface, the crew will change lifting slings. Usually lower connections are with trunnion and top connections with lifting padeye. Fig. 2.32 shows a lifting padeye and Fig. 2.33 shows a trunnion.

1. This padeye is connected to a jacket leg. Heavy welding is clear. At one point padeye weld line becomes near brace to leg weld.
2. Horizontal members and lateral braces are shown. Their weld lines also have a small gap. Overlapping permanent welds are avoided.
3. Weld line connecting joint CAN to leg is also clear. In this case the external diameter of tubular members is kept constant. To facilitate pile driving in most conditions, the internal diameter is kept constant.
4. Jacket leg elevation marks are shown a little bit below padeye and joint CAN weld line.

As is clear, the lifting padeye includes a main plate and two cheek plates with stiffening ribs at both ends. A hole is driven in the middle that keeps

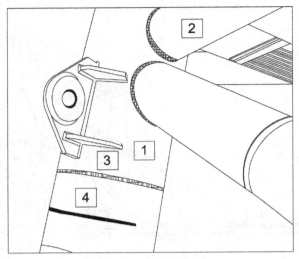

Figure 2.32 Lifting padeye.

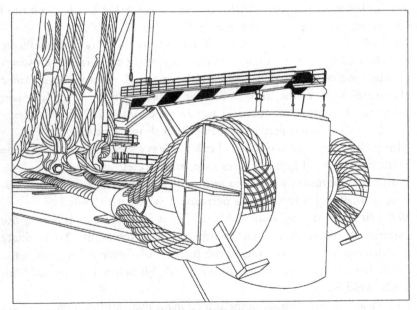

Figure 2.33 Trunnion.

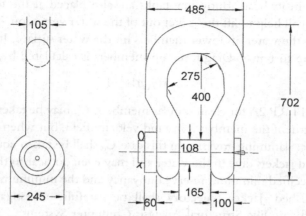

Working Load Limit 150 Ton Weight 160 kg

Figure 2.34 Sample green pin heavy duty shackle dimensions.

the pin of the shackle. Lifting sling is connected to shackle. To connect sling to a lifting padeye, lot of effort is needed. It has to be noted that a shackle suitable for lifting 200 tons weighs about 235 kg. Fig. 2.34 shows typical dimensions. It cannot be installed without using a crane.

A trunnion consists of two tubular members welded to a lifting point that may become the extended part of the topside main leg. Since during operation the platform weight is transferred via main legs to the jacket, it will be more beneficial if lifting is also done through the same legs. To connect the lifting sling to the trunnion it is only required to lower the sling nearby it and allow it to pass one of the smaller tubulars. In Fig. 2.33 the interwoven part of slings are embedded in a concrete block for better protection. Due to sling length limitation, an intermediate shackle connects two slings from trunnion to crane barge main hook.

For jackets, normally the trunnion is placed near the bottom of it and a padeye is used for above-water locations. For topsides, trunnion or lifting padeye is located at the highest elevation.

In padeye design several parameters shall be controlled. They include:
- Shackle pin bearing on main and cheek plates
- Main and cheek plates' shear stress
- Bending strength of padeye body in connection to the leg
- Shear/tensile stress of the weld connecting padeye to leg

Launching can only be performed if jacket has launch trusses and barge is equipped with a rocker arm. Launch trusses are visible in Fig. 2.25. Barge rocker arm allows jacket tilting before final drop to water. Jacket is launched from its head. Buoyancy tanks are also placed at the top of the jacket. This will help to lift the jacket out of the water very soon. When the jacket enters the water, its lower members hit the water surface. Its velocity may reach up to 6 m/s. Drag force on members is calculated by:

$$F_d = 0.5 C_d \rho V^2 A$$

As per API RP 2A for clean tubular members, C_d may be taken equal to 0.7. This is true if the member is in fluid velocity field, but when it hits the water surface, slamming governs. In this case C_d shall be taken equal to 5.5.

Launched jackets dive in the water and may even go near to the seabed. Code has specified minimum reserve buoyancy and the minimum tolerance from the seabed. Jacket trajectory shall be carefully investigated using specialist software like Structural Analysis Computer Systems.

For upending, a crane barge lifts the jacket from two points. Then while the main hook is lifting, its head auxiliary hook releases slings, which allows the jacket to become vertical. When it is placed on the seabed, mud mats shall provide temporary stability. Code has specified minimum safety factors against overturning, sliding, and settlement.

After upending and before pile driving has finished is a very critical stage. Immediately after the jacket has become vertical, pile driving shall start. First segments are placed in all four legs. First segment height shall be at least equal to jacket leg length plus a minimum extra piece. Pad ears are welded to the top of the pile segment, which does not allow pile penetration. The second segment (first add on) length shall consider an estimated first penetration under pile segments plus the weight of the hammer.

The number of pile segments shall be minimized to consume the least offshore welding time. Pile driving may proceed in a diagonal pattern. After all segments at each stage are driven, jacket levelness will be checked. If one side of the jacket is lower than others, using lifting jacks it will be lifted. Lifting one leg of the jacket in small steps is possible. Pulling it to a large distance requires a large pulling force and may damage the leg itself.

After jacket elevation becomes satisfactory, next pile segments in all legs are placed and welded. Assuming Ø1500 × 50 pile with 45 degree bevel, weld volume is calculated to be 4630 cm³ equivalent to 36.3 kg electrode rod. Assume a welder is using 5-mm-diameter electrodes and can melt each 40-cm-long electrode within 2 min. For this volume he has to melt about 590 electrodes. If two teams are working, one leg welding will take about

10 h. To this time shall also be added the duration needed for jacket leveling, placing next segment, welder/fitter replacement, idle time, and NDT duration.

Time in offshore activities is a very precious commodity. Therefore a maximum number of welders will be mobilized. It is important to note that in some clayey soils, a setup phenomenon happens. During pile driving soil will be pressurized. Especially in piles that have experienced plug conditions, soil will be displaced and therefore stressed. When driving is stopped, water will migrate from soil voids due to increased pressure. This will increase soil effective pressure and consequently its resistance to driving. This phenomenon is called setup. If sufficient time has elapsed between driving of two segments, setup may gain values such that driving of the next segment becomes impossible. In this case two approaches can be taken.

The first solution (and maybe the most expensive) is to drill inside the pile and remove internal soil. Drill bit will penetrate much below the end of the pile. This will reduce internal skin friction and part of the end bearing and therefore will reduce soil resistance to pile penetration. After soil removal, the pile is redriven. After breaking the static resistance, soil strength will decrease and the pile can be more easily driven.

The second approach is to find a much heavier hammer and start redriving. Pile refusal criteria will be defined in the geotechnical section.

2.10.2 Topside Analysis

Topside analyses are again divided similar to the jacket. In topside, permanent analyses include in place and seismic. Fatigue is not a determining factor in topside members' analysis. Temporary analyses include load out, transportation, and lifting. Miscellaneous analyses include helideck design and equipment/piping support design.

Although number and topics of topside analysis are less than jacket analysis, considering the variety of loads, members, and interfaces with other disciplines, the work load may even be higher.

Topside structure consists of truss-like members in rows. Primary elements connect these columns and braces. Secondary members covered with steel plate or grating covers the space between primary members. Columns and braces are tubular members while primary and secondary members are normally I beams. Fabricated plate girders are also used when adequate-size rolled sections are not available. Acceptable tolerances in girder fabrication like end-to-end dimensions, curvature in vertical and horizontal elevation, twisting, etc. are very low and shall follow specified limitations. Flange and web welding shall be based on predefined procedures with suitable fixtures. Using I beams is at the same time faster.

Topside operational loading is more complex than jacket while environmental loading is easier. Environmental wave and current loads have less impact on topside. Their action induces bending in the platform, which has some sort of secondary impact on topside stresses. Wind load in global analysis causes shear which shall be taken by braces. For single elements, instantaneous gust speed is used.

2.10.2.1 Piping/Packages/Vessels Loads

Piping/equipment (packages/vessels) supports on the topside exert heavy loads on individual members. Small reaction forces may be applied as uniformly distributed load but heavy forces (above 50 knots) shall be individually applied in the model and locally checked. In some projects this limit may be taken as 100 knots. This will cover piping supports, packaged equipment, and vessels' loads on secondary members.

Vessels may have either large dead weight or large volume that causes large operational load or both. High-pressure vessels have a thick wall. Therefore their dead weight is very high compared to atmospheric vessels or storage tanks. Assume a 3.0-m-diameter vessel containing crude at 120 bar internal pressure. If a steel plate with 350 MPa yield stress is used in its fabrication, a minimum thickness equal to 86 mm will be necessary. This is only for structural strength of the shell due to hoop stress. Stresses due to operational loads like weight of the inside fluid, thermal stresses, etc. shall also be considered. In addition, a corrosion-resistant layer and local-strengthening thickness shall be added. For a guide please refer to Section 3.1.3 "First Phase Separator."

Every meter length of the shell of this vessel weighs more than 64 kn. Weight of nozzles, flanges, internal fixtures, stiffening plates, corrosion allowance, etc. shall be added to this weight. If T/T^1 length of this vessel is 10 m and headers are assumed to be semispherical with the same thickness, total dry weight of this vessel may be well over 900 kn.

---

[1] T/T = Tangent-to-tangent length. Distance from tangent point of one header to the other in vessels. All headers have an additional piece from the tangent point to the weld line to the cylindrical portion. This portion that is fabricated with the header is included in the cylindrical portion length.

Vessels are supported on two saddles. If geometry of the platform or equipment layout permits, each vessel may be supported by two or three secondary members. Saddle width may be between 2.0 and 2.5 m and normally less than vessel diameter. Saddle spacing shall be carefully selected considering the following items:

- It shall match (as far as possible) with platform structural members.
- Assuming vessel to act as a uniformly loaded member, saddle location shall distribute internal loads evenly:

  Distance from each end to saddle centerline $= L_1$

  Distance between saddles $= L_2$

  Total vessel length, $L = 2 \times L_1 + L_2$

  To equalize bending moment of cantilever portion with midsection bending moment shall have: $L_1 = 0.21 \times L$

- Saddle weld lines shall not interfere with nozzle locations and adjacent plates weld lines.

  Not all three criteria can be satisfied simultaneously. A compromise between these criteria shall be reached. Saddle weld line away from nozzle welds is always followed. In addition there should be enough space for connections to nozzles.

  For installation purposes, normally a vessels' saddle rests on a small pedestal 20–30 cm height. One pedestal is fixed while the other is sliding support. This will take care of vessel expansion due to temperature changes. On the pedestal supporting the sliding saddle, Teflon pads may be installed. They reduce steel-to-steel friction factor and prevent rusting. Fig. 2.35 shows a vessel resting on a pedestal support.

1. Two pedestals, one in front and the other in the back, are fabricated. Only the front pedestal is clear. Pedestal and saddle final coating cover are different. One reason may be that vessel is painted in vendor fabrication shop and painted as equipment. But pedestal is fabricated in EPCIC contractor yard and painted as part of platform structure. Structural configuration of all pedestals is similar. They consist of top and bottom flanges supported by a middle web. Stiffeners are located at specific intervals to prevent web buckling. Pedestal usage facilitates bolt tightening. In addition, fixed or sliding support can be easily provided.
2. Level gauge is visible near front saddle. It consists of a pipe connected to top and bottom of the vessel. Two separate indicators with a small overlap cover bottom and top portions. To protect the glass tube, metallic plates bolted to each other enclose it. For clarity, support of the top section gauge is deleted. Bolt spacing on the cover plates is in 10-cm steps.

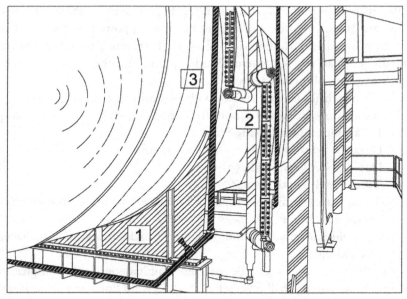

Figure 2.35 Vessel saddle resting on a pedestal.

3. Fire ring nozzles are visible at front and middle section. The third ring covering the back section is not shown. Their main intention is to cool down the vessel and its containments to prevent fire spreading.

Other than operational loads, hydrostatic test load of vessels is also a critical case for local design. In a hydrostatic test it is filled with water (higher–density fluid) to a pressure 50% above design pressure, which itself may be more than 20% above operating pressure. This ratio is defined by allowable pressure based on piping class rating.

For many vessels (which are normally filled with gas) hydrostatic test conditions may govern local design.

For the above vessel, ID = 3.0 m and T/T = 11.0 m with semispherical heads. Fluid and gas weight in operating condition will be about 32.0 ton while water weight during hydrostatic test will be more than 85 ton. In hydrostatic case, each saddle will transfer about an 87-ton load.

One of the global load cases that may be used in lieu of more refined load distribution is 10-kn blanket loading. This is applied all over the deck area to represent structural dead load, equipment load and piping, electrical, instrument, safety, and architectural bulk loads. This load condition is useful for jacket design when topside load distribution is not yet clearly defined. Although this blanket loading is more or less sufficient for skid packages less than 100 kn, it is not useful for large vessels that may weigh over 1400 kn and are supported on two saddles.

In topside modeling partial or full jacket may be modeled. With modern computers it has become very easy to include a full jacket model with piles and include soil interaction against lateral environmental loading. To reduce computer output, jacket members' stress check will be skipped.

In addition to vertical loads as explained previously, horizontal loads due to expansion shall also be considered. Section 2.7.5 has provided some insight on this subject. During shutdown, platform reaches ambient environmental temperature. Platform operating temperature is much higher than this value. For large vessels temperature distribution may not be completely uniform. For piping it is uniform. It is seen that 50°C temperature difference may increase stresses up to 110 MPa. This only happens if both supports prevent sliding. By multiplying this unit stress change to a large vessel cross-section area, huge loads are calculated. The only solution is to allow one sliding support. Careful detailing shall be constructed to ensure minimum friction factor. This may be accomplished by attaching Teflon pads on stainless steel surfaces. Axial load due to change in temperature is so high that it will even overcome steel-to-steel sliding friction for normal weights. Based on the selected detailing, a sliding force using a suitable friction factor shall be used in vessel-support pedestal design.

2.10.2.2 Blast Design

Blast design is a special case. Very high loads act on a small area for a very short period. Peak blast load is applied within 0.2 s. Typical positive pressure duration may range from 50 to 200 ms. Shorter duration is normally associated with higher overpressure values. This very rapid load application has a reduction effect due to inertia of the structure. Therefore in simplified analysis using an elastic approach allowable stress is taken equal to yield strength. In fact the structure experiences plastic behavior in which material stresses pass yield strength level. After yield stage, mild steel has a strain-hardening behavior that can accept stresses higher than yield strength. In some cases, for simplicity elastic analysis and design is considered for blast design. In this approach we shall consider that plastic modulus (Z) is larger than elastic modulus (S).

- For strain hardening, code allows 1.10.
- For I beam sections, code allows $Z/S = 1.15$

For a rectangular section having width "b" and depth "d" referring to M_e and M_p, the following definitions can conclude $Z/S = 1.5$. For I beams this value reduces:

$$M_e = \frac{bd^2}{6}\sigma_y$$

$$M_p = \frac{bd^2}{4}\sigma_y$$

For I beams with flange width b, flange thickness t_f, total depth d, and web thickness t_w:

$$M_e = \frac{I}{C}\sigma_y$$

$$M_p = \sigma_y\left[bt_f(d - t_f) + (d/2 - t_f)^2 t_w\right]$$

In the previous formula and in order to simplify calculations, we have neglected curvature in connecting web to flange. For HEA beams, M_p/M_e ratio may be a little less than 1.15 and for HEB beams it may be larger than 1.15.

Using software that calculates allowable stresses based on working, stress methods have to include modification factors to increase allowable stresses to yield strength.

- For bending along weak axis, $0.75\ F_y$ is used. Therefore modification factor becomes $1.1 \times 1.15/0.75 = 1.69$.
- For bending along strong axis in compact sections, $0.66\ F_y$ is used. Therefore modification factor becomes $1.1 \times 1.15/0.66 = 1.92$.
- For axial tension, $0.6\ F_y$ is used. Therefore modification factor becomes $1.1 \times 1.15/0.6 = 2.11$.

Blast pressure reduces with distance from blast center. Therefore for design of a wall in which blast pressure is applied on a large area, unit pressure is reduced with factor 0.8. This is not a code-specified value and has to be agreed on in a rational basis.

Blast overpressure study shall be performed to determine expected value. In several projects 0.3 bar pressure has been used. This is in lieu of the fact that the actual calculated overpressure study showed it to be less than 0.2 bar.

Dynamic amplification factor shall also be included in this analysis. Normally blast wall consists of I beams welded to top and bottom secondary members and placed at 1-m intervals. The distance between them is covered by steel plate. Size of I beam, spacing, and steel plate thickness shall be calculated in each case. The blast wall structural system affects its dynamic behavior. A rigid structure will have a low period of vibration that is near to the blast-loading period. This will induce a high DAF. On the other hand, if a flexible structure is selected, it may show large deformations and impair platform operational conditions.

Frequency of vibration of a uniform beam simply supported or fixed at both ends with uniform loading is taken from Pope [4] and Young and Budynas [32].

$$f_n = \frac{K_n}{2\pi} \sqrt{\frac{EIg}{wl^4}}$$

For the first mode of vibration and simply supported condition $K_n = 9.87$ and for fixed support condition $K_n = 22.4$. Assuming HEA200 I beams placed at 1-m intervals and covered with 10-mm plate. Distance between top and bottom fixity point of this wall is 4 m. Combined moment of inertia is calculated to be 7196.4 cm^4. For a simply supported beam the first mode period of vibration becomes 0.15 s and for fixed end condition it is 0.07 s. This is a relatively stiff configuration. For an effective blast loading with a period of 0.2 s, DAF for these conditions are, respectively, 2.18 and 1.12. It is very difficult if not impossible to calculate an accurate DAF. Boundary conditions of the wall configuration, behavior of the bottom and top secondary or primary members, effects of other structural elements like braces, etc. cannot be accurately modeled. When effective force period is near to the structural system, DAF varies considerably. In theory, when force period is equal to structural period, DAF becomes infinite. It is noted that assuming a very high DAF, applied load will increase and it is necessary to design a big structure. This does not lead necessarily to a safer platform. In many cases DAF = 1.5 has been used.

2.10.2.3 Temperature Impact
Structures near the flare have a special load condition due to excessive temperature. During emergency flaring flame length may reach several tens of meters. Members near the flare tip may be heated up to 400°C. Carbon steel elements undergo large changes in modulus of elasticity and yield stress due to high temperatures. API RP 2A has provided a table showing percentage reduction due to high temperatures. At normal room temperature mild steel reaches yield stress at 0.2% strain. Table 2.14 shows reduction factors in temperatures from 100°C to 600° C.

Table 2.14 Yield stress and modulus of elasticity reduction versus temperature

Temperature °C	100	150	200	250	300	350	400	450	500	550	600
API % reduction F_y	0.940	0.898	0.847	0.769	0.653	0.626	0.600	0.531	0.467	0.368	0.265
CTICM % reduction F_y	0.961	0.932	0.898	0.857	0.811	0.758	0.699	0.632	0.557	0.472	0.377
CTICM % reduction E	0.979	0.962	0.941	0.916	0.885	0.847	0.802	0.748	0.683	0.603	0.505

This table shows at 250°C more than 25% and at 450°C about 50% yield stress is reduced. In temperature impact analysis, only deadweight and applicable operational loads are included in the model. Temperature isothermal curves have spatial variation. This temperature affects flare tower and bridge or nearby structures. Although temperature may vary at beginning and end of a structure, the worst temperature is included in analysis.

API table provides yield stress changes in different strains, ie, 0.2%, 0.5%, 1.5%, and 2.0%. At low strains yield stress and modulus of elasticity are assumed to vary in proportion. However, engineering judgment is required to assign correct value for F_y and E.

Other institutes like CTICM[2] show separate relations for F_y and E.

$$\frac{F_{y\theta}}{F_y} = 1 + \frac{\theta}{900 \times \ln\left(\frac{\theta}{1750}\right)}$$

$$\frac{E_\theta}{E} = 1 + \frac{\theta}{2000 \times \ln\left(\frac{\theta}{1100}\right)}$$

θ: Temperature in °C (for $0°C \le \theta \le 600°C$)
E: Initial Young's modulus
E_θ: Young's modulus at temperature θ
F_y: Initial yield stress
$F_{y\theta}$: Initial yield stress at temperature θ

The thermal coefficient of expansion is $12.0 \times 10^{-6}/°C$. This formula will relieve the engineer of deciding which strain shall be used in defining Young's modulus.

2.10.2.4 Topside Fabrication
Topside elevations are first fabricated on the ground. This is to avoid welding in elevation. Temporary supports are placed under main beams to prevent excessive deformations under own dead load. Before installation of columns and braces each elevation is a flexible plane structure. Some of the piping, electrical, instrument, and vessels supports may be fabricated with the primary and secondary members. Plating is also placed. Sometimes deck penetrations including piping, electrical, instrument, gutters, drains, etc. are punched and actual cutting is left during installation of related crossings.

[2] CTICM = Centre Technique Industriel de la Construction Metallique.

Topside fabrication requires a very good coordination between engineering, procurement, and yard teams. The best condition is when equipment of each elevation arrives at the time that elevation is ready. In this case first cellar deck is fabricated, its equipment placed, columns and braces are erected, and then lower deck structure is lifted. Similarly, lower deck equipment is placed and fabrication proceeds to the next level. Fig. 2.36 shows main deck being lifted in place. It seems all or a majority of lower deck equipment has already been placed. Six crawler cranes (three from each side) are performing this yard lifting.

In this figure six cranes (three on each side) have been used for lifting. It is also seen that they have not used temporary lifting padeyes on top of the deck.

In practice some of the equipment in cellar or lower deck are long lead. High-pressure separators, dehydration, regeneration, turbo generators, turbo compressors, etc. are all long-lead packages. If construction group wants to place them before installing the next deck, it will delay fabrication schedule. Yards may adopt several rectifying measures.

Figure 2.36 Main deck lifting.

- They may place the next deck with some temporary supports and leave some of the columns and braces for a future time. Whenever related equipment arrives, they may be skidded in. A very detailed procedure shall be prepared for this approach. This procedure shall include at least:
    - All required temporary supports and analysis to ensure next deck is stable under its own weight plus equipment.
    - The skidding way shall be carefully checked for the applied loads. Structural members in the skidding path may not be able to withstand applied heavy loads.
    - Supports, piping spools, cable trays, and other minor items may not be placed in the skidding path, or they shall be temporarily installed to enable easy dismantling.
- They may not fabricate a portion of next deck to enable lifting equipment from above it and placing it on the deck below.

Both methods have some degrees of rework and risk. But sometimes schedule constraints require to select one of them.

2.10.2.5 Topside Load Out

Topside load out has some similarities to jacket [1,5,14]. Small topsides may be loaded out using trolley or lifting. However, majority of topsides from one side are heavy enough and from second point exert concentrated loads on their supports. Therefore they are normally loaded out using skidding, which distributes loads over a relatively long span. Before load out, topside shall be weighted. This gives an accurate evaluation of total weight and its distribution on each support. Loads on supports located under one row are normally unequal. After their transfer on barge deck area this may cause a heel rotation, which shall be immediately rectified by uneven ballasting of tanks in the same row. Otherwise a continuously increased heel may lead to barge instability and topside failure in water.

At start of construction a decision shall be made about which platform side shall first enter the barge deck. This edge shall be placed toward the quay wall. Although placing skid beams from the first date are not necessary, after some stages of job progress it will be extremely difficult to lift topside and place the skid beam. Therefore skid beams may remain under topside during a majority of the construction phase. Normally wood is placed under skid shoe to distribute applied loads uniformly. Although grease is applied on the surface between skid shoe and skid beam, after a long time static friction factor may become very high. This is the reason for very high jacking capacity required at start of load out.

With topside approach on barge, deck tanks shall be deballasted under it and ballasted in the opposite direction to keep the barge surface in a horizontal elevation. Ballast plan at each step shall be determined carefully based on tide-level variation and loads transferred from topside movement. Majority of ballast plan refers to tanks at extreme ends. Due to long rotating arm to barge COG, they are very effective in barge longitudinal rotation. Small adjustment is possible with ballasting tanks located near to barge COG.

To countereffect support of unequal loads in the same row, tanks in starboard and portside may be ballasted or deballasted unevenly.

Fig. 2.37 shows a side view of a Load out support frame (LSF) after returning from offshore installation. This LSF has been used in several successful floatover installations. Designer had to match topside framing with LSF supporting span.

1. Barge skid beams with side stiffeners are installed at barge deck and LSF moves on top of it. To be able to transport different structures, moving barge skid beam over an area on deck is possible. To do this, temporary welding shall be cut.

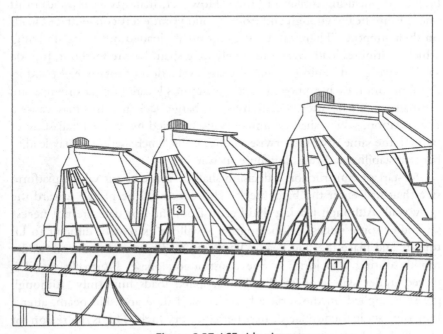

Figure 2.37 LSF side view.

2. Wooden timber to distribute topside loads is installed between LSF and skid beam top flange. It is bolted to LSF and moves with it. Hard durable wood is used.

3. Numerous lateral stiffeners are installed. Small ones placed at the edge of large stiffeners are mainly intended to prevent plate buckling. The main function of large stiffeners is to prevent vertical load concentration and improve its uniform distribution at least in the 45 degree sector. Topside load is applied on the elastomers. Normal plate girder action may not be sufficient for its uniform distribution. Although very high loads can be tolerated at barge bulkheads, acceptable loads on transverse frames and between them is much lower.

4. LSF general dimensions are shown in Fig. 2.38. Maximum overhang length from skid beam centerline is 2.0 m. Elastomer can be installed at 1.5 m. Center to center of LSF longitudinal skid beams is 20.0 m. This means LSF structure can be adapted to transport structures with support spacing from 17.0 to 23.0 m.

5. Transverse girders are placed at 15.5 m span. They also can be adjusted from 13.5 to 17.5 m spans. For each case local stiffeners shall be adjusted. Even larger differences in both longitudinal and transverse directions can be handled with suitable strengthening.

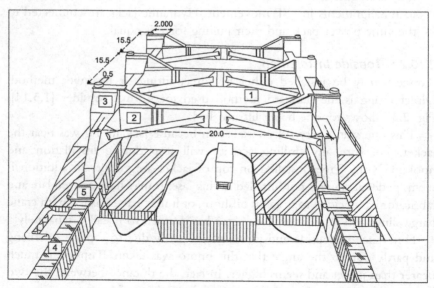

Figure 2.38 LSF top view.

Fig. 2.38 shows top view of the same LSF. Transverse beams and horizontal braces are visible.

1. In the original design, X braces were used. Later it was changed to K brace configuration to support transverse girders laterally. Although a majority of push pull jacks are equipped with a system to ensure equal volume of hydraulic oil is injected on both lines, horizontal braces will protect LSF against unequal movements at each side.
2. Side girder height is 3.5 m and rests on 30.0-cm wooden planks. Transverse beam height is 2.0 m. These large girders with thick flanges and stiffened webs enable transportation of heavy topside loads in different projects.
3. Elastomer support height is 1.8 m. This excludes 30.0 cm elastomer height. After each installation, elastomer shall be carefully examined to ensure plastic deformations have not happened. If in doubt, elastomers shall be changed after each operation.
4. In this sketch LSF is being transferred to fabrication yard. Grease applied on yard skid beam to reduce friction is visible. Land cranes and tugs have been removed for clarity.
5. Hydraulic jacks installed at each side pull LSF toward yard. Pulling action pushes barge to quay wall face. Therefore although two wires connect transport barge to yard bollards, they are not needed. However, they are tightened. Tightened wires prevent barge yaw movement due to misalignments in LSF movement. Hydraulic jacks are connected to the same power pack and their pulling force is equal.

2.10.2.6 Topside Installation

Topsides may be installed either by direct lifting or floatover method. Direct lifting is the standard method used for normal topsides [1,5,14]. Fig. 2.39 shows topside being lift installed.

This one was a very risky installation because jack-up rig was near the jacket. For sure any drilling activity will stop during installation and hookup. Xmas Trees installed on top of jacket are visible. In addition it seems pedestal crane is not installed in this case. It may have arrived late and fabricator has decided to install it offshore, or it may have clashed with crane barge slings and installation contractor has decided to install it separately.

Difference in topside and jacket size is partly due to topside overhang and partly due to the angle that this photo was taken. Topside is much nearer than jacket and seems bigger. In fact, the distance between the two jacket rows is 20 m. Topside outermost dimension is 31 m.

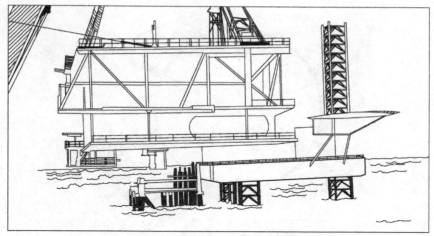

Figure 2.39 Topside lift installation.

Floatover installation is normally used for very heavy platforms. In this method topside is placed on a special cargo barge. After arriving near installed jacket location it is deballasted to have maximum clearance with top of jacket legs. Using tugboats, topside and barge are maneuvered to inside jacket legs. Then barge is ballasted so that topside legs rest on the jacket leg contact points. After topside is installed, barge ballast is increased to provide sufficient clearance with topside members before pulling it out of jacket. To enable this installation method, several requirements are necessary:

- Distance between jacket legs shall be wide enough to allow cargo barge entrance.
- Jacket shall not have any horizontal member from top to an elevation below cargo barge's lowest elevation to enable exit with sufficient tolerance.
- Bumper set is placed on jacket legs' internal surface. Normally the first row in entrance path will have one in 45 degree angle and another one in perpendicular direction. The second row may have only bumpers placed perpendicular to barge wall.
- Bumper placed inside jacket leg or topside leg.
- In some cases internal leg bumpers may be removed and long strike jacks are used. This way barge ballasting/deballasting will also be minimized.

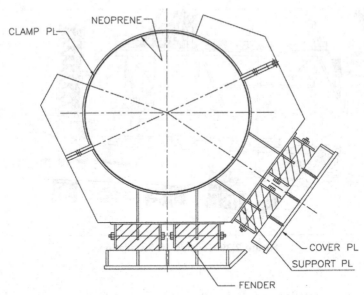

Figure 2.40 Jacket leg bumpers.

Fig. 2.40 shows a bumper configuration on jacket legs. Rubber fenders and spacing limits are visible. Some main features are described:
- Bumpers are placed at entrance row in 45 and 90 degree and in internal rows only at 90 degree perpendicular to cargo barge side wall.
- Bumpers are connected with a removable clamp to jacket leg. Clamp is fabricated in two half-circle pieces. Bolts connect the two pieces. A small gap (~6 mm) is kept between the two connecting plates. This enables to increase clamp grip on the jacket legs by tightening the bolts.
- Clamp internal surface is covered by neoprene, which has a high friction factor in contact with steel. In addition, it has a small modulus of elasticity compared to steel. Therefore applied pressure is uniformly distributed on the jacket leg.
- Large radial force induced by tightening bolts creates a considerable friction force against longitudinal and rotational slippage.
- Number of fenders at connecting point to barge wall depends on impact energy and tolerable uniform load by barge wall.
- Fenders can deflect up to 50% of their thickness and still perform elastically. Installation bolts and front panel arrangement shall allow this much deformation.

Leg contact is a very critical moment. Before this stage topside load is taken by barge. Total barge and topside weight is very heavy and has a dynamic motion due to environmental conditions. Tide variation is important in adjusting barge level via ballasting but does not have an important dynamic impact. If rapid tide level variation is expected, heavy-duty ballasting pumps shall be utilized with sufficient contingency. Entrance is usually done in rising tide and exit in falling tide.

Before transferring topside load to jacket legs, swell may have a considerable influence. Heavy barge and topside have a long vertical period of motion. With long period swells this may induce a large DAF.

Fig. 2.41 shows a topside leg during contact with jacket leg in a floatover installation. Bumpers are placed inside topside leg. Jacket leg contains bearing plate and guide cone. Horizontal (Δ_h) and vertical (Δ_v) tolerances between topside and jacket legs are shown. Lateral bumpers are not visible. Barge is connected to jacket legs with some wire ropes.

Safe weather window for floatover installation shall be clearly defined. Before entrance, the cargo barge shall be properly anchored. Anchor pattern shall be carefully selected to enable:

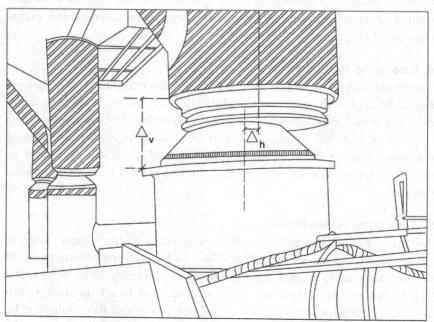

Figure 2.41 Topside/jacket leg contact.

- Pulling cargo barge in and out of jacket legs without any necessity to change anchors.
- Anchors shall not have any clash with jacket legs during any operational phase.
- Anchor stiffness shall be selected considering environmental loads and jacket stiffness.

This method does not require very high capacity crane barges and is therefore cost-effective in installation of heavy topsides.

After topside load is transferred to jacket legs, barge period of vibration will change considerably. The reason is that a considerable weight is relieved while barge stiffness remains more or less unchanged. Therefore it may show rapid movement and oscillations against environmental loads. At this stage barge shall be ballasted rapidly to provide sufficient tolerance from topside members and then pulled out of jacket legs.

2.10.3 Hurricane Impact

Platform design against hurricanes has impact on several disciplines. The most important one is structural group. Calculation of induced stresses in members and platform integrity check against environmental forces follows normal structural procedures. In addition, some items need careful examination. They include:

A. Forecasting Method

There are several statistical theories to predict magnitude of a natural event. An applicable statistical model shall be selected to ensure forecasting does not introduce considerable errors. With recent developments in data acquisition systems, available data banks and progress in forecasting methods selecting the most appropriate statistical method is no longer an unknown factor. We recommend that the reader refer to related books in this field.

B. Appropriate Return Period

An appropriate return period shall be selected to ensure design values of environmental forces are neither too low to have an inadequate platform nor too high to have an overdesign platform. Intensity of environmental factors like winds, waves, and currents is defined based on their return period. Risk matrix approach (using likelihood or probability multiplied by consequence) provides quantitative evaluation of the subject.

Forecasting environmental forces based on return periods several times longer than platform design life have been widely accepted by industry. Design life of a platform does not mean that immediately after that period it will be decommissioned.

Probability of occurrence of an event with equal or higher intensity than event with T years return period, within platform design life (N), is calculated by [13]:

$$P = 1 - \left(1 - \frac{1}{T}\right)^N$$

Normal platforms design life is 25–30 years. This is in spite of the fact that many platforms after 40 years of operation have been refurbished for further usage. Table 2.15 shows some typical occurrence probability value based on platform design life and environmental force return period.

Note that statistical studies never give an exact date or certainty for occurrence of any phenomenon. They only predict its probability of occurrence based on certain assumptions that may sometimes be inaccurate. In a bridge project to design foundation temporary retaining walls, water level for the Karun River (with 10-y return period) was calculated to be +10.0 CD. This study was based on more than 50 years of collected data using several statistical models. Construction of this part took a little less than a year. Within this period water level twice exceeded +11.0 CD, causing some financial loss and project delay. As can be seen from Table 2.15, selecting a 10-year return period for a 1-year activity has less than half probability of occurrence compared to a 100-year storm in a 25-years platform design life. In spite of this fact, in that particular project an event with much higher return period happened twice. The author checked with residents in that area and they confirmed that for several years before starting and later after finishing that critical project stage, water level did not even reach +6.0 CD. This is just to emphasize that even when using a quantitative risk management approach it has inherent uncertainty factors that somehow are not predictable.

Table 2.15 Probability of occurrence

Design life (N year)	1	25		30	
Return period (T year)	10	50	100	50	100
Probability %	10.00	39.65	22.22	45.45	26.03

It seems that the recent climate change phenomenon has increased the intensity of some events. For example, Table 2.3.4.1 [31] has recommended wind speed (1 h at 10 m) to be 80 knots (43.24 m/s) and wave height for 1000-ft water depth to be 70.5 ft (21.50 m) for high–consequence platforms (category L–1) to be built in the Gulf of Mexico. In 1992 hurricane Andrew had wind speeds above 50 m/s and waves more than 72 ft (21.95 m).

API approach is based on quantitative risk assessment. Category L–1 is a manned-nonevacuated platform with high-failure consequence. This means that even for cases for which the code tries to provide a higher safety margin, unexpected values may occur.

C. Platform Configuration

Proper platform configuration (like lowest deck elevation or platform orientation) may avoid heavy wave impacts. Wave forces hitting the platform deck exert considerable forces. Code recommendation is to avoid these forces by increasing lowest deck elevation. It is understood that this approach increases platform first mode shape period of vibration. As per API [31], the bottom of the lowest deck shall be above the following:

- Recommended air gap 5 ft (1.5 m) to cover platform settlement, water depth uncertainty, and possibility of extreme waves
- Design wave crest height (normally 100-year return period)
- Water level at HAT
- Storm surge (sea level rise due to storm)
- Long-term seafloor subsidence

It is worth keeping in mind that for Airy linear wave theory crest and trough heights are equal but for nonlinear waves like Stokes' fifth order usually the crest is larger than the trough.

D. Member Response

Critical member responses shall be evaluated. For example, due to increased wind velocity during hurricanes, vortex–induced vibrations may become critical for slender topside elements. Vortex shedding for elements subject to wave action is usually more critical than members subject to wind velocity. Seawater density is about $(1025/1.2) \sim 850$ times heavier than air density. In normal conditions jacket members like riser, J tube, and slender braces (which are hit by water waves) are more susceptible to vortex-induced vibrations compared to topside wind impacted members. In hurricanes wind velocity increases considerably. Therefore in addition to structural strength check, for decks' slender elements like lighting poles, pedestal crane boom rest, antenna masts, drilling derrick members, and other similar elements response to vortex-induced vibrations shall also be examined.

E. Other Disciplines

Since there is no section to explain hurricane-related issues in other disciplines, some of its impacts are explained here.

Piping: Small-sized pipes shall be appropriately supported at shorter intervals to ensure that they possess suitable strength against induced vibration. At the same time support of large-diameter pipes shall withstand larger applied hurricane loads and movements. All shutdown and blowdown valves are normally motor operated. In addition, major isolation valves shall also be motorized. This is to ensure that when operators are unable to walk outside quarters areas, closing of all piping systems is possible.

Instrument: Operating range of instruments shall match new conditions. This includes both threshold measuring value and actual position of installed instruments on supporting structures.

Electrical: Electrical power generation shall be able to continue working in abnormal situations. This covers all items like exhaust, fuel transfer, ventilation, etc.

In case personnel evacuation within a short notice is not possible, sufficient supply of fuel, food, drinking water, etc. shall be stored in suitable conditions. These requirements apply both to mechanical and architectural disciplines.

HVAC ducts, louvers, fans, and similar items shall be able to withstand hurricane winds. Air conditioning systems for control room and quarters area shall be able to work at very high wind speeds and rain storms.

Telecommunications: Radio and satellite communication means fail during hurricanes. In this condition, data transfer (including signals, voice, video, fax, etc.) with subsea cable is a suitable backup.

2.10.4 Installation Interface

Normally each engineer will do his or her job quite satisfactorily. The neglected point is the interface items. If proper coordination is not done by the lead engineer, they may work as separate islands and the outcomes may not fit to each other. The art of lead engineer is to put separate pieces of this puzzle together in a timely manner. A list of some major points follows. Although some of them do not belong to installation, the majority are related to it. Only topics are listed in this book. For more detailed descriptions, please refer to structural books.

A. Pile
- Leg extension compared to pile self-penetration depth
- Possibility of rubber diaphragm rupture due to seabed material and leg penetration
- Ease in rubber diaphragm rupture due to pile first segment plus first add-on weight
- Pile leg spacer clearance in the lowest seabed level and higher elevations
- Pile leg clearance for grouted and nongrouted legs
- Pile cutoff point and its distance with leg end point
- Pile leg connection with crown piece, dimensions, bend radius, and Z material type
- Pile penetration depth based on maximum load and maximum uplift
- Pile segment cut point especially in mud-line elevation and splash zone
- Driving segment dimensions
- Selecting first segment length
- Selecting each add-on piece length to avoid termination in setup prone layer

B. Mud Mat
- Jacket initial settlement during on-bottom stability
- Selecting mud mat dimensions and elevation
- Placing compression-only gap elements to define pressure distribution under mud mat. Elements directly below or nearby stiffened points will be highly stressed
- Placing skirt plates around mud mat to prevent mud overflowing. In case mud reaches above mud mat level, jacket uplifting to correct elevation becomes very much difficult

C. Conductor Guide
- Placing lowest conductor guide in an elevation higher than the lowest jacket level
- Conductor guide dimension is selected based on conductor driving method. If conductors are driven by hammer, only 1-in. clearance from each side is needed. If conductors are driven by drilling, then conductor guide shall have at least 2″. clearance from 32″ drilling bit. Jacket vertical installation tolerance shall also be considered in clearance.

- Conductor guide to accommodate predrilled appraisal well shall have a very big bell-shape configuration.
- Selecting conductor guide configuration and intermediate supports
- Conductor penetration depth

D. Jacket Design
- Jacket leg thickness to be designed to tolerate hydrostatic pressure during installation and loads based on temporary and permanent analysis results. However, sometimes it may be economical to reduce thickness and use hydrostatic collapse ring stiffeners.
- Leg CAN thickness is selected to tolerate fatigue loads. Its dimension is such as to enable taking in all connecting braces.
- Grouting pile leg internal space increases leg fatigue strength and reduces leg thickness
- Selecting brace dimensions and their impact in providing jacket buoyancy
- Brace design and control based on buckling criteria in two dimensions (X or K brace)
- Ductility level design for rare intense earthquake acceleration level
- Leg CAN design against fatigue and weld profile control or grinding
- In-place design in maximum and minimum water elevation
- Load out design criteria for one leg/row settlement or failure and combination

E. Boat Landing and Barge Bumper
- Selecting elevations and dimensions of boat landing to enable access in different tide levels
- Selecting boat landing configuration to avoid vessel clash with cellar deck members
- Selecting and design of shock cell criteria
- Possibility of a vessel hitting boat landing at one or two points
- Barge bumper configuration, dimension and design

F. Temporary Supports
- Sea fastening supports are used during transportation. They shall be designed/placed in such a way that their cutting is done at the minimum required time and after cutting leave a minimum of debris.
- Rigging platform is used to accommodate lifting slings during offshore transportation. To minimize offshore work, normally slings are

connected onshore. Onshore crane access is also much better. They shall be secured in such a way as to be safe against accelerations and movements during offshore voyage, while easily accessible to connect a crane barge hook.

- Barge bumpers are installed but boat landings are transported separately. After jacket is fixed via piling, boat landing is lifted and placed in location.
- After jacket is installed and riser is connected to the underwater pipeline and topside piping, it becomes stable. These end connections act as additional supports. However, during transportation and installation jacket may require temporary supports. They may be left in place after installation. Normally riser clamps do not provide any torsional restriction. During sea transportation, lifting, or launch it may experience torsion. Riser bend connected to end blind flange is a heavy item. The bend may protrude several meters away from the jacket face and the blind flange will be located at the farmost end point. This may induce a considerable torsion that may only be taken with heavy support elements connected to jacket members.
- J tubes normally have small loads and are closed with a small, lightweight end flange. A diver will remove it and connect the pulling cable attached to it to a subsea cable.

2.10.5 Contingency Factors

Contingency factors are needed to procure bulk material. They shall take into account several factors like:

- Modeling accuracy: Accuracy both in analytical modeling/calculation and drafting affects bulk item quantity. Some of the small stiffeners may not be accurately modeled in detail design and only exactly modeled in work orders. Using 3D models increases the accuracy.
- Cutting plan accuracy and waste: Preparing an optimized cutting plan and usage of remaining portions will reduce required additional material. This necessitates proper material tracing procedures. Before start of cutting, all small-sized stiffeners shall be identified. Although available software provides a preliminary proposal, it can be further improved by careful examination.
- Availability of specific items: Some items like thick Z-quality material with higher grades may require long production duration and higher costs. Therefore the yard shall handle them very carefully to avoid any damage or rework.

- Vendor accuracy in packing and delivery: Final packing list and loading may have some inherent mistakes. It shall be carefully examined by vendor or purchaser representative before dispatching.
- Damage during transportation: Bundling material, lifting by cranes and loading/offloading, and sea transportation may damage some of the material in the manufacturer mill or storage area or during voyage and offloading. Normally one of the purchase order terms and conditions for vendors is to provide sea worthy packing. However, some damage is inevitable. Even if vendor accepts to replace damaged goods, this takes considerable time. Therefore EPCIC contractors prefer to include some margin for these damaged items.
- Damage during storage and handling in yard: Bulk items are stored in open area. They are transported in yard by trailers pulled by trucks. Normally each yard has specific lifting and transportation procedures. In spite of this storage, handling in yard for relocation or transfer to cutting/welding site may damage some portions of steel material.
- Yard practice in fabrication: Although API RP 2A in some cases has approved usage of pieces up to 90 cm, some fabrication yards are reluctant to use them due to higher man power and welding/fabrication costs unless they face material shortages.
- Damages/mistakes in cutting/fabrication: Normally pieces from special materials like joint CANs, etc. are fabricated very carefully. Therefore less damage is expected during cutting and fabrication. However, it is not nil. A portion is damaged during cutting and prefabrication operations. Cutting to size may have mistakes or bevels may be wrong or fit up may not be correct.
- Rejected material: Some of the material that has arrived in the yard may not be acceptable.
- Some fabricated items may not pass the quality control tests and client may instruct to cut. This requires additional material of the same quality.
- Some welds may not pass nondestructive Tests. Based on the extent of weld defects, repair action may vary from removing part of the weld and rewelding to removing the welded piece and connecting a new one.

An exact figure cannot be quoted for each item. Yards may have developed specific figures from passed projects. This will be the best reference. If nothing is available then for small size and low-grade material higher contingencies may be used. Different factors may be used for tubular, rolled sections and plates. As a general practice, figures of 1—10% may be used.

2.10.6 Material of Construction

Structural steel for platforms shall be suitable for offshore environment. This includes structural mechanical property (allowable and yield stress besides ultimate tensile strength), deformation property, ductile behavior, and weldability. The author does not have much experience with American standards and is not describing ASTM requirements in this book. Two major standards normally used to define steel material suitable for offshore construction are:

- Det Norske Veritas (2000). "Metallic Material," DNV-OS-B101.
- European Standard (2004). "Hot Rolled Products of Structural Steels," EN10025.

Although they use different nomenclature and have differences in chemical composition and other criteria, more or less they have a similar approach. Steel material is divided into three categories:

- Normal strength: Yield stress is about 240 MPa, tensile strength about 360 MPa.
- High strength: Yield stress is about 340 MPa, tensile strength about 470 MPa.
- Extra-high strength: Yield stress is about 450 MPa, tensile strength about 550 MPa.

DNV has also defined materials with higher yield stress, but they are not normally used. As a general criterion, increasing steel material strength reduces its ductility and weldability. This is due to higher carbon rate.

It is understood that with the same chemical composition rolling to thinner plates increases yield stress. Changes are of the order of 10 MPa, which will be defined based on thickness range. For example, high-strength steel may have yield stress equal to 355 MPa if less than 16 mm thick. For 16−40 mm yield stress may reduce to 345 and for 40−63 mm it may reduce to 335 MPa. For each type of material, it is necessary to refer to the latest applicable code.

Normal strength steel is used for secondary and tertiary members. Stairs, handrails, and some equipment supports are in this category. Those secondary beams that support major equipment and all primary members are fabricated with high-strength material. In special joints, extra-high-strength material may be used. But its fabrication and welding requires specific procedures to ensure that its response to cyclic loads present in offshore environment follows code requirements.

Material strength is checked with tensile test. A sample with specific dimensions is cut from the material. It is fixed in one end and pulled from other end. Specimens shall not have any curvature. This ensures only tensile stresses are developed. Code specifies test sample dimensions and number of tests per specific tonnage of material. Yield stress, ultimate tensile strength, and percent elongation besides failure section contraction are reported.

Material ductile behavior is checked with a Charpy test. In this test a specimen is cut from parent metal. A notch is made almost in mid-length of specimen. Then it is hit with a free-falling object in a swinging motion with approximately 290 J energy. Total energy needed to break the specimen and percent of brittle surface at the broken area is an indication of material behavior. A Charpy test for DNV materials that are normally used in cold northern areas is performed at 0, −20, −40, and −60°C. The same test for EN (Europaische Norm/European Standard) material is performed at +20, 0, −20, and −50°C. Minimum energy specified by both standards is approximately equal to 27 J.

Materials that may experience stresses perpendicular to their thickness shall be tested for lamination. Braces connected to joint CANs induce this type of stress. TTP test for plates less than 25 mm thick is not performed. The reason is repeated rolling to reduce original thickness has already removed all gaps. In addition, since joint CANs are constructed from thick plates, checking TTP criteria is crucial.

Extra-high-strength material shall be checked for weldability. This is done by measuring either of two parameters, carbon equivalent value (CEV) and P_{CM}. These two parameters are also used for lower grade steel material.

$$CEV = C + \frac{Mn}{6} + \frac{Cr + Mo + V}{5} + \frac{Ni + Cu}{15}$$

$$P_{CM} = C + \frac{Si}{30} + \frac{Mn + Cu + Cr}{20} + \frac{Ni}{60} + \frac{Mo}{15} + \frac{V}{10} + 5B$$

Table 2.16 shows sample results for three specimens tested to check their conformity with S355 material. Additional digits in steel grade define its manufacturing process.

Each project shall define acceptable CEV or P_{CM} values. Although for high strength steel code may accept CEV < 0.43, in some project specifications CEV may be limited to 0.41. Other than that, project material specification may assign different requirements for similar material used in

Table 2.16 Sample material test

Sample	C %	SI %	MN %	P %	S %	Cr %	MO %	NI %	Cu %	AL %	Ti %	Nb/Cb %	V %	B%	CEV	P_{CM}
321546	0.135	0.376	1.550	0.013	0.001	0.039	0.017	0.161	0.168	0.038	0.003	0.030	0.001	0.0003	0.427	0.241
321694	0.135	0.386	1.530	0.011	0.001	0.031	0.016	0.154	0.157	0.038	0.004	0.025	0.001	0.0004	0.420	0.240
322006	0.134	0.359	1.580	0.011	0.001	0.026	0.013	0.051	0.032	0.034	0.003	0.030	0.001	0.0004	0.411	0.232

different platform members based on importance of that member in platform structural integrity. This is additional to more strict criteria in construction QC procedures.

2.10.7 Architectural Interface

Structure has lot of interfaces with architecture. An explanation was given in Section 2.1.3. All wall panels, doors, windows, chairs, beds, HVAC units, etc. shall be properly supported.

- Doors and windows are supported via hollow rectangular section profiles to the floor and roof beams.
- Wall panels may be supported via angles to the floor and roof at both ends.
- Panel surface between two ends can be connected to corrugated plates or supported by flat bars installed at short spans.
- False floor support is normally provided by its manufacturer. It has to be adjustable in height.
- A false ceiling can be supported from secondary beam elements. Connecting it to the roof plate shall be avoided.
- False ceilings shall be properly supported from roof to withstand their own weight and possible loads by HVAC ducts, diffusers, etc.

The governing factor in some cases may be deflection instead of stress. Excessive wall panel deformations (especially if permanent) may impact operators' sense of security. During heavy storms and rough weather, an inside accommodation area shall provide a safe haven for operators.

Cabin areas shall have fire-rated walls with exterior areas and have comfort-rated panels with other interior accommodation areas.

2.10.8 Environmental Conditions

Site-specific environmental data shall be provided by the client. Nowadays with progresses in satellite measurements and the huge available database and improvements in forecasting methods, accurate data for almost every point can be obtained with a reasonable cost.

An environmental conditions report shall provide data about:

- Maximum wind speed with specific return periods like 1, 10, and 100 years from eight directions with proper averaging time.
- Maximum wave height and period with the same return periods from eight directions.
- Tide levels compared to chart data in the area showing at least LAT, HAT, and MSL.

- Maximum current velocity with the same return periods as wind and waves in the same directions from water surface to seabed.
- Average wind occurrence in velocity groups per year for wind fatigue calculations.
- Significant wave height groups occurrence per year for deterministic or spectral fatigue calculations.

Each category of data will be used in certain conditions. For example, 1 and 10-y return period data are used for operation and transportation conditions. Storm conditions are used with minimum 100-y return period events. Normal wave occurrence data is used for fatigue calculations. Earthquake intensity is defined based on ratio of gravitational activity. For low-seismic risk (less than 0.05 g) storm conditions may govern. Increased seismic levels shall follow special ductility considerations.

2.10.9 Pile Lateral and Axial Capacity

Site-specific geophysical and geotechnical data shall be provided by the client. These data shall give water depth, expected scour depth, mud mat load-bearing capacity for different mud mat shapes, soil lateral and vertical response, and pile load-bearing capacity [1,5,7,9].

Piles are used for stabilizing fixed jacket structures on the seabed. Both lateral and vertical loads are transferred via piles to seabed.

Pile lateral capacity is governed by soil lateral response, which is generally given by $P-Y$ curve data. These data show the mobilized soil strength at each lateral movement for each layer against cyclic loads. This relation is nonlinear and normally after a specific movement soil fails, which means the resistance drops considerably. Based on soil type, its resistance will change, which may have weak layers between stiff layers. Table 2.17 shows typical $P-Y$ curve values for $\varnothing 1220 \times 38.1$-mm pile. In each geotechnical report, $P-Y$ values for several expected pile diameters are obtained. If pile sizes do not differ considerably, the change is almost linear based on pile diameter size. In this table, soil layer thickness is given based on distance from seabed. Soil mobilized lateral resistance P is in kn/cm and soil lateral movement is in cm. Distance from seabed to 1.72 m is considered as soil scour depth and soil resistance for it is not given.

Consider soil resistance at 4.2 m depth. When pile pushes soil, it will show resistance. As is clear, soil resistance (almost linearly) increases to 3.93 kn/cm up to a lateral movement of 2.05 cm. After this value until 3.07 cm displacement, its resistance decreases negligibly to 3.79 kn/cm.

Table 2.17 Soil lateral resistance

Soil layer depth (m)	P(1)	Y(1)	P(2)	Y(2)	P(3)	Y(3)	P(4)	Y(4)	P(5)	Y(5)	P(6)	Y(6)	P(7)	Y(7)	P(8)	Y(8)	P(9)	Y(9)
1.72	0	0	0.3645	0.305	0.523	0.915	0.7924	3.05	1.1411	9.15	1.5849	24.4	1.5849	36.6				
2.00	0	0	0.4317	0.305	0.6194	0.915	0.9384	3.05	1.3514	9.15	1.8769	24.4	1.8769	36.6				
3.00	0	0	0.5425	0.305	0.7784	0.915	1.1794	3.05	1.6983	9.15	2.3588	24.4	2.3588	36.6				
4.00	0	0	0.6125	0.305	0.8788	0.915	1.3315	3.05	1.9173	9.15	2.6629	24.4	2.6629	36.6				
4.20	0	0	1.4978	0.154	2.1147	0.307	2.731	0.512	3.8104	1.54	3.9323	2.05	3.7928	3.07	0.6202	9.22	0.6185	10.2
5.00	0	0	1.6622	0.154	2.3468	0.307	3.0307	0.512	4.2286	1.54	4.3639	2.05	4.209	3.07	0.6882	9.22	0.6864	10.2
6.00	0	0	1.8448	0.154	2.6047	0.307	3.3638	0.512	4.6933	1.54	4.8434	2.05	4.6716	3.07	0.7639	9.22	0.7618	10.2
7.00	0	0	2.0275	0.154	2.8626	0.307	3.6968	0.512	5.158	1.54	5.323	2.05	5.1341	3.07	0.8395	9.22	0.8373	10.2
8.00	0	0	2.2101	0.154	3.1205	0.307	4.0299	0.512	5.6227	1.54	5.8025	2.05	5.5966	3.07	0.9151	9.22	0.9127	10.2
9.00	0	0	2.3928	0.154	3.3784	0.307	4.3629	0.512	6.0874	1.54	6.2821	2.05	6.0591	3.07	0.9908	9.22	0.9881	10.2
10.00	0	0	2.5754	0.154	3.6363	0.307	4.6959	0.512	6.5521	1.54	6.7616	2.05	6.5217	3.07	1.0664	9.22	1.0636	10.2

Continued

Table 2.17 Soil lateral resistance—cont'd

Soil layer depth (m)	P(1) / Y(1)	P(2) / Y(2)	P(3) / Y(3)	P(4) / Y(4)	P(5) / Y(5)	P(6) / Y(6)	P(7) / Y(7)	P(8) / Y(8)	P(9) / Y(9)
11.00	0 / 0	2.7581 / 0.154	3.8942 / 0.307	5.029 / 0.512	7.0168 / 1.54	7.2412 / 2.05	6.9842 / 3.07	1.142 / 9.22	1.139 / 10.2
12.00	0 / 0	0.5125 / 0.122	1.459 / 0.61	2.2894 / 1.22	3.5925 / 2.44	4.6758 / 3.66	6.5171 / 6.1	10.226 / 12.2	29.111 / 61
12.40	0 / 0	2.573 / 0.154	3.6329 / 0.307	4.6915 / 0.512	6.5459 / 1.54	6.7553 / 2.05	6.5156 / 3.07	1.0654 / 9.22	1.0626 / 10.2
13.00	0 / 0	2.5958 / 0.154	3.6651 / 0.307	4.7331 / 0.512	6.6039 / 1.54	6.8152 / 2.05	6.5733 / 3.07	1.0748 / 9.22	1.072 / 10.2
14.00	0 / 0	2.6243 / 0.154	3.7053 / 0.307	4.7851 / 0.512	6.6764 / 1.54	6.89 / 2.05	6.6455 / 3.07	1.0866 / 9.22	1.0837 / 10.2
15.00	0 / 0	2.6528 / 0.154	3.7455 / 0.307	4.837 / 0.512	6.7489 / 1.54	6.9648 / 2.05	6.7176 / 3.07	1.0984 / 9.22	1.0955 / 10.2
16.00	0 / 0	2.6813 / 0.154	3.7858 / 0.307	4.889 / 0.512	6.8214 / 1.54	7.0396 / 2.05	6.7898 / 3.07	1.1102 / 9.22	1.1073 / 10.2
17.00	0 / 0	2.7098 / 0.154	3.826 / 0.307	4.9409 / 0.512	6.8939 / 1.54	7.1144 / 2.05	6.8619 / 3.07	1.122 / 9.22	1.1191 / 10.2
18.00	0 / 0	2.7383 / 0.154	3.8662 / 0.307	4.9929 / 0.512	6.9664 / 1.54	7.1892 / 2.05	6.9341 / 3.07	1.1338 / 9.22	1.1308 / 10.2

This point shows an abrupt change in soil behavior. After this point, soil resistance drops to 0.62 kn/cm at a displacement of 9.22 cm. This is only about 16% of the maximum resistance. After this stage, resistance is fairly constant.

Soil resistance at 18 m has a maximum mobilized resistance of 7.19 kn/cm against 2.05 cm displacement. This is about twice the resistance at 4.2 m depth. At the same time, its absolute value loss of strength is also 1.83 times loss of strength for 4.2 m depth.

Soil layer at 12 m is an exceptionally stiff soil. Its resistance increases even to 61 cm displacement. It is clear that no platform can tolerate such a displacement at 12 m below the seabed. Normal displacements at the seabed for fixed platforms are on the order of a few centimeters. Based on jacket stiffness and water depth, a few centimeters at the seabed may be more than half a meter at water level. Soil at 1.72—4 m also shows resistance increase up to 24.4 cm and after that is constant.

Fig. 2.42 shows different soil behaviors as explained herein. For chart clarity, soil response for displacements above 250 mm is not shown.

Software for offshore platform design shall be able to incorporate this nonlinear behavior. Nonlinear behavior is normally modeled by stepwise load increase and linear behavior in each substep. For single piles it is

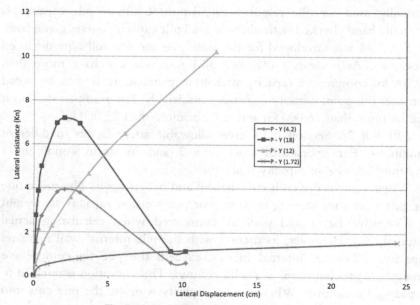

Figure 2.42 Different lateral soil behavior.

relatively easy to use 5—10% load increments and adjust the spring factor along the pile length to the calculated displacements. It shall be understood that large load increments may diverge iterations. In addition, springs shall be modeled relatively nearby each other.

At different sections a major portion of lateral loads is taken by soil resistance and the remaining is taken by pile shear. In a sample example for this, for a pile with 20 m depth more than 70% lateral force was taken by soil resistance and the remaining tolerated by pile shear. It is understood that for a full-length pile lateral soil resistance shall be equal to applied lateral loads.

Vertical soil response is divided into two categories: pile tip and pile shaft. The first is shown by $Q-Z$ and the second is given by $T-Z$ data. Each gives the soil mobilized response with specific displacement. Although $Q-Z$ values will be given through pile length, only a single value corresponding to pile tip will be utilized. $T-Z$ values for the same soil in Table 2.17 are given in Table 2.18. In this table, T values are in Kpa and Z in cm.

Pile axial capacity is mobilized after some movement occurs. General accepted criterion for shaft friction is to move at least 1% and for the end bearing is 10% pile diameter. For granular soils, less movement is required to mobilize soil resistance.

An important point is that pile lateral capacity/response are entirely calculated and governed by $P-Y$ curve while $T-Z/Q-Z$ curves are normally used for pile response calculation. Ultimate pile capacity is normally hand checked with ultimate axial pile capacity curves given later.

Fig. 2.43 was developed for the same pile size and soil type discussed previously. As is clear, a Ø1220 × 38.1-mm pile will have more than 10,000 kn compressive capacity after 40 m penetration. It is to be noted that after this point capacity increases considerably. For example, at 50 m it will be more than 16,000 kn and at 60 m more than 22,000 kn.

API RP 2A Section 6.3.4 gives allowable safety factors in different conditions. For operational cases SF = 2 and in storm conditions a one-third increase in capacity is accepted.

When a pile is driven it cuts the soil and penetrates it. The soil inside the pile continues moving upward. For compression capacity its weight is a negative factor and shall be considered when calculating actual external load. However, its friction with the pile internal wall increases capacity. Whenever internal friction exceeds the pile tip compressive strength, it penetrates soil as a solid column. This condition is referred to as plugged condition. Whenever tip capacity is more, the pile cuts into soil below.

Table 2.18 Soil axial resistance along pile shaft (T–Z)

Soil layer depth (m)	T(1)	Z(1)	T(2)	Z(2)	T(3)	Z(3)	T(4)	Z(4)	T(5)	Z(5)	T(6)	Z(6)	T(7)	Z(7)
1.72	3.9	0.195	6.49	0.378	9.74	0.695	11.69	0.976	12.99	1.22	9.09	2.44	9.09	3.66
2.00	4.68	0.195	7.8	0.378	11.71	0.695	14.05	0.976	15.61	1.22	10.93	2.44	10.93	3.66
3.00	5.81	0.195	9.69	0.378	14.53	0.695	17.44	0.976	19.38	1.22	13.57	2.44	13.57	3.66
4.00	6.47	0.195	10.77	0.378	16.16	0.695	19.4	0.976	21.55	1.22	15.09	2.44	15.09	3.66
4.20	7.02	0.195	11.69	0.378	17.54	0.695	21.05	0.976	23.39	1.22	16.37	2.44	16.37	3.66
5.00	8.0	0.195	13.34	0.378	20.0	0.695	24.0	0.976	26.67	1.22	18.67	2.44	18.67	3.66
6.00	9.08	0.195	15.14	0.378	22.7	0.695	27.24	0.976	30.27	1.22	21.19	2.44	21.19	3.66
7.00	10.15	0.195	16.92	0.378	25.37	0.695	30.45	0.976	33.83	1.22	23.68	2.44	23.68	3.66
8.00	11.21	0.195	18.68	0.378	28.03	0.695	33.63	0.976	37.37	1.22	26.16	2.44	26.16	3.66
9.00	12.27	0.195	20.45	0.378	30.67	0.695	36.81	0.976	40.9	1.22	28.63	2.44	28.63	3.66
10.00	13.32	0.195	22.2	0.378	33.31	0.695	39.97	0.976	44.41	1.22	31.09	2.44	31.09	3.66

Continued

Table 2.18 Soil axial resistance along pile shaft (T–Z)—cont'd

Soil layer depth (m)	T(1)	Z(1)	T(2)	Z(2)	T(3)	Z(3)	T(4)	Z(4)	T(5)	Z(5)	T(6)	Z(6)	T(7)	Z(7)
11.00	14.38	0.195	23.96	0.378	35.94	0.695	43.13	0.976	47.92	1.22	33.54	2.44	33.54	3.66
12.00	13.92	0.254	13.92	10.0										3.66
12.40	14.56	0.195	24.27	0.378	36.41	0.695	43.69	0.976	48.55	1.22	33.99	2.44	33.99	3.66
13.00	15.14	0.195	25.24	0.378	37.85	0.695	45.42	0.976	50.47	1.22	35.33	2.44	35.33	3.66
14.00	15.85	0.195	26.41	0.378	39.62	0.695	47.54	0.976	52.82	1.22	36.97	2.44	36.97	3.66
15.00	16.53	0.195	27.55	0.378	41.33	0.695	49.59	0.976	55.1	1.22	38.57	2.44	38.57	3.66
16.00	17.2	0.195	28.67	0.378	43.01	0.695	51.61	0.976	57.34	1.22	40.14	2.44	40.14	3.66
17.00	17.86	0.195	29.76	0.378	44.65	0.695	53.58	0.976	59.53	1.22	41.67	2.44	41.67	3.66
18.00	18.5	0.195	30.84	0.378	46.26	0.695	55.51	0.976	61.68	1.22	43.18	2.44	43.18	3.66

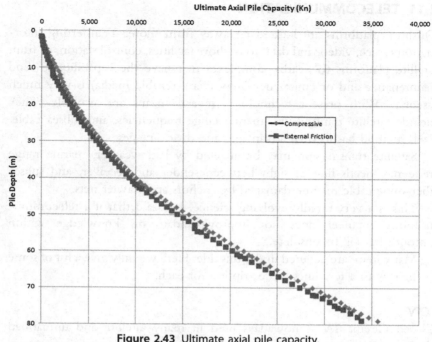

Figure 2.43 Ultimate axial pile capacity.

Clayey soils show increase in capacity after driving is finished. This is considered as Soil Setup. Increases up to 60% have also been observed. Based on soil type it may happen within a few days to a few weeks after driving is finished. In fact, pile axial capacity is governed by soil effective pressure. Water particles will migrate gradually due to the additional stresses induced by driving. This will increase effective stress portion, which in turn will increase soil effective stress, which itself results in increased pile capacity.

API RP 2A Section 6.10.6 specifies 250 blows per foot as refusal criterion. Pile refusal is not considered as the end of driving. The consultant specifies penetration depth target based on the geotechnical report. If design penetration is not achieved, remedial action may vary including:

• Check hammer performance or use stronger hammer.
• Test pile capacity. Dynamic pile testing is a very rapid nondestructive and cost-effective test method. In several cases actual capacity is higher than calculated.
• Drill inside pile, remove internal friction, and redrive.

2.11 TELECOMMUNICATION

Offshore platforms are located far away from shore. Transferring information (voice, video, and data) to onshore facilities, control station, or from satellite platforms to nearby complexes that have the capability to send maintenance and or emergency crew (via a reliable media) is very much essential. With present technology, several means are available. They include satellite transmission, marine range frequencies, and subsea cable. Each method has its own advantages and disadvantages.

Satellite transmission may be affected by bad weather, marine range frequency needs line of sight between sender and receiver, and subsea fiber-optic cable may be damaged by anchors and trawler nets.

This is a very rapidly evolving science. It is said that if a telecommunications engineer does not improve/update his knowledge within 6 month he will be obsolete.

Many items are covered under this title. Here we only give a list of some of them with a few lines of description for each.

CCTV

Closed Circuit TV is nowadays used in many satellite and unmanned platforms. They may be connected to an anti-intrusion surveillance system. In unmanned platforms normally they cover the wellhead area, main process equipment (if any), technical room, boat landing, and other access to the platform. The intention is to give proper information to the controller. In case of any event in the wellhead and process area, the ESD and FGS system will automatically function as planned. Video transmission of the event will give the operator a real-time understanding of what is happening.

The cameras shall have pan, tilt, and zoom capability. Based on their installation location, they shall fulfill required ingress protection or hazardous zone classification. They are also equipped with washing tanks. Since most of them are placed outdoors, they shall be able to tolerate harsh environmental conditions.

VHF/UHF/FM

Very-high frequency amplitude modulation (VHF AM) radio system provides communication between platform personnel and the approaching aircraft. VHF frequency modulation (FM) radio system can provide communication with marine vessels within 40 nautical miles.

Ultra-high frequency FM handheld radio system provides communication between operation personnel and platform control room. They may be connected via platform telephone system.

PAGA

Public address and general alarm (PAGA) system consists of loudspeakers/beacons placed in different locations of the platform. This enables operator in the control room to communicate with and issue immediate warnings/instructions to the personnel present in the field, either automatically or manually.

Each operator may have his own specifications for alarm tone and its meaning. Following is a typical list of signals:

- Boat landing alarm (BLA): This is continuous ding sounds for 1 s duration that is repeated with yellow light only on boat landing.
- Alarm level 1 has single "ding" sounds for 1-s duration with yellow light action everywhere.
- Alarm level 2 has continuous "ding" sounds for 1-s duration repeated with yellow light action everywhere.
- LSD has continuous "sound — silence" sound with yellow light action everywhere. The sound frequency is 400 Hz with 1-s duration. It is followed by 2 s of silence. Yellow light is acting during sound cycle.
- PSD has a continuous sound silence sequence with yellow light action everywhere. The sound frequency is 400 Hz with 2-s duration. It is followed by 1-s silence.
- Alarm level 3 has continuous sound silence with yellow light everywhere. The sound is standard YELP. The sound with 600-Hz frequency with 0.25 s duration is followed by sound with 1250 Hz-frequency for 0.25 s duration, then there will be 1-s silence, and this pattern is repeated.
- Alarm level 4 has continuous sound silence with yellow light everywhere. The sound is standard YELP. The sound with 600-Hz frequency with 0.25-s duration is followed by sound with 1250-Hz frequency for 0.25-s duration then both tones will be repeated again with the same duration. After this there will be 0.5-s silence, and this pattern is repeated.
- YESD has continuous sound with yellow light everywhere. The sound is standard YELP. The sound with 600-Hz frequency with 0.25-s duration is followed by sound with 1250-Hz frequency for 0.25-s duration, and this pattern is repeated.

- Alarm level 5 has continuous sound silence with red light everywhere. The sound is standard (slow whoop). The sound with 500-Hz frequency with 4-s duration is followed by sound with 1200-Hz frequency for 1 s. Then there will be 3-s silence, and this pattern is repeated.
- RESD has continuous sound with red light everywhere. The sound is standard (slow whoop). The sound with 500-Hz frequency with 4-s duration is followed by sound with 1200-Hz frequency for 1 s. This pattern is repeated.
- MOB has continuous sound with blue light everywhere. The sound is standard (YEOW). The sound with 1200-Hz frequency with 1.5-s duration is followed by sound with 600-Hz frequency for 1.5 s. This pattern is repeated.
- ASD has continuous sound with blue light everywhere. The sound is standard (HORN). The sound with 450-Hz frequency is released continuously.

Alarm levels as per their priority include:
- Emergency Speech
- ASD (abandon platform shutdown)
- MOB (man overboard)
- RESD (red emergency shutdown)
- YESD (yellow emergency shutdown)
- Level 5
- Level 4
- Level 3
- PSD (process sdown)
- LSD (local shutdown)
- Level 2
- BLA (boat landing alarm)
- Level 1
- Speech

Three colors are used for alarms. They include:
- Red for fire case
- Yellow for GPA general platform alarm, toxic, explosive gas detection
- Blue for PAPA (prepare to abandon platform)

For manned platforms 2 × 100% independent PAGA system shall be installed. This means as per safety requirements full redundancy shall be provided. Wiring to connect field instruments like loudspeakers and flashing beacons shall be done via different routes. This will ensure if one of the routes is cut the system is still able to broadcast necessary alarms.

Living Quarter rooms and enclosed office spaces will have their dedicated loudspeakers. Flashing beacons are most effective in areas with high-noise levels. In these locations operating sound levels may already be high. Adding a new sound source may make alarms or instructions incomprehensible. Architectural or structural walls with heat and sound insulation reduce sound level considerably. To find an accurate sound reduction level vendors technical catalog shall be reviewed. Insulations with 50-mm thickness may reduce sound level up to 45 dB.

The human ear can accept a considerably large range of sound energy. Each sound has an energy that can be expressed in watt per m^2. Decibels are defined as logarithms in base 10 of sound energy. The threshold of hearing is taken as zero dB with an energy level $I_o = 10^{-12}$ W/m^2.

Sound Intensity Level, SIL (dB) $= 10 \log \frac{I}{I_o}$

Since sound has energy and can move air it has pressure. Hearing threshold pressure is 2×10^{-5} Pa.

Sound Pressure Level, SPL (dB) $= 20 \log \frac{P}{P_o}$

Table 2.19 shows dB definition and their energy value.

Physical observation shows sound intensity reduces with distance. The question is in what ratio? Sound is distributed in a planar area, therefore its intensity varies inversely proportional to the square of the distance away from the sound source.

Table 2.19 Sound level definition

Sound level (dB)	Energy intensity (W/m^2)	Pressure level (Pa)	Description
0	0.000000000001	0.000020	Threshold of hearing
10	0.00000000001	0.000063	Rustle of leaves
20	0.0000000001	0.00020	Rural ambient
30	0.000000001	0.00063	Quiet unoccupied office
50	0.0000001	0.0063	Average office
70	0.00001	0.063	Speech at 3 ft
90	0.001	0.63	Shouting at 5 ft
110	0.1	6.3	75-piece orchestra
130	10	63.2	Painful
140	100	200.0	Near jet engine

Table 2.20 Exposure duration to sound levels

Maximum exposure duration (h)	15 min	0.5	1.0	1.5	2.0	3.0	4.0	6.0	8.0
SIL (dB)	115	110	105	102	100	97	95	92	90

$$\frac{I_1}{I_2} = \frac{R_2^2}{R_1^2} \text{ or } SIL_1 - SIL_2 = 10\log\left(\frac{R_2}{R_1}\right)^2$$

Only energy intensity in W/m^2 can be algebraically added. This means that if we have two noise sources each producing 70 dB in a location $(0.00001\ W/m^2)$, the combined intensity level will produce $0.00002\ W/m^2$, which is equivalent to 73 dB.

Sound intensity control is needed for operators' safety. Permanent hearing loss can result if loud noises are experienced for too long periods. The US Occupational Safety and Health Administration has established criteria dictating exposure limits for various environments, as shown in Table 2.20.

A widely accepted condition for sound level limitation is to use 85 dB at 1 m distance from any package. This can be achieved in many packages. For generators, either diesel or gas turbine or compressors, this may not be easily achievable. In this case installing sound barriers or using headphones in the work area is recommended. For these conditions using loudspeakers for emergency alarms is useless. For these cases normally visual aids are used. Flashing beacons shall be located in a place visible in all conditions. Their light shall not be so severe as to cause temporary blindness. In fact during emergency conditions operators shall act immediately. Temporary blindness due to extra-bright lights may hamper their reaction.

NDB

A nondirectional beacon is placed on the helideck to provide the helicopter pilot with bearing information of the platform. It may be remotely controlled to start/stop. Its' operating frequency, power, and duration to be operative after platform is shut down shall follow International Civil Aviation Organization regulations.

Meteorological System

A meteorological system is necessary for all platforms that have a living quarter. Before transporting crew with either tugboat or helicopter,

weather conditions at the platform shall be known. The gathered information may include:
- Wind speed and direction
- Temperature
- Barometric pressure
- Humidity
- Sea current speed and direction
- Wave height
- Tidal level variation

This information may be stored in a central unit and at the same time displayed online to the operator.

RACON

A radar beacon is a transmitter receiver device. Platform navigational identifications are programmed in RACON. When it is triggered by radar (ie, helicopter radar) it will automatically return signals, which appear on the display of the aircraft. These signals provide bearing, range, and identification information of the platform.

PABX

Private automatic branch exchange telephone system provides voice communication facilities for operators in the platform. This will include both official and private communications. This system is connected to onshore and normally to a country's telephone system. Fax machines, hotline telephone lines, and indoor telephones are all connected to this system.

2.12 VENDOR DATA

Vendor data plays a major role in offshore platform engineering. It is understood that some data are part of vendor proprietary design. They may not be revealed to the purchaser. Other than that, remaining parts of information that affect platform construction, commissioning, and operation shall be submitted by vendor in a format as specified in MR. Each vendor may have its own way of presenting technical data, which may differ in some aspects from other vendors. However, operators on a certain platform tend to use specific routines. In daily workload the necessary data shall be easily accessible. Nowadays computerized search engines have facilitated this task very much. A single document can be easily selected from a large database.

Some packages have software that can check working conditions. Built-in functions can send alarms and provide guidelines and procedures easily. These alarms can start from routine maintenance activities like lubrication oil refilling, cooling water volume, filter cleaning or replacing, washer change, etc. to major overhauls and troubleshooting/fault-finding hints. In addition, each platform may have its own job specifications and routines. To match these formats (as much as possible) vendor data is preferred to be given in a uniform format.

For example, general arrangement drawing in all packages is better to be presented in the same section of final data book. The same is applicable for other technical documents like PID, consumption list, maintenance manuals, and equivalent lubrication oils list, which for rapid access are preferred to appear in the same section of FDB for all packages.

Vendor data review, commenting, and implementation of tasks is applicable to all disciplines. Therefore it has been discussed under a separate topic. After issuing MR for purchase, vendors start generating their own documents. This data has a detrimental influence on platform design. Three subjects can be identified for this task:

A. Conformity With Project Specifications

At first the consultant shall carefully review vendor data to ensure manufactured equipment (packages, vessels, pumps, generators, compressors, valves, etc.) match with project specifications. This shall cover all aspects like electrical power rating, fuel consumption, required utilities, weights, materials of construction, type of signals, general arrangements, package capacities, etc. International manufacturers produce equipment for a worldwide market. Different environmental conditions, crude composition, operating pressure and temperature, required capacity, etc. impose certain specifications that if not followed by vendors may endanger the platform and operators' lives. HSE regulations in different regions may also impose certain restrictions. Some examples are explained herein:

- Using carbon steel material for highly corrosive crude will result in rapid corrosion and deterioration of package. Either leakage of toxic material or blast of high-pressure containment may result in disastrous consequences. Exotic material as specified in project specifications ensures design life and safe operation.
- Conformity with hazardous zone classification and IP rating for the supplied instrument within package shall be checked and confirmed.

Assume package and its UCP are installed at different locations. IP rating for outdoor equipment is different from indoor units.
- Sparing of electromotors, detectors, analyzers, etc. shall match project specifications.
- Type of cables used in package manufacturing shall be according to project requirements.
- Alternating current frequency in North America is different from Europe. This is applicable both on the electrical generator itself and on consumer electromotors. Consultant shall check to ensure supplied generator frequency matches equipment input power. Otherwise frequency convertor is needed, which itself affects the value of input power.
- Type of signals from instruments installed on the package (which are to be sent to ICSS) shall match project requirements. Otherwise, connecting it to marshaling cabinet and ICSS may face difficulty.

B. Review of Deviations

Vendor design may have several deviations from project requirements. This may be due to availability of certain materials, limitations in manufacturer machinery, time constraints, etc. Consultant shall review them carefully and identify major and minor deviations. Those that cannot be accepted shall be rejected by the consultant. Some items may not be classified as deviation. They are design faults. Two examples follow:
- If a crude composition includes considerable H_2S, then using carbon steel material not complying with NACE requirements or sour service conditions shall be rejected.
- If a generator (which is intended to be installed in a hot environment) is designed with low-capacity fans for its cooling system and vendor expects to have large temperature exchange with cooling media. This is not a deviation. It is a design fault and shall be rejected.

Minor deviations are those that the consultant can accept and close the issue. For example, project painting topcoat material may not comply with vendor painting specifications. This is in case either vendor or independent evaluation confirms painting system itself complies with the expected environmental conditions.

Some major deviations may be technically acceptable. However, they have either some cost or time or operational impact. In that case, EPCIC contractor or client approval is required before permitting vendor to proceed with fabrication. Widely accepted criteria for defining major and minor deviations have not been established. To the author's understanding, any difference with project specifications that can be technically accepted can be categorized as a deviation. The decision to classify it under minor or major may be decided on a project basis.

C. Implementing As-built Vendor Data

Implementing final as-built vendor data is important for all remaining parts of the project including construction, commissioning, installation, hookup, and operations.

- For example, if package footprint and supporting system has changed from original design, consultant shall obtain data on time and revise secondary (or even primary) members to adapt with (will be supplied) package supporting system.
- If signals from package are different from those identified in project I/O list, they shall be identified on time to enable control system vendor to foresee required modifications in his system. If these differences are revealed after dispatching the control system, necessary modifications can be performed during commissioning. It may require both hardware and software changes. As much as possible this exercise shall be done before offshore installation. It has to be kept in mind that any offshore modification is several times more costly than a similar action when the platform is still in the EPCIC contractor yard.
- If package operation or maintenance has changed, all the necessary steps shall clearly appear in package manuals to enable the operation team to manage their day-to-day duties.

As is clear, vendor data review and implementation starts from engineering and extends to platform operating life.

Cost-wise these tasks may be a considerable portion. If vendor data arrive on time (during main course of the engineering contract) this is to the benefit of all. They can be efficiently reviewed, commented on, and implemented in project documents with less cost. If they arrive late after the main engineering team is demobilized or assigned to other tasks, it may have an adverse impact. Mobilizing new engineers and necessary learning curve lost time add to costs. The EPCIC contractor may argue that vendor data review and implementation costs had already been included as part of the main contract. The consultant argues that a specific man power was assigned for the project for a certain period and is no longer available to perform the task. This way the two sides may not reach an agreement. Specifying a certain timetable in the main contract for supplying vendor data is helpful.

CHAPTER 3

Systems and Equipment for Offshore Platform Design

Systems and equipment installed in an offshore platform depend on its intended functions and crude type/composition. Whether the platform is intended for oil or gas plants? Does it perform any processing or is it only a wellhead platform? Does it include any enhanced oil recovery (EOR) package? Is it a living quarter platform? Change in any of these criteria may need different types of equipment.

This chapter introduces various equipment functions and gives a brief description of their specifications. Some items may be common in different platforms. Some equipment may become necessary after several years of production in a field. Although platforms are divided into gas or oil type, they have many common features.

Oil platforms may have a considerable gas yield (associated gas) but normally gas platforms have a small condensate ratio. In fact light hydrocarbons, which are dissolved in fluid under high pressure, may escape fluid after pressure is reduced. Stabilizing means bringing oil to a condition that under normal/atmospheric pressure and ambient temperature gas does not escape from liquid to create a dangerous situation.

Designing platform vessels and machinery to operate under very high pressure increases costs in addition to introducing safety concerns. Each high-pressure vessel is a potential explosion source. Normally immediately after Xmas Tree crude pressure is reduced and flow is regulated by choke valves. This reduction is only to the extent to ensure its transfer to onshore plant or for export does not need additional pumping.

Both oil and gas reservoirs may contain water either as moisture or liquid. With high pressure and temperature inside the reservoir, the saturation limit is much higher than platform operating conditions or in the export pipeline. Saturation water may condense and add to existing free-water volume.

Two important ratios in each crude composition include water cut and gas oil ratio (GOR). They are necessary for designers to determine required process facilities and material selection.

Practical Engineering Management of Offshore Oil and Gas Platforms
ISBN 978-0-12-809331-3
http://dx.doi.org/10.1016/B978-0-12-809331-3.00003-X

Water cut is the ratio of water compared to the volume of total liquid in an oil field. For gas fields water volume is expressed per standard cubic feet of gas. Of course normally water daily volume shall be so low as to be expressible only per thousands of BOPD or millions of SCFD. Otherwise, production from that field (with current oil prices) may not be economical. These two ratios not only change from reservoir to reservoir but also during the lifetime of a reservoir. Their future changes may impact offshore complex design.

Actual crude composition is very important in defining a reservoir development scheme. The author has encountered a case in which the client gave year 1972 crude composition in a bid in 2007. The original crude had only 2% water cut without sand trace. After 40 years of production the same reservoir had a large water cut ratio with considerable sand. In fact, the water portion volume in the first stage separators was filled up to 60% with sand. Approximately every 3 months operators had to shut down a train and clean up sand. Finally they were forced to define a new desanding project. The original separators were not designed for desanding. Inevitably, sand settled in the first space in which its momentum was reduced. However, some small particles escaped downstream and caused erosion in pipes, valves, instruments, etc. They had this information in hand. Well testing was performed regularly. Only correct coordination was not performed between operators and main office. This discrepancy was highlighted during bid meeting. Owner original cost estimate and scope of work (SOW) definition did not include anything about this subject. They had to stop bidding and go back to redefine their studies.

Water is heavier than condensate, oil, and gas. In a reservoir, gas and oil are located above water. But it has to be noted that the situation is not like a static pool with clearly distinguishable layers. In fact it is like a sponge. The three elements are available in layers of reservoir with different ratios. In higher elevations the ratio of lighter material is considerably more and in lower elevations the ratio of heavier material is much higher. It has to be understood that all oil volume inside a reservoir is not recoverable. Oil viscosity, reservoir pressure, bedrock characteristics, drilling techniques, etc. all impact reservoir yield.

A study from the United States Geological Survey (USGS) indicates that under natural flow heavy oil recovery factor may vary from 12% to 19% around the world. Another study again by USGS specifically performed for certain American fields indicates using EOR methods may increase this factor from 25% to 65% (Ref. [33]).

In an oil reservoir, at the early stages of production oil is the major part of the crude. After some years two phenomena affect the crude. First, oil volume in the reservoir decreases. Second, due to crude extraction reservoir pressure drops. Due to pressure reduction some hydrocarbons are separated from liquid phase. This separation increases gas volume. At the same time water that is under higher pressure at lower layers finds a way to seep to higher elevations and fill the empty porous media above. Therefore both GOR and water cut increase. Decrease in total volume of reservoir yield is another issue.

For gas reservoirs a similar phenomenon may happen. After some years of production and decrease in reservoir pressure some condensate/water that under higher pressure had been forced to lower depths of the reservoir may find their way to upper layers and hence increase related portions in total crude composition.

As an example, see curve presented in Fig. 3.1 for a gas field. Values are changed from actual data. As shown in this figure, after 10 years' flat production plateau, reservoir yield reduces to less than half in about 15 years. Pressure drop on the other hand is more or less gradual. Within 25 years reservoir pressure drops to about one-fifth of the original value.

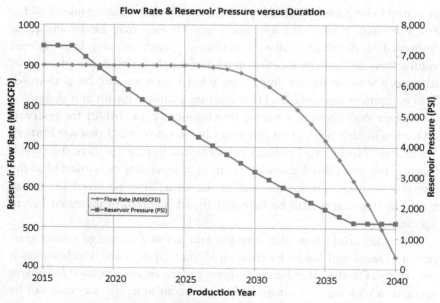

Figure 3.1 Sample gas field production and pressure decline.

This field is understood to be relatively dry at the start of production. However, after several years it will yield water to a rate of 2.5 B/MMSCF of gas. With the specified production capacity this means it will produce about 15 m^3/h water. Considering high H_2S and CO_2 content, this field has very high corrosive crude and requires either corrosion resistant alloys (CRA) or injection of considerable volume of corrosion inhibitor or a combination of both. Dry gas at first stage does not mean that carbon steel (CS) piping can be used for that duration.

This same reservoir has free elemental sulfur. Under high pressure and temperature (inside reservoir) sulfur is dissolved in gas. Immediately after the choke valve (with considerable pressure and temperature reduction) it starts to settle. Part of free sulfur will move with gas due to high speed and gained momentum. This portion will also later settle in bends and/or after impacting baffles or other barriers. Settled sulfur may plug the gas route and induce corrosion in platform piping and equipment. A solution to this problem is to use either physical or chemical solvents. Packages for solvent transfer, storage, and injection shall be foreseen. In the onshore plant sulfur separation from solvent shall be included. Based on required solvent injection pressure and volume, different schemes shall be investigated. In case very high pressure is needed, injection pumps can be installed on the platform. They may require considerable electrical power, which may be preferred to be generated on the platform itself instead of using subsea cables. For very high power demand, using gas turbines may be advantageous compared to diesel generators. Continuous supply of very large solvent volume may be done by pipeline instead of supply boats. A complete set of laboratory tests for solvent impact on other chemicals shall be performed. This equipment has considerable impact on platform layout and design.

Owner shall provide a master development plan (MDP) for reservoir utilization in each field. Long ago associated gas in oil platforms was burned. Nowadays "No Flaring" philosophy is adapted in many parts of the world. Gas is either transported onshore via crude or separately or is injected to the reservoir to boost production and at the same time be saved for future usage. In this case it shall be dehydrated and its pressure increased before injection.

For main production platforms gas may act as a source of power generation. Diesel fuel has to be transferred via supply boats, which makes it very costly. On the other hand, platform gas can be directly used for power generation with minor treatment. The treatment in many cases may not be more than scrubbing the moisture and heating to ensure its temperature is above the dew point such that water will not condense. New turbogenerators can accept a variety of gas compositions. If H_2S ratio in gas is too high, turbogenerators may not be able to burn it directly. In addition, after burning with exhaust line in a nearby location, its sulfur content may endanger operators' health. In this case sweetening may become necessary before burning gas as a fuel source for electric power generation. Extracted H_2S may be burned in flare. Factors discussed previously change platform layouts and designs.

Normal water volume in oil and gas reservoirs is only treated to remove oil to allowable limits. Several years ago the allowable oil ratio in water discharged to sea was 40 ppm. Later it was reduced to 15 ppm and now is 10 ppm in some locations. For each specific case, local regulations shall be carefully studied before establishing limiting criteria. This is the regular allowed oil level discharge to sea. The actual oil discharge for a specific time span may be much higher. Unwanted spillages, accidents, and emergency conditions will discharge more oil to sea. If a platform is producing 2000 BWPD with 10 ppm oil discharge limit (if all packages work properly and 10 ppm limit is not exceeded), it will discharge 3.2 L per day oil to sea. It is important to note that this amount of water is for a considerably large platform. Assuming water cut to be 5%, this water volume is equivalent to 40,000 BOPD crude oil production. It has to be highlighted that these figures are given only for information. The author in no case wants to underrate oil discharge to sea problems. In spite of specified limitations, unwanted spills inevitably exceed these levels.

Similar to associated gas, produced water can also be injected into a reservoir instead of discharging to sea. Normally volume of separated water is much less than required for injection. A rule of thumb for enhanced oil production is to inject three to four times extracted oil volume. For example, if 10,000 BOPD is target oil volume, at least 30,000 BWPD shall be injected. This is approximately equivalent to 200 m^3/h.

Before injecting water or gas several treatments shall be done. They may include:

- Gas separation: This will ensure H_2S and CO_2 are removed. Therefore corrosive acid will not be produced in injection lines and in the reservoir.
- Removing algae from water: Normally this is for pumped seawater. Water separated from oil does not include algae.

- Deoxygenizing: This will ensure chemical composition is not formed. In addition, oxygen–consuming living organisms will not be transferred to reservoir.
- Biocide injection: Will ensure living organisms are killed before injection. This prevents bacterial and biological growth.
- Sulfate removal: Prevents scale buildup in the well and reservoir.
- Based on project specifications, other chemicals may also be injected.
- After these treatments water pressure shall be considerably increased before injecting to a depth below present layers.

Including all these packages on a platform changes its configuration completely.

Both gas injection and gas lift are enhanced recovery methods. However, there is a major difference in their effect. Gas or water injection increases reservoir pressure and therefore moves oil upward. This is used where reservoir pressure has dropped to a considerable value. Gas lift is done in reservoirs where some pressure is still left but crude is heavy and viscous. Therefore head loss from reservoir to wellhead is too high. It consumes all available pressure head. In this case reservoir static head is not enough to drive oil to the well surface. Here gas is mixed with oil to reduce its specific gravity and viscosity. Therefore absolute value of head loss is reduced and existing pressure becomes sufficient to drive oil inside the well column to the surface. Selection of gas injection or gas lift is based on reservoir studies in MDP. Basic and detail design engineering only select necessary equipment.

Similar to export gas, injection gas is also treated to remove water. This will save compressor life. Gas may be taken from different points in process train. Based on their pressure difference, two- or even three-stage compressors may be used. Normally turbo compressors are able to increase their suction pressure several times. Therefore if the required injection pressure is very high or pressures of obtained gas from different tie-in points are very different from each other, they may be directed to different points of compressor train. The lowest pressure gas is directed to the suction of the first-stage compressor. Pressure increase will increase temperature and decrease volume and consequently increase specific gravity. Therefore a cooler is provided at the discharge of the first-stage compressor before it is directed to the suction of the second-stage one. Cooling will bring gas closer to dew point temperature. At temperatures equal and below dew point, droplets are generated, which may harm compressor blades. To avoid this, a suction scrubber is placed to remove water particles and reduce water content below condensing ratio at the reduced temperature. The same approach may be followed at the discharge of second stage and suction of the third stage.

Based on the platform location (including distance from shore, water depth, environmental conditions, etc.) the company may decide to have some treatment facilities on the platform or transfer untreated crude to onshore plant. This decision determines platform design. As stated several times, systems in each category of platform are more or less similar. The main difference is the operating/flowing/shut in pressure and temperature, crude composition, GOR, Water Cut, etc., which determine the material selection, design class, and platform equipment/layout.

One example of changes in company decision and therefore design approach happened in one of the gas development projects. The reservoir was divided into several phases. The first phase included production and living quarter platforms installed offshore about 100 km away from shore. Production platform had dehydration and glycol regeneration package. Other than this phase, all other phases only included water removal. Gas and condensate were transported with a pipeline to onshore. In later stages reservoir pressure dropped. Therefore design had to foresee future compression platform with necessary future tie-ins.

Many disciplines are involved in the design of an offshore complex or platform. Based on different practices, the platform packages and systems may be categorized in different groups. In addition, different users may categorize some of the packages in another category. The author uses five main categories. This categorization is only for presentation purposes and does not imply any specific rigid meaning.

- Process: The systems affecting crude production, treatment, and transportation.
- Utility: The systems providing required power and facilitating the function of process systems without direct impact on the crude.
- Instrumentation and Control: The systems enabling monitoring and control of all platform equipment and packages.
- Safety: The systems providing hazard mitigation, fire extinguishing, and personnel rescue.
- Accommodation: The systems enabling operators' life and work on the platform.

It has to be emphasized that based on platform usage and owner/developer decisions, some of the systems listed below may not appear on the platform.

Grouping between process, utility, etc. systems as given in this chapter is not a universal agreement. In many platforms all systems not affecting process are simply classified under utility. In others only equipment that impacts crude treatment is classified under process. Of course due to its importance and standard regulations, safety category items are recognized everywhere. The previous groupings may change between different projects.

In this book I have tried to explain process equipment as they first appear in the production line after Xmas Tree. For other systems their relation to production is considered.

3.1 PROCESS SYSTEMS

Process systems are located either on production or wellhead platforms. Their main intention is to separate gas, oil, and water or guide crude flow by regulating pressure, temperature, and volume. In this process certain crude components may be extracted. This process will facilitate transportation to onshore or to floating loading facilities and will reduce potential corrosion of the pipeline. In addition, with some minor treatment it may enable crude export via single buoy mooring system to a very large crude carrier (VLCC). This will certainly increase its commercial value.

Separation is achieved by a change in pressure, temperature, and momentum of the crude. Consider a soda bottle that bubbles when shaken and gas separates from the liquid. When you open the soda bottle to atmospheric pressure a considerable pressure reduction happens. This is the major reason to separate dissolved gas by bubbling. More or less the same phenomenon happens for oil and gas separation.

3.1.1 Xmas (Christmas) Tree

Xmas (Christmas) Tree is the first equipment in any well, regardless of being onshore/offshore/subsea/gas/oil and injection wells. Since it does not impact crude composition or separation, in a broad sense it is not process equipment. The main function of Xmas Tree valves is to protect topside facilities against reservoir pressure surges and abnormal conditions. It is the first equipment connected to the drill pipe. Regardless of whatever is done on the production platform, it shall have a Xmas Tree on each well.

This valve closes or opens the flow in and out of the well. This means it is not only used for production wells but is also used in water/gas injection wells. Xmas Tree mostly consists of a block of steel containing several valves. These valves can be remote or manually operated.

Automatic operation of Xmas Tree valves are governed by distributed control system (DCS) or emergency shutdown (ESD) system. Tie-in points have been foreseen for the utility systems like chemical injection, well intervention, and well monitoring (such as pressure, temperature, corrosion, erosion, sand detection, etc.). A connection is kept for injecting cement when it is wanted to close the well. Main Xmas Tree valves include:

- Subsurface safety valve (SSSV) or down hole safety valve (DHSV)
- Surface safety valve (SSV) or master valve (MV)
- Wing valve (WV)

Xmas Trees may have a wide operational pressure range up to 15,000 psi (1034 bar)[1].

SSSV is placed inside well up to 100 m below seabed (mudline). Its function is to close flow during shutdowns. Due to high pressure, it is hydraulically operated. This and other safety valves are fail-close type valves. In order to enable immediate closure of the well flow, the hydraulic pressure usually keeps a spring open. Therefore as soon as the hydraulic pressure is removed, the valve closes flow. Theoretically, pressure of noncompressible fluids will drop to zero as soon as one drop of fluid is removed. A sketch of Xmas Tree is shown following in Fig. 3.2. This was installed on a gas platform. Master valve (MV) and wing valve (WV) large actuators are visible. Both of them have a handwheel for manual operation.

SSV (MV) is also used to close the flow and is sometimes placed in double configuration. The lower one is left open unless there is a need to repair the upper one.

If due to any reason well shall be permanently closed, a kill wing valve is used. Through it cement is injected to close the wellbore.

As explained before, valves placed in the Xmas Tree in particular and valves in flow line are fail-close type. This is a safety precaution to ensure that during power failure only production is stopped and operator/platform safety is not endangered.

Xmas Trees are designed for well shut-in pressure. This is equal to a pressure that when the flow is stopped builds up upstream of the flow line. In fact, shut-in pressure is equal to the reservoir pressure minus static weight of crude column (gas or fluid) in the well. During production, pressure drops to well flowing pressure. This is much less than shut-in pressure due to head losses. Platform equipment shall be designed for the operating pressure, which is much less than shut-in and flowing pressures. If they are

---

[1] 1 psi = 0.069 bar = 6903.3 Pa.

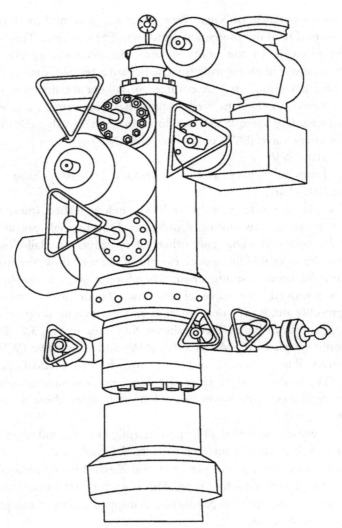

Figure 3.2 Xmas tree.

to be designed as per well shut-in pressure, all the equipment will become very heavy and costly.

Connections to Xmas Tree may include a variety of chemical injections. Methanol is injected to prevent hydrate formation or freezing. This is especially needed during start-up operations. Corrosion inhibitor may also be injected to well or in the piping system.

Xmas Tree installation is done by drilling contractor. Conductor pipe diameter limits horizontal installation tolerance. Drill pipe will also have vertical tolerance. Xmas Tree will have several branches with connecting valves and gauges. Operator shall have access to them. In some cases intermediate levels may become necessary. Covered area shall be as close to Xmas Tree body as possible. Xmas Tree may undergo vertical movements during operation. The main cause is temperature variation. Movement range shall be carefully determined in each platform. Values up to several centimeters are expected. In a gas platform values of 30 mm squat and 90 mm growth have been used.

As an indication, assume seabed temperature to be 13°C while air normal temperature may be around 35°C. Operating temperature of a platform may reach 90°C. With 70 m water depth and Xmas Trees located in elevation +15.0, this well will experience an average temperature change of ~70°C in 85 m pipe length. Nickel and carbon steel have approximately similar expansion coefficient. Assuming 1.1×10^{-5} as expansion coefficient, this yields about 6.5 cm total deformation.

Location of structural members shall be carefully selected to avoid clash during these movements. To cater for lateral installation tolerances, other than a few primary members all other members for this access platform may be offshore installed.

Space needed for Xmas Tree depends on its dimensions. With larger flow volume and higher pressure, rating dimensions will become larger. For center-to-center spacing of conductors, squares with a side length of 6.0 ft (1830 mm) to 8.0 ft (2400 mm) have been used.

3.1.2 Production/Test/Flare/Export Header

Headers collect fluid from all wells and route them to necessary equipment. Production/test header configuration in wellhead and production platforms may be slightly different. Each well is connected to both production and test headers. If it is located on a production platform, then lines from different satellite wellhead platforms may also be connected to production header. Normally test headers/test separators for satellite platforms are located on them. Installing a dedicated test line to transfer satellite platform crude to production platform is very costly. It is worth to note that testing is done only to obtain a measurement of each well production. Gas, oil, and water ratios are obtained in this manner for total reservoir. This enables reservoir engineering team to prepare a production plan for each well. They may even shut down or reduce production of some wells to have maximum production from reservoir.

Each header consists of a pipe capped at one end and connected to the flow line at the other end. As many branches as required are connected to it. Branches are normally connected with an angle from an elevation above header centerline to prevent fluid backflow during shutdown. This will save space. To get better hydraulic conditions, in some cases 45 degrees bends may be used.

For large production platforms two separate trains may be provided. This is to ensure that when one train is shut down the next one can continue production. In this case half the wells or lines are connected to one header and the remaining lines to the other header. Header size is determined by the volume of crude oil or gas, considering the following items:

- Maximum erosion velocity
- Minimum slugging in separation equipment
- Suitable pressure drop

Each line shall be connected to the header with sufficient isolation valves. Flow line valves are fail-close type. In addition, check valves are installed to ensure in case of any failure reverse flow will not occur. In many cases double-block and bleed valves are necessary based on H_2S content. In this case a drain valve shall be connected to drain header.

To enable separate flaring of each individual well, it may be directly connected to flare header. Increased number of wells and connection lines makes heavy congestion in the well bay area. A very accurate 3D modeling is required to ensure each well is connected to all required lines and at the same time sufficient access area to all valves is provided.

Line from Xmas Tree wing valve to choke and from there to production/test headers and from pressure safety valve (PSV) to flare header shall have sufficient support and flexibility. Guided supports may be used to allow movement in one direction and prevent other directions. Spring supports are used to limit force value exerted on the line. In support design several factors shall be considered:

- Installation: Normally wells are drilled and Xmas Trees are installed after jacket installation. This is to prevent cement, mud, oil, and other contaminants during drilling damage topside equipment. To cater for installation tolerances, line from mating flange to choke and from there to header is offshore installed during hookup activities. Therefore supports for this segment are only designed for platform in-place condition and not transportation.
- Well vertical movements: Some tolerance for well squat and growth during different phases of production shall be considered.

- Temperature deformation: This line experiences high temperatures during production and very low temperatures during blowdown or emergency flaring. It is better to allow header movement instead of exerting large forces to keep it in location. For an estimate of required forces, refer to chapter "Disciplines Involved in Offshore Platforms Design," Section 2.7.5.

- Choke and PSV operation: Choke is in continuous operation and PSV may pop up when reservoir pressure exceeds set point. Both of them change direction of large volume high-pressure fluid. In gas, flow density is low and pressure is high. In oil platforms fluid density is high and pressure is normally lower than gas platforms. Both of them have large thrust force that shall be counterbalanced by support system.
Sample 24″ production header and 8″ test header are shown in Fig. 3.3.

- Since headers may become very hot, personnel protection insulation is installed around them.

- The aisle between the two headers is for operators' access to valve handles. As can be seen, handles are turned toward each other to enable access to both headers through a single passage.

- All valves are actuated and can be automatically controlled. Valves are connected with one in between. Free flange is blinded. This may be either for future lines or two trains are constructed. Half of the wells are connected to each header with a tie-in flange for future access.

- Area is covered by grating. Header supports are fabricated from normal profiles and welded to secondary members.

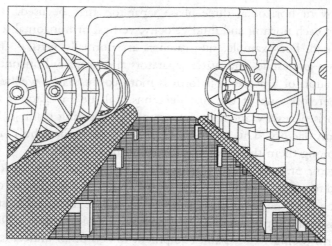

Figure 3.3 Production, test, and flare header.

3.1.3 First-phase Separator (Free-water Knockout Drum)

Water is heavier than hydrocarbon. Any crude contains gas, hydrocarbon, and water. Due to reservoir pressure and temperature, these are mixed. After reaching platform surface and based on the changes in fluid pressure and temperature, three phases tend to separate. If separation is done in the export pipeline, it may block pipeline and inhibit transportation.

Extracted crude from reservoir contains several materials/contaminants. They may range from sand, bitumen, wax, heavy hydrocarbons, light (gas) hydrocarbons, H_2S, CO_2, nitrogen, water, etc. Several methods may be used for separation:

- Gravity: Large vertical or horizontal (slug catcher/two- or three-phase separator) vessels may be used to separate sand, slugs, water, gas, etc.
- Centrifugal action: This is normally used to separate very small percentages of oil from water, or to separate two mixed chemicals with different specific gravities, for example, diesel fuel separation from other impurities, like water.
- Physical absorbents/filters: Used for separation of small quantities of impurities.
- Vibration and pressure reduction: Are used to separate dissolved gas from fluid.
- Gas bubbles: In essence this is also a gravity-based method for small-size particle separation.

In an offshore platform hydrocarbon separation after Xmas Tree almost always means gravity type. Either slug catchers to remove slug/sand from crude flow or horizontal three-phase separators may be used. Based on volume of each portion and fluid velocity, vertical or horizontal separators may be used. For example, slug catchers are normally vertical. Two- or three-phase gas, crude, and water separators are normally horizontal.

Fabrication of a process platform is more costly than a wellhead platform. Therefore one of the first decisions shall be to check if offshore separation is necessary and to what extent. Engineering team shall investigate many factors. This decision shall be made before the conceptual phase. Among them the following items can be highlighted:

- Reservoir pressure: Reservoir pressure shall overcome head loss from seabed to surface, inside platform piping, and in the pipeline to onshore. At the same time (if possible), pressure at onshore inlet shall be sufficient to pass onshore plant systems with lowest possible pumping. Designing platform piping, vessels, and equipment to accept an unnecessary high pressure will increase costs due to greater thickness of material and stringent code requirements for high-pressure equipment.

• Reservoir temperature: Very high/very low temperatures require higher quality material. In addition they require personnel protection insulation. Allowable stress ratio to yield stress is reduced with increasing temperature. On the other hand, material brittleness increases considerably with very cold temperatures. This means both high/low temperature conditions require higher quality material.

• Combination of temperature and pressure also has a similar impact. With the same temperature higher design pressure may require a higher piping class. For example, for carbon steel A105 material, class 300 piping at 100°F can tolerate up to 740 psi, but at the same temperature for 1480 psi pressure a class 600 pipe is required (Ref. [34]).

• Increased temperature and pressure increases gas solubility in crude or water vapor saturation point in gas. Therefore along the pipeline as both pressure and temperature are reduced, three-phase separation tendency increases. Due to considerable differences in specific gravity, this phenomenon may block pipeline.

• Water content: Very high water volume in the crude will cause two problems. First, in the export pipeline it may separate forming a layer at the bottom of the pipeline. Two- or three-phase flow condition changes hydraulic parameters and may prevent proper fluid export. This will both obstruct gas passage and increase pipeline corrosion. H_2S or CO_2 may dissolve in separated water and form acid, which increases corrosion considerably. Second, a lot of energy is wasted in transporting it to onshore or during its separation and treatment for offshore disposal.

• GOR ratio shall be carefully studied. In some cases it is more economical to separate gas and crude oil and transport them via separate pipelines. Gas can also be used for power generation in the offshore platform. This may be economically beneficial because in spite of its higher capital expenditure (CAPEX) it has considerably less operating expenditure (OPEX).

• Crude composition shall be carefully studied to adopt suitable mitigation measures for corrosion or sedimentation prevention.

Considering these previous factors a thorough study shall be performed during MDP stage to determine whether it is economical to separate gas/oil/water offshore and to what extent.

A decision shall be made on the degree of separation including gas—liquid and liquid—liquid separation. It is understood that with temperature and pressure reduction of the product its gas solubility reduces. Therefore some gases tend to separate. Very large pressure differences at each separation stage shall be avoided. Crude stabilization necessary for export is very far from offshore operations, which still has high pressure. Therefore it may be performed in several stages.

High-level cost evaluations are normally done in MDP and conceptual studies. Basic and detail design stages focus on practical issues to enable rapid procurement and fabrication.

Based on the previous explanation, first-stage separation includes two- or three-phase separation with limited pressure drop (to platform operating pressure). Terminology used in some oil and gas platforms may be somewhat different. In oil platforms, "production separator" is used, in gas platforms "free-water knockout drum" is used, because it separates free water from gas. The term "knockout" is used for removal of liquid from gas. The same applies for "flare knockout drum" (Section 3.1.10). This liquid may be a combination of water and hydrocarbon. Separation process is done inside a pressure vessel to enable using reservoir pressure in crude export. This will save energy necessary for pumping.

Separator functions in oil and gas platforms are more or less similar. Of course, the number/type and function of internal elements, vessel design dimensions, volume of different compartments, thickness of walls/heads (due to differences between volume of gas, water, oil, and platform pressure) are different. In the following discussion they are treated more or less the same. The first step in a production line is to separate water and then condensate/gas.

Both production and test separators are intended to separate the three-phase gas, water, and hydrocarbon. Their function is more or less similar. Production separator does it for the combined crude product of the platform. Test separator is only intended for one well at a time to give an understanding of how the reservoir is acting. Only the size is different. Separators consist of a cylindrical vessel with end caps. Majority of separators are installed in horizontal position. Internal parts may consist of but are not limited to (Ref. [35]):

- Inlet deflectors used to reduce flow momentum, perform an initial bulk separation of gas/liquid, and distribute gas uniformly.
- Baffles with perforation to increase pressure drop and accelerate separation. They reduce turbulence and at the same time smaller droplets adhere to each other. This effect may be accelerated by injecting

demulsifier. Eventually it leads to smaller length and shorter retention time for larger droplet settling. The Gas Processors Suppliers Association "Engineering Data Book" [35] gives some typical retention times for gas—liquid and liquid—liquid separation.

- Weirs to allow lighter crude separation from top of water.
- Mist collectors at gas outlet to reduce humidity from saturation point. They mostly consist of knitted-wire mesh pads.
- Vortex breakers at liquid outlet.

Internals have no rotating or electrical parts. They mainly consist of plates in different shapes and configurations connected to each other. Those internals that may need future maintenance for cleaning, clog removal, etc. shall be fabricated in several sections. These sections shall be bolted to vessel wall and to each other.

Internals design shall enable their removal through manholes. This means large plates shall be divided into several smaller sections and bolted to each other. At least one dimension of a plate shall be suitably smaller than the free distance between manhole walls.

At preplanned platform shutdowns, the separator is isolated, drained, washed, and purged. This will ensure toxic and explosive gases/hydrocarbons are removed. Then manholes are opened. Even with these precautions, all safety measures shall be followed. It means detectors shall check gas presence inside vessel, personal breathing apparatus shall be worn, inside vessel shall be properly lit up to make every corner visible, lights or other electrical devices taken in shall have suitable explosion-proof rating, etc. After that operators may enter inside vessel, open internals one by one, and transfer them out of the vessel for cleaning.

Construction material of internals shall be from exotic material. This prevents them from rusting in vessel design life and prevents slug buildup. Supports welded to vessel walls and holding internals against crude flow shall also be from the same material.

Chemical injection including demulsifier and antifoam are used to accelerate separation and prevent foaming. Their name indicates their function.

Excessive foaming may trigger false signals in high and high-high level alarms (HHLA). HHLA will cause process shutdown. This may lead to loss of production.

Demulsifier helps physically dissolved material to coagulate. This will add particle size/weight and help rapid settlement. It reduces settling time, which leads to smaller vessel size.

Water separation is based on gravity settling. Water is heavier than crude oil and gas. Gas separation is done by reducing its solubility in crude. This is achieved through agitation and pressure reduction stages, change in temperature, and providing enough stilling time. Therefore the size (length/diameter) of the vessel, location of deflectors, and location/elevation of the weirs shall be carefully selected. The calculation shall be done by specialized engineering companies.

Normally in a production platform, crude from several wells driven in a reservoir are mixed for transport. Test separators are used to check each single well and obtain a description of well performance. Their action is exactly similar to that of the main separator with a smaller volume. The reservoir engineering shall continuously update their database and reservoir model regarding water cut ratio, GOR, and the product volume. This is achieved by well testing. If production separation is not done in the platform, the separated gas and crude will be remixed and directed to export flow line. Water is normally treated and discharged to sea, otherwise it will cause corrosion, deposit as slug, or both.

As stated the only difference between test and production separator is their size. In wellhead platforms, test separator is only designed for one well testing at a time, but production separator shall have enough capacity for all the wells connected to that train. Based on the number of wells, platform capacity, and crude composition taken from different layers of a reservoir, one or two trains can be used. Considering connected piping, valve, and other equipment, a platform with two trains will be more costly than a platform with one train with the same capacity. But operators know that other than unexpected problems each year, they have to do maintenance and repair work to ensure continuous operation. Selection of two trains enables routine maintenance without total platform shutdown.

Production separator vessel function is to separate gas, liquid, and condensate. This may serve several intentions.

Gas may be needed as fuel gas or for injection to the reservoir to enhance reservoir yield or for dedicated gas export line.

Condensate and water are also normally separated. Separated water is discharged to sea. Before discharging to sea, it shall be treated sufficiently to reduce the oil ratio to acceptable limits. The acceptable oil limit is around 10−15 ppm. For very high water cut ratio the oily water treatment package may become so costly that pumping the two-phase liquid to the export line becomes economical. In this case, a production separator may not be used. The separated gas and crude in test separator are also mixed again and

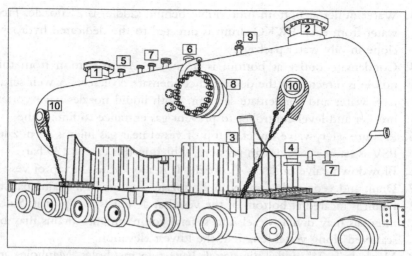

Figure 3.4 Photo of an free-water knockout (FWKO) drum.

injected to the export header. Piping and instrumentation diagram (PID) of a sample separator vessel was given in chapter "Disciplines Involved in Offshore Platforms Design," Section 2.8.5. Separator sketch is shown in Figs. 3.4 and 3.6. This free-water knockout (FWKO) was fabricated for a gas field. Fig. 3.4 shows it in the manufacturer shop on the transportation carriage ready for dispatch to engineering, procurement, construction, installation, and commissioning (EPCIC) contractor yard. In the original photo, between wooden planks and separator body a rubber mattress was placed to prevent damage to vessel paint. The main paint system shall be applied in manufacturer plant. Paint touch-ups are permitted in EPCIC contractor yard.

As the figure shows, this vessel has several piping and instrumentation nozzles. Protection caps are placed on nozzle ends to prevent air entrance. Connecting flanges are not in manufacturer SOW. Design pressure of this vessel was 139 bar and design capacity was 600 MMSCFD. Main piping nozzles include:

1. Crude inlet at top (gas, condensate, and water) is a 24″ nozzle. Immediately at entrance it hits baffles and changes direction to reduce flow momentum.
2. Gas outlet at top with a mist collector is also 24″ nozzle. It has a drain line directing liquid to water compartment before weir.

3. Water outlet at bottom (not visible behind saddle) is 2″ nozzle. Free water from each FWKO drum is directed to the dedicated hydrocyclone in oily water package.
4. Condensate outlet at bottom is 6″ nozzle. Hydrocarbon from this nozzle is directed to the dedicated condensate coalescer. A weir separates water and condensate sections. Both liquid nozzles have vortex breaker and level controller to prevent gas entrance to liquid line.
5. Pressure safety valve outlet at top of vessel near gas inlet is 3″ nozzle. PSV is set at vessel design pressure, which in this case is 139 bar.
6. Blowdown valve outlet is 4″ nozzle. It is also located at top of vessel.
7. Drain and vent outlets are both 2″ nozzles and used for maintenance. Drain is located at bottom of the vessel and vent is at top of it. Since internals may divide vessel into several compartments, drains may be scattered along vessel length at the lowest elevation.
8. Manhole is 24″ wide. Big vessels have two manholes. Manholes are placed at different sides of the vessel.
9. Temporary gauge to check internal pressure. Two steel plates were welded to protect gauge from damage during transportation, which are not shown here.
10. Temporary lifting ears are used for movements in the fabrication shop and transportation.
11. Some nozzles are kept to inject purging gas during maintenance. After washing/draining and before operator entering inside vessel, purge gas is injected to ensure all combustible or toxic gas exits from inside vessel. Instrumentation nozzles are used for installation of:
- Pressure gauges or transmitters
- Temperature gauges or transmitters
- Level measuring gauges or transmitters

Separation may be done at different stages. The first stage is at relatively high pressure. This will separate water and hydrocarbons, which need a very high pressure to remain dissolved. Second-stage separators operate at much lower pressures. This is normally used to stabilize oil. Stabilization means removing so much of hydrocarbons from liquid phase that in normal conditions (ambient temperature and atmospheric pressure) no gas is discharged from stored oil. This is necessary to enable its export via tankers, pipelines, or trucks to consumers. Stabilizing requires large vessels and is normally done onshore.

First-stage separators, which operate at very high pressures, may have large wall thicknesses. Even using simple hoop stress formula can give an

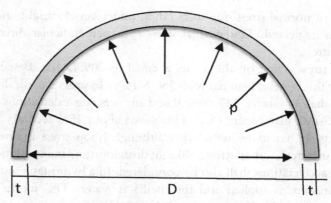

Figure 3.5 Pressure vessel internal pressure.

answer within ~15% accuracy. To get a feeling of the values, an example is given. Refer to Fig. 3.5.

Vessel design pressure (p) = 140 bar

Vessel internal diameter (D) = 3500 mm

$$\sigma = \frac{P}{2t} = \frac{Dp}{2t} \leq 0.6F_y$$

For rapid calculations, always remember 1 bar is equivalent to a column of water 10 m high. This is equivalent to a pressure of 1 kg/cm². In order to get a rapid answer with less than 15% error, we have made several simplifications:

- A vessel consists of shell, heads, and nozzles. The critical point may not always be the shell. Usually the nozzle-to-shell connection is locally stiffened. Heads are rolled from thicker plates to cater for thickness reduction during bending. The reason hoop stress is used in this text is that the shell is the major part of the vessel and hoop stress calculation for the shell is very straightforward and rapid and can easily be calculated by hand.
- Hoop stress is due to internal pressure. It is not always the most critical case. Additional loads due to environmental factors and platform operation shall also be considered.
- Limiting allowable stress to 0.6 yield stress is another simplification. For a detailed design, the American Society of Mechanical Engineers' recommended values shall be used. In some cases, it may be below 0.6. For example, increased temperature reduces allowable stress.

- Effect of normal stress due to bending under vessel weight and inside fluid is neglected. This normal stress is in perpendicular direction to hoop stress.

Hoop stress acting on this vessel is equal to 209.4 MPa. Based on the above, for this vessel if material with 350 MPa yield stress is used, shell plate thickness shall be above 117 mm. Based on accurate calculations for this vessel, 135-mm plates were used. This shows about 15% accuracy in hand calculations. It has to be noted that although hoop stress is one of the controlling factors, other stresses like hydrostatic test, transportation, and operation accelerations shall also be considered. In a hydrostatic test, higher internal pressure is applied and the media is water. This means that in addition to normal hoop stresses due to internal hydrostatic pressure, impact of bending stresses due to vessel's own weight and water shall also be added.

Heads are normally elliptical. Due to its longer required space, spherical shape is normally not used. One of the common elliptical shapes used is the one that its long diameter is twice its short diameter. Head plate thickness shall be selected a little bit higher than shell thickness. During head forming, plate thinning happens. The additional thickness caters to it.

Besides, at the head—shell connection a stress perpendicular to hoop stress acts. This is due to capping effect. From the following formula it is seen that its value is half the hoop stress:

$$\sigma = \frac{(\pi D^2)/(4p)}{\pi D t} = \frac{Dp}{4t}$$

Hoop stress acts in radial direction, while this stress acts in longitudinal direction. At a specific point one may act as normal stress and the other as shear while in other location their role may change. In addition to their separate checking with allowable normal or shear forces their combined effect shall also be checked with von misses criteria.

During head forming a tangent section is also considered. This will facilitate cap welding to shell. Volume of the vessel is governed by retention time. The major part of oil and water are separated when crude coming from the well impacts on entrance baffles, deflectors, etc. Lower half of the vessel is intended for oil/water separation. Typically, a 5-min retention period is selected. Gas volume changes with pressure, but water (completely) and oil (partially) are considered incompressible. For ideal gas, a linear equation relates pressure, volume, and temperature. In reality a compressibility factor shall be introduced in this equation.

Table 3.1 Expected oil and water volume

Fluid flow	Oil volume		Water volume	
	Lean start	Rich end	Lean start	Rich end
Mass flow rate (kg/h)	70,570	81,061	6452	6369
Density (kg/m³)	676.7	662.9	973.9	961.4
Flow rate (m³/h)	104.3	122.3	6.6	6.6
5 min flow (m³)	8.7	10.2	0.6	0.6

At beginning and end stages of reservoir life, the water/fluid ratio changes. Process calculations will show expected volumes. Table 3.1 shows expected volume of oil and water at the two extreme conditions for a gas reservoir. Here changes are very small.

The volume of this vessel dedicated to fluid part shall be at least 10.8 m^3. The sketch in Fig. 3.6 shows different compartments of a first stage separator. This sketch is taken from Devold [3] with some modifications. At inlet a slug catcher may remove large liquid plugs. Inlet baffles reduce momentum considerably. Slugs are combination of heavy hydrocarbon and water. The turbulence helps separate gas that exits from the top outlet. Before the gas discharge nozzle, a demister absorbs parts of saturated water vapor. Since oil is lighter than water and stays at the top, a weir separates the water and oil outlet. The two main internals are inlet baffle to change fluid momentum and demister to remove water mist. Other internals may also be installed, which are not shown here.

After fabrication and for transportation, inside vessel is filled with inert gas. The intention is to prevent humid air entrance. Internal pressure is a little above atmospheric. All nozzle ends are capped and will be removed

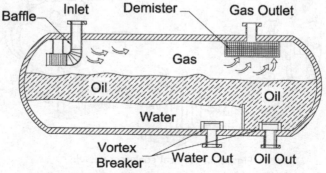

Figure 3.6 First-stage separator compartments.

after installation in final location and before connecting piping tie-in points. For lifting in manufacturing yard or transportation or for placing in its final location on the platform, four lifting lugs are welded to vessel sides. This shall be done before heat treatment.

The condensate coalescer is a liquid—liquid separator. It is located downstream of a first-stage separator. Its main function is to separate the remaining water portion in the liquid phase of the product. Therefore its design is completely different. Here water/gas volumes are considerably less than first-stage separator. Therefore normal gravity separation is not efficient. Lower operating pressure is caused by head loss from first-stage separator to this vessel. In addition, this lower pressure facilitates further gas bubbling out and separation from the fluid. In some platforms condensate coalescer may be placed after second- and even third-stage separation. Gas and water are extracted from condensate.

Fig. 3.7 shows a typical coalescing vessel. Nozzles are more or less similar to FWKO drum.

1. Manholes (24″) are located at both heads. This is for two reasons. First of all, this vessel is much smaller than FWKO drum, therefore installing a large hole at sides may hinder other nozzles' installation. Second, vessel internals are located along its length. Placing manholes at ends facilitates future internal removal.

2. Inlet and outlet nozzles are both 10″. In fact, piping line connected to condensate coalescer is 6″ and near the vessel a reducer changes the size to match nozzle.

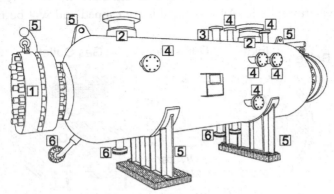

Figure 3.7 Typical coalescer vessel.

3. Similar to other pressure vessels a PSV protects the vessel from overpressure. For this vessel, operating pressure was 124 bar and PSV is set at 139 bar, which is vessel design pressure.
4. Other nozzles for installation of level/pressure/temperature transmitters are also connected.
5. Miscellaneous items like davit for holding manway door, saddle location, lifting padeyes, etc. are also constructed.
6. Connections for drain and purging gas for maintenance are also installed. Water drain is both for normal operation and after washing (maintenance). For efficient draining of vessel compartments at each area, a water drain is installed. Washing drain is directed to closed drain vessel, while operating water drain is directed to dedicated hydrocyclone.

Each vessel shall have entrance hole(s). This will be used for cleaning and maintenance operations. Similar to large rooms in the platforms, for large vessels separate entrance/exit holes shall be considered. Manholes are used for this purpose. Normally they are 24″ openings covered with bolted doors. A moveable davit shall bear its weight, such that rotating manhole door along davit vertical axis enables easy operator access. It shall be gas/water tight during design life. Door for a 24″ clear opening may have a diameter greater than 1 m. Its thickness shall be more than vessel wall. For a vessel constructed from 140-mm-thick plates, davit shall handle a very heavy load including cover plate, fastening blots, gas-tight gasket, and handles. Davit shall rotate freely while holding this load to enable free access.

3.1.4 Slug Catcher

Slug catcher is used when large content of slug water and sand are expected in the crude. In addition to damaging downstream systems, instruments, pumps, etc., sand may plug platform piping. Slug catcher may be placed upstream of the production header.

Slug catcher is normally a vertical vessel with some deflectors to impose an abrupt change in crude moving direction. Horizontal vessels can also be used. Deflectors will help in separating large-size slug and sand. They will be gathered at the bottom of the vessel to be removed later. After several years of production in a reservoir, slug or sand content may increase. Some reservoirs may have large contents of sand production even from the beginning. Other than sand (which is trapped), other material like gas, condensate, and water can continue flowing. Gas can be directed to dehydration package or gas export header. Condensate is directed to condensate coalescer to be dewatered. Water is directed to oily water treatment package.

In one of the gas fields, slug catcher with 3.0 m internal diameter and 9.5 m T/T distance has been installed. This platform has two trains, each designed for 190.5 m^3/h of hydrocarbon (HC) condensate and 9.3 m^3/h water, assuming 8 min residence time gives ~ 26 m^3 for total slug volume. Space between HLA (6.75) and Normal Liquid Level (NLL; 2.6) provides this necessary volume (~ 29 m^3). Vessel internal volume is much more than this value. Fig. 3.8 shows typical slug catcher.

Its main process nozzles include:
- Crude inlet 24″
- Vapor outlet 24″
- Condensate outlet 12″
- Water outlet 3″
- Pressure safety valve 6″
- Vent valve 2″
- Manway 24″ for internal access

Design pressure for this slug catcher was 139 bar. Wall thickness for this vessel was calculated to be 139 mm. As can be seen, head thickness is about half shell thickness. Refer to Section 3.1.3 for an explanation. A pressure transmitter continuously checks vessel internal pressure. Three-mm cladding thickness as corrosion protection had to be added. Vessel weight with internals, nozzles, manways, support, etc. was above 150 ton.

This vessel has several utility nozzles that are omitted from this figure for clarity.

In some complexes in which several satellites are connected to the main platform (to clean infield pipeline), periodic pigging between satellites and main is performed. Based on size and length of the infield pipeline, an expected amount of slug is calculated. It may be stored in a temporary vessel and later transported offshore by supply boats or may be directed to closed drain drum or to a treatment system. Slug catchers shall also have suitable manholes.

3.1.5 Sand Management System

Sometimes in oil or even gas platforms sand may be transferred via crude to the topside. Sand consists of small-size particles around 150 μ or less. Due to higher momentum and viscosity, sand in oil platforms is normally produced in much higher volumes.

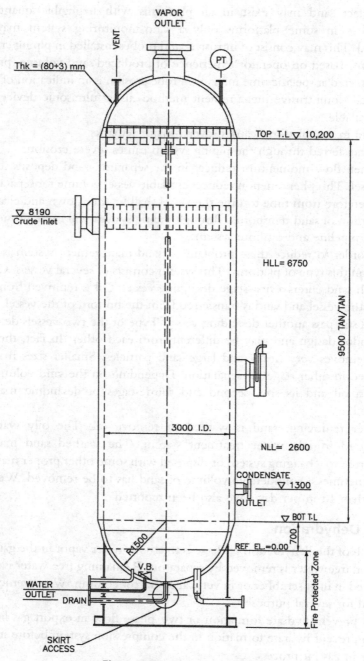

Figure 3.8 Typical slug catcher.

In fact sand may exist in all platforms with negligible quantities. Therefore in some platforms only a sand-monitoring system may be installed. This may consist of intrusive sand probes installed in piping critical locations. Based on operator experience of produced sand volume, probes are extracted at specific time intervals. Probe erosion is an indication of sand volume. Nonintrusive measurement methods using ultrasonic devices are also available.

Sand in any platform causes several major setbacks:

- If transferred through the piping system causes severe erosion.
- When flow momentum reduces in the separator, sand deposits in the vessel. This phenomenon reduces available vessel volume for separation. Therefore from time to time the vessel shall be shut down and cleaned.
- Portion of sand transported through piping system will cause erosion in the pipeline and onshore systems.

In order to reduce these problems, a sand management system is provided on this type of platform. This system consists of several vessels. Crude oil with sand enters a first-stage desanding vessel. Oil is removed from the top of the vessel and sand is transferred from the bottom of the vessel. This oil will still pass another desanding vessel. Type of the two vessels depends on vendor design and may be different from each other. In fact, the first stage removes very heavy and large sand particles. Smaller sizes may be removed in other stages of separation. Depending on the sand volumetric ratio to oil and its size, second and third stages of desanding may be required.

After removing, sand is washed to remove oil. The oily water is transferred to oily water treatment system. The washed sand may be transferred to a bagging system or disposed with some other proper manner.

Sometimes a considerable volume of sand has to be removed. Weights more than 1 ton per day have also been reported.

3.1.6 Dehydration

The job of the dehydration package is to absorb water vapor in the gas. It is assumed free water is removed by separators. Remaining free water is again removed in inlet scrubber or in vertical contactor column. Water removal is needed for several purposes:

- To prevent hydrate formation or two-phase flow in export gas line.
- To prevent hydrate formation in the compression system before injecting in gas lift process.
- To enhance gas quality as fuel gas for selling to customers.

- To reduce possibility of CO_2 or H_2S combination with condensed water. This phenomenon produces acid and induces corrosion.
- For flaring normally dehydration is not done as it is considered too expensive for gas, which will be flared. However, other means of water vapor reduction are employed like passing through a flare knockout drum.

It has to be noted that water saturation volume in gas is reduced with decreasing temperature. Gas composition has a detrimental impact on water ratio. Gas Processors Suppliers Association "Engineering Data Book" [35] provides a thorough description of the phenomenon. Different curves showing water saturation in different gases (temperature–pressure) are given.

Dehydration package normally consists of two sections, namely dehydration and regeneration. Process flow diagram (PFD) of this package and sketch of the vessel are shown in Figs. 3.9 and 3.10. The process of absorbing water vapor from gas is called dehydration, while the process of removing water from the absorbing media/catalyst to be reused is called regeneration.

Dehydration is normally done using triethylene glycol (TEG). TEG has high water absorption tendency. It is poured from the top-most level of the contactor tower. During descending to the floor, it contacts gas that is moving upward. When in contact with wet gas in the contactor column it

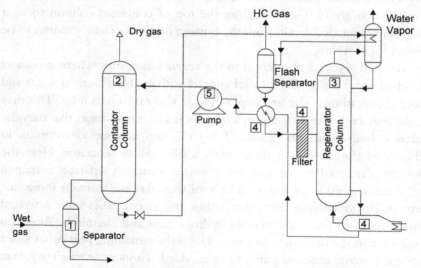

Figure 3.9 Dehydration package process flow diagram (PFD).

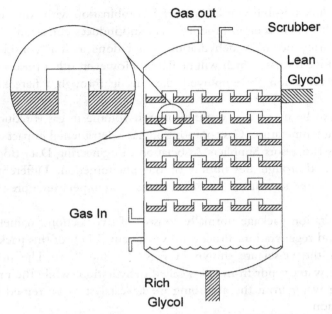

Figure 3.10 Contactor column internal (Ref. [3]).

absorbs water vapor in the gas. This vessel is working at high pressure. A small portion of TEG may be entrained in the gas, which shall be separated. Dry TEG without water is called lean glycol and after saturation with water is called rich glycol. Gas exits from the top of contactor column to next stage. Saturated glycol exits from the bottom of the contactor column to be treated for recycling.

The rich glycol is transferred to the regeneration unit where it releases absorbed water vapor. Glycol after contact with gas has absorbed water and some hydrocarbons. The first step to release water is to heat it up. This may be done in a reboiler. In essence, a reboiler is a heat exchanger that may use different heat-exchanging media. Fired reboilers are not so common in offshore platforms. Glycol then enters a three-phase separator. Here the pressure is reduced to atmospheric. Pressure reduction separates entrained gas, a major portion of water and glycol that may still contain some impurities. Before entering the regeneration unit glycol passes some activated carbon filters to remove entrained hydrocarbons and chemicals. After this stage it enters the stripping column and loses the remaining portion of water vapor and attains required purity to be recycled. To increase efficiency, lean and rich glycol may pass the same heat exchanger.

In theory water absorption and glycol regeneration is a closed cycle. In practice total glycol volume is not regenerated. Some portion of it is lost with the gas. The lost glycol shall be replaced from the glycol storage tank. A periodic supply of glycol to the platform is needed to ensure continuous platform operation. Normally a 2-week supply is stored in the tanks. The following main items are observed in this PFD (Fig. 3.9):

1. Before the contactor column an inlet scrubber vessel is installed. This is mainly for condensate, slug, and free water collection to prevent consuming a large volume of absorbing media. Cooler before scrubber reduces gas temperature. This will reduce amount of water that can be held in gas at vapor stage to the saturation point and increases scrubber efficiency. In fact, direct cooling itself is a dehydration method. Liquid collected at inlet normally contains a large volume of hydrocarbon as condensate. Therefore it may be first directed to condensate coalescer. Liquid from outlet scrubber is mainly water. Water from both scrubbers is directed to oily water separator package.

2. Contactor column is a vertical vessel that gas enters from its bottom and in moving upward contacts with water-absorbing media. Fig. 3.10 shows sketch of its internals. In general it has several levels of perforated plates. Caps are installed on perforations. Therefore gas has to lift each cap to move upward. After lifting cap, it has to pass a pool of water-absorbing media. Funnel-shape internals can also be used. The intention is to maximize gas/absorbent media contact for better efficiency. Means shall be provided to discharge rich glycol to regeneration package.

3. Rich glycol that has absorbed water is regenerated to be reused. This process mainly involves heating it to separate water. Glycol boiling point depends on its type. Minimum boiling point for monoethylene glycol (MEG) is about 200°C. Water boils at 100°C.

4. Heating can be done by reboilers or heated glycol. As stated before, to increase water/slug separation efficiency carbon filters can also be used. This process is mainly performed at low atmospheric pressure.

5. It is not economical to reduce contactor column pressure to atmospheric. To transfer gas to other stages or to export line, its pressure is needed. If dehydration is located at onshore plant, lower pressures can be used. For dehydration packages installed on offshore platforms, pressure is not very much reduced. Therefore lean glycol shall be pumped to increase its pressure and enable it to enter contactor column.

Part of dehydrated gas can be burned to generate necessary power or provide necessary heat for reboiler unit. In some cases it is possible to use excess heat of turbogenerators' exhaust as heat exchanger. It requires additional piping with suitable insulation and protection to prevent operators' or flammable crude unwanted contact.

Fig. 3.10 shows schematic representation of internals inside a contactor column. Glycol holding pots and caps to ensure gas will bubble up from glycol pools are shown.

3.1.7 Sweetening System

Sweetening package is very similar to dehydration. The main difference is in absorption media and final outcome. Sweetening duty is to absorb H_2S/CO_2 molecules in the gas. It may be done offshore or onshore but is normally more cost-effective if placed onshore. In some cases to sweeten gas for power generation this package is placed offshore. In this case its capacity will be limited to volume of gas necessary for power generation.

As long as the acid gas ratio is not very high (to induce corrosion in the pipeline) or is not required for a specific purpose (like power generation), it is preferred to perform sweetening in onshore plant. One of the factors increasing cost is that in addition to sweetening package a system for safe disposal of H_2S or CO_2 shall also be considered. Before injecting gas to a domestic pipeline, sweetening may be needed for several purposes:

- To prevent release of sulfur components after burning in houses.
- To prevent accidental release of toxic gas. Refer to Section 2.9.4, which shows even small percentages are lethal.
- To enhance gas quality as fuel gas.
- To reduce possibility of CO_2 or H_2S combination with water. This phenomenon produces acid and induces corrosion.
- To extract sulfur for commercial purposes.

Similar to dehydration, sweetening package normally consists of two separate sections, namely sweetening and regeneration. The process of absorbing H_2S/CO_2 molecules from gas is called sweetening, while the process of removing acid gas molecules from the absorbing media to be reused is called regeneration.

Several catalysts can be used for sweetening. One of the most commonly used catalyzers is amine. Sweetening is normally done using aqueous solutions of various alkylamines. Different amines are used in gas treating like monoethanolamine, diethanolamine, and methyldiethanolamine.

In its simplest description, amine is poured from top of contactor column to contact sour gas. While descending to floor it contacts sour gas, which is moving upward. In contact with sour gas in the contactor column, it absorbs H_2S/CO_2 molecules in the gas. Then it is poured from top of another column to be stripped of these components. Regenerated amine is recycled. Both columns are long vertical cylinders. They can be operated with low or high pressure. Design sizes like column dimensions, wall thickness, etc. will be different. A typical PFD for amine unit is presented in Fig. 3.11.

1. To improve gas sweetening efficiency at inlet and outlet of contactor column a scrubber or separator can be installed. This will work as water vapor or other liquid contaminants separation. Sour gas passes from a prefilter (inlet separator) before entering to contactor column. This is a gas–liquid separator to absorb water and heavy hydrocarbons. Gas is directed to bottom of contactor column. Similar to a dehydration package, water from both inlet/outlet separators may be directed to oily water treatment. Since H_2S has a high solubility ratio in water for onshore plants, which may contaminate surface water, further limitations in H_2S level in discharged water may be enforced. For offshore platforms regional restrictions for allowable H_2S level in seawater shall be investigated. Removing H_2S requires stripping units, which impacts platform layout, equipment, and power consumption.

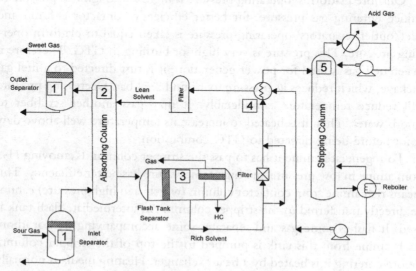

Figure 3.11 Natural gas sweetening.

2. Lean amine is poured from top of absorber column. Amine moving downward absorbs H_2S/CO_2 of gas moving upward. Now it is called rich amine. It exits from the bottom of this column. Gas exiting from the top of column has some impurities. It passes a postfilter (outlet separator). This will absorb amine and water. This vessel is working at high pressure. Dry amine without impurities is called lean amine and after saturation with sour gas is called rich amine. Gas exits from the top of contactor column to its next stage. Saturated amine exits from the bottom of the contactor column to be treated for recycling.
3. Rich solvent, water, and liquid hydrocarbon enter a flash tank. Hydrocarbon and part of gas separate in this vessel. Pressure reduction is the main means of separation.
4. To improve regeneration, filters and heat exchangers are used. Similar to a dehydration package for heating energy, either part of the sweetened gas can be burned or excess heat of generators' exhaust can be used. This may require some additional piping if the two units are not located nearby each other.
5. Regeneration is done in stripping column and reboiler units. H_2S, CO_2, and water vapor that exit from the top of the column may still contain some solvent, which can be collected in an accumulator and returned. The final portion of H_2S and CO_2 can be burned or directed to the next stage of processing like sulfur recovery units.

Offshore platforms' operating pressure is high. Although it is possible to reduce entering gas pressure, for better efficiency contactor column and inlet/outlet separators' operating pressure is taken equal to platform operating pressure. This pressure is very high for burning in GTG. Therefore if sweetened gas is used for power generation, it is first directed to a fuel gas package, which reduces its pressure to around 3–4 bar. Decreasing pressure will reduce temperature considerably. It may pass another scrubber to absorb water. Then it is heated to increase its temperature well above dew point before being directed to GTG combustion.

To regenerate amine it has to pass the stripper column. Removing H_2S from amine in low pressure and high temperature has better efficiency. This means rich amine from contactor column (which is in high pressure) cannot be directly transferred to the stripper column. An intermediate flash tank is used. It flashes some gas and separates some accompanying hydrocarbon. Rich amine from this tank is pumped to the top of the stripper column. Before entering it is heated by a heat exchanger. Heating media is normally lean amine from the stripper column. Gas is released from top of stripper column. It may contain some amine. Therefore before any action it is directed to a reflux unit to return the liquid part.

Rich amine is transferred to the regeneration unit where it releases the absorbed molecules (mostly H_2S/CO_2). In offshore platforms it is difficult (if not impossible) to send removed H_2S to a sulfur recovery unit. In fact, installing sulfur recovery in offshore platforms is too costly. Therefore sour gas will be burned and sweet gas is used for either power generation or directed to the export line. As an example, assume a platform has 1000 MMSCFD gas production. H_2S mole percentage is 4.164%. Gas molecular weight is 19.73 g/mol. Use Avogadro's law, which states that "Equal volumes of ideal gases at the same temperature and pressure contain the same number of molecules regardless of their chemical nature and physical properties. One mole of ideal gas occupies a volume of 22.414 L at standard temperature and pressure condition." In addition, from basic chemistry recall that H_2S molecular weight is 34.08 g/mol and sulfur is 32.06 g. This means that if gas from this reservoir can be categorized as "ideal gas" and if all sulfur is removed from sour gas on a daily basis, more than 1680 ton sulfur is produced. It is understood that natural gas from a reservoir is not ideal gas and this calculation has some error. In spite of that, this calculation gives a very good approximate value. This clearly shows why sulfur recovery unit installation on offshore platforms is not possible. A sweetening package installed in offshore platforms is with small capacity. Released H_2S gas can be burned.

In brief, the sweetening process can be expressed in several stages. Amine after contact with sour gas has absorbed H_2S and CO_2. Rich amine is routed to a flash tank to remove hydrocarbons and other contaminants. It may further pass a filter. Then it will be heated. Heating may be done in a heat exchanger using lean amine media or electrical heating or reboiler. Normally in offshore platforms installation of fired equipment is avoided as much as possible. Heated rich amine is poured from the top of a stripping column to lose sour gas and attain required purity to be recycled. Rich amine exits from the bottom of stripper column.

Although sour gas absorption and amine regeneration is a closed cycle, in this process, total amine volume is not regenerated. Some portion of it is lost with the gas. The lost amine shall be replaced from storage tanks.

A sweetening package installed offshore shall operate in higher pressure. This means that vessels may have smaller volume but thicker walls. Sweetening installed onshore operates at much lower pressures. One of the plants installed more than 40 years ago had 75 psi operating pressure. It sweetened associated gas released from second- and third-stage separators.

H_2S and CO_2 released from the sweetening process shall be properly disposed. Based on the volume they may be flared. If the portion of the sweetened gas is low compared to total platform gas volume, then separated H_2S and CO_2 can be injected to the export line. This way they are transferred to onshore for further treatment and suitable disposal.

Part of lean amine from the stripper column may be directed to a reboiler unit. As stated, fired boilers are avoided in offshore units. Electric power may be used to generate required heating. High temperature of GTG exhaust system can also be used effectively for this purpose. In this case, for start-up conditions in which sweetened gas and running GTG are not available, a dual-fuel generator shall be used.

In this process part of amine will be lost. A backup amine tank is used to supply lost amine. It is periodically filled in by supply vessel.

Required sweet gas volume can be calculated using basic chemical concepts. Assume for an offshore platform 1000 kW power generation is required. Gas from that platform is calculated to have net heating value equal to 28 MJ/m^3. Note, 1 watt-hour $= (1 \text{ J/s})/3600$. GTG efficiency may be taken as 0.3. Required volume of sweet gas is determined by dividing required power to net heating value. It is converted to hours by multiplying to 3600, then, divided by GTG efficiency. For the above, case we need about 430 m^3 sweet gas per hour in standard conditions.

Further assume this platform gas composition is as shown in Table 3.2. As per Avogadro's law, in standard conditions "equal volume of gas

Table 3.2 Typical gas composition

Typical gas composition	Mol Wt. g/mol	Mole %	Mol Wt. g/mol
N_2	28.01	6.346	1.777515
CO_2	44.01	7.578	3.335078
H_2S	34.08	4.164	1.419091
C1	16.04	81.601	13.0888
C2	30.07	0.233	0.070063
C3	44.10	0.056	0.024696
i-C4	58.12	0.008	0.00465
n-C4	58.12	0.005	0.002906
i-C5	72.15	0.003	0.002165
n-C5	72.15	0.002	0.001443
C6	84.00	0.001	0.00084
Benzene	78.11	0.002	0.001562
Toluene	92.14	0.001	0.000921
		100.00	19.72973

contains same number of molecules." Therefore this sweetening package releases about 31.6 m³ H_2S per hour.

Typical gas composition shown in Table 3.2 is for a highly sour gas. Second column shows chemical molecular weight of each gas. This is constant and can be derived from chemical tables. Third column shows mole percent of each gas obtained from analysis. In some reports weight percent may be shown. The engineer shall be careful to understand what has been reported. Although this gas is very sour, it contains more than 80% CH_4, which after treatment can be used directly for domestic usage.

Refer to Section 3.2.2 for air composition. Multiplying each gas mole percent to its molecular weight we get:

Nitrogen = 0.78 × 28.02 = 21.86 g/mol
Oxygen = 0.21 × 32.00 = 6.72 g/mol
Others = 0.5 g/mol
Sum Standard Air = ∼29.08 g/mol

Specific gravity of above gas is obtained by dividing 19.73/29.08 = 0.678.

Avogadro's law (hypothesis) is a principle stated in 1811 by the Italian chemist Lorenzo Romano Amedeo Carlo Avogadro (1776—1856). This number (Avogadro's number) is 6.022×10^{23}. It is the number of molecules of any gas present in a volume of 22.414 L at 0°C (273.15 K) and 1 atm (101,325 pa) pressure for ideal gases. Avogadro's number is the same for the lightest gas (hydrogen) as for a heavy gas such as carbon dioxide.

If a well is producing 100 MMSCFD of the above gas, total volume of H_2S is calculated as shown below:

H_2S Volume = 4.164% × 100 MMSCF × 0.3048^3 = 117,911.3 m³
H_2S Weight = 117,911.3 M³ × (34.08/1000)/(22.414/1000) = 179,281.6 kg
Sulfur Weight = 179,281.6 × 32/34.08 = 168.3 ton

Sweetening plant and sulfur handling equipment capacity shall be designed considering the above production rate. Handling this much sulfur on a daily basis requires huge facilities. This shows why installing a large sweetening plant in offshore platforms is not recommended. Small packages can be installed to generate sweet gas necessary for offshore power requirements. Normally in offshore platforms extracted H_2S is burned without sulfur separation.

3.1.8 Oily Water Treatment System

Extracted crude/gas from the reservoir may contain a lot of water. This is other than free water that is extracted in separators. Water ratio varies considerably from location to location and in a specific reservoir increases after several years of production. Water presence has several adverse impacts.

- Water content may reach up to 10%. In this case the simplest disadvantage is that 10% of the export power is wasted. This wastage requires lot of effort in separating. Therefore crude production cost increases considerably.
- Water combines with gases like CO_2 and H_2S and produces acids that are highly corrosive to the platform piping, equipment, valves, and export pipeline. To prevent corrosion, exotic material shall be used and/or corrosion inhibitor material shall be injected in the line.
- Water is heavier than oil. After cooling down in the seabed and pressure reduction due to head loss, water vapor condenses. After separation, it settles down in the lower part of the pipeline, reducing the available area for crude/gas transfer. Laminar water flow increases corrosion rate in the pipeline.
- Water accumulation may be so much as to completely block the export pipeline. In this case a blocked out signal may initiate platform total shutdown. Then cleaning pigs shall be launched to remove water/slug from the export line and allow for crude/gas passage.

Oily water treatment system may consist of two separate skids. The first skid contains a deoiling unit and the second skid contains a degassing unit. In a majority of techniques for oil separation, difference in oil and water specific gravity (light oil and heavy water) concept is used. Density difference can be used in approaches like gravity settling, centrifugal or gas induced separation. Oil absorption filters use the difference in adhesion tendency between materials.

At this stage and after first separation, oil ratio in water is very low, hence more efficiency is needed compared to first-stage separation units. When oil particle size is in the order of 50 µm gravity-based methods are not efficient. Therefore instead of large gravity-based stilling vessels, hydrocyclones are used. Hydrocyclones have a cylindrical inlet connected to a tapered end. Based on their internal conical shape and helical flow path, hydrocyclones induce a swirling motion to fluid, which exerts heavy centrifugal forces. This swirling motion is achieved by fluid high pressure and flow rate. Its separation efficiency is highest at a specific combination of oil—water ratio, uniform flow, and pressure. Therefore hydrocyclones are placed near to outlet of first-stage separators. This ensures sufficient pressure exists in the system. To ensure uniform flow, a recycle pump at outlet may inject part of the flow to upstream of hydrocyclone.

To increase centrifugal action, reverse demulsifier is injected at inlet of deoiling skid to enhance oil, water separation. The static concept used in first stage separation cannot be used here. Considering the very low oil volume, it requires very long retention time, which in turn requires very big vessels. Hydrocyclones provide better separation. Due to centrifugal action water is forced to the sidewall and oil is kept at the middle. Oil from hydrocyclones is directed to a closed drain system. From there it will be pumped to the export line. Water from hydrocyclones is directed to degassing skid.

To increase gas separation efficiency, degassing drum operating pressure is reduced to near atmospheric pressure. Gas outlet is directed to low pressure (LP) flare and water outlet is directed to open drain system via a last-stage filter.

Water is separated at several points. In each stage, volume of separated water and its oil content varies. Therefore oily water from different points of the platform may be directed to different tie-in points of the water treatment system. Treated water will be finally routed to an open drain system and the oil will be pumped to the production line via the closed drain system.

PFD in Section 2.8.4 shows the same concept. Water from each first-stage separator is directed to deoiling skid (dedicated hydrocyclone) while water from both condensate coalescers is directed to a single hydrocyclone in deoiling skid. The reason is the volume of water extracted in first-stage separator is much more than water from condensate coalescers. In addition, first-stage separator water contains higher oil ratio. Therefore it requires further treatment process.

The points at which contaminated water is expected to flow to either of oily water treatment skids may include:
- Production separator (FWKO drum)/test separator
- Condensate coalescer
- Drip pans of any package
- Pig launcher/pig receiver

- Slug catcher
- Sand-collecting vessels
- Drain from any vessel/equipment during maintenance
- Drain from any valve/piping section during maintenance

Water from maintenance and drip pans may also be routed to the closed drain system. There a second-stage separation is achieved. The oil is pumped back to the export line and the water is sent to an open drain system.

The oily water treatment package receives water mainly from test/production separator and gas water separation system. This water contains high levels of hydrocarbon material. Before this stage, the process of water separation is based on the philosophy of changing fluid momentum by providing baffles in fluid passage, adding demulsifiers to vessels. This causes hydrocarbon molecules to disintegrate and become lighter. Given sufficient settling time, the fluid water, which is heavier than oil, occupies the bottom level of the vessel and hydrocarbon flows above it.

In the oily water treatment, higher degrees of purity have to be achieved. Several years ago the acceptable level of oil in the water to be discharged to sea was 40 ppm. Now it has been reduced to 15 ppm, and in some cases 10 ppm is used. The scale is nonlinear. Therefore the change in equipment used to reduce oil content from 15 to 10 ppm may be much bigger than the change in package from 40 to 15 ppm.

For the first case (40 ppm limit), a series of hydrocyclones and degassing drum may be sufficient. For further reduction in oil ratio several scenarios can be followed. The first solution, which requires more space and may not be so efficient, is to use a recycling pump to transfer water from discharge of the last hydrocyclone to the inlet of the stream. Another solution may be to add a new hydrocyclone. The third and maybe more efficient method is to pass water through filters. A combination of all three solutions can also be used. Of course each vendor may have its own solution to achieve established oil discharge limits.

The water satisfying acceptable discharge criteria is then routed to an open drain system. Normally open drain consists of a caisson where the water is discharged to sea from the bottom of the caisson and the oil is pumped to a collecting vessel. Separated hydrocarbons may be routed to a closed drain system to be pumped to the export line.

Fig. 3.12 shows three hydrocyclones on a skid from quotation stage.

1. Here three hydrocyclones with different inlet nozzles are located on a skid.

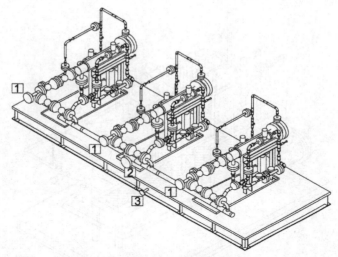

Figure 3.12 Hydrocyclone.

2. Their outlet is connected to the common header.
3. Only main items of the package are shown. Skid contains many instruments that have been removed from the sketch for clarity.

Fig. 3.13 shows degassing drum skid of the same package.

1. Degassing vessel is located at edge of the skid to facilitate access.
2. Before discharging water to the open drain caisson, it may pass a filter vessel.
3. Internal access platform may be installed for operator access to gauges and instruments.
4. Although in the sketch manholes are installed at the side of the vessel, in the final fabricated package it was installed at the head.

Separation can also be achieved by introducing a large number of small-sized gas bubbles in the oily water. Gas bubbles attach themselves to the particles. This way unit density of the combination further reduces compared to water. At the same time, its size increases. This reduction in density and increase in size increases its tendency to separate. Gas bubbles with contaminants rise to the surface and are collected by skimming the froth. Clean water is discharged from the bottom outlet. Droplet settling velocity (rising/falling) is determined by Stokes's law.

$$V = \frac{d^2 g (\rho_w - \rho_o)}{18 \mu_w}$$

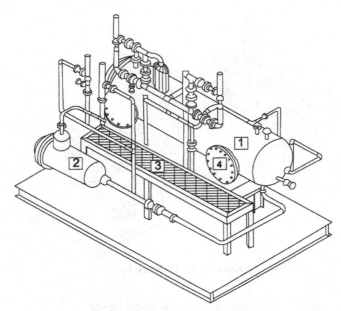

Figure 3.13 Degassing drum.

V = droplet settling velocity, rising/falling (m/s)
d = droplet diameter size (m)
g = gravitational acceleration (9.81 m/s^2)
$\rho_w - \rho_o$ = difference in density between water and oil (kg/m^3)
μ_w = dynamic viscosity of water (Pa s)

Dimensional check of this formula shows conformity. Other units can also be used. Other than gravitational acceleration, which is a universal constant, other factors can be manipulated to increase efficiency. For example, assume a droplet has 150 μm diameter, water density is 1000 kg/m^3, oil density is 850 kg/m^3, and water dynamic viscosity at 20°C is 0.001 Pa s, particle rising velocity is calculated to be 1.84 mm/s. If all parameters remain constant and particle diameter reduces to 50 μm, settling velocity decreases considerably to 0.2 mm/s. With 50 μm size, settling velocity will be nine times less.

With these velocities vessel sizes become so large that it is practically impossible to use this technique in offshore platforms. To facilitate calculations, assume a spherical bubble (150 μm) attaches to a spherical particle (50 μm) and the combined shape remains spherical. This increases bubble size very little to 151.8 μm. Oil particle specific gravity is 850 kg/m³, gas bubble weight is negligible. Combined sphere density becomes 30.36 kg/m³. If water temperature is increased to 50°C, water dynamic viscosity reduces to 0.000547 Pa s. With these values, now the particle rising velocity becomes 22.3 mm/s, which is 110 times more. This way package dimensions become manageable.

Gas bubble introduction can be done by two methods. The first one induces bubbles by gas flow through a tube and continuous movement of an impeller. This method is called "induced gas flotation (IGF)." The second method dissolves gas bubbles in the liquid at a separate chamber under pressure. Gas bubbles are released when pressure is reduced. This method is called "dissolved gas flotation (DGF)." Fig. 3.14 shows both schemes. In both methods uniform distribution of gas bubbles and their size play a major role in package efficiency.

3.1.9 Chemical Injection

Chemicals are injected to crude in several locations and for different purposes. It may be used to prevent hydrate formation, to increase or decrease coagulation rate, to prevent foaming or corrosion, etc. In this text, some chemicals are explained. For each specific case based on availability of chemical and its compatibility with crude composition, a suitable trademark shall be selected. Before selection, a thorough study shall be made to investigate the chemical's impact on hydrocarbon composition, required rate to perform the necessary action, its availability in the long run, etc. Operators shall normally have a large quantity of extra chemicals. A chemical injection package may consist of several chemicals such as:

- Hydrate inhibitor (MEG/low-dosage hydrate inhibitor (LDHI)/ Methanol)
- Demulsifier
- Reverse demulsifier
- Antifoam
- Corrosion inhibitor
- Scale inhibitor

Hydrate Inhibitor

In certain conditions crude may experience an abrupt reduction in pressure like during start-up of well. In others, long transportation distances and head loss reduces crude pressure. In addition, long-distance transport in

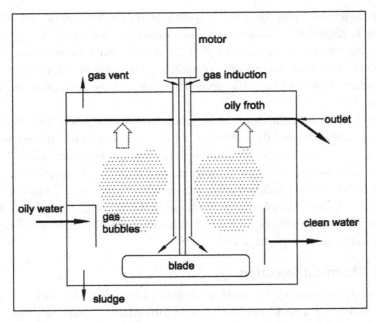

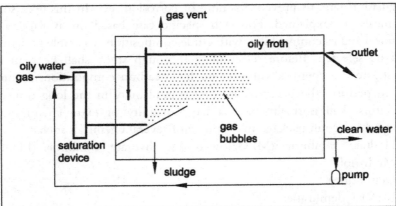

Figure 3.14 Induced gas flotation (IGF) and dissolved gas flotation (DGF) schemes.

underwater pipeline has heat dissipation with surrounding water. Both pressure reduction and head loss will cause considerable temperature reduction. Then water particles may separate from the crude. If done inside well or along blowdown line, since pressure/temperature difference is very high ice particles may form. Along underwater pipeline, since already free water has been separated and only saturated water in vapor or water mixed with crude exists, it may separate without ice formation. This phenomenon is called hydrate formation.

They contain a hydrocarbon nucleus. If untreated, these nuclei may connect to each other and block the passage. Methanol/MEG or any LDHI may be used to prevent hydrate formation. Methanol is normally injected at the Xmas Tree upstream choke valve in start-up condition to prevent freezing. MEG or LDHI is injected to the export line during normal operation. If an additive to control PH is included in MEG, it will have corrosion inhibitor effect as well. LDHI is much more expensive and needs corrosion inhibitor chemicals to be additionally injected.

Injection rate depends on several factors like crude composition, seawater temperature, and transport distance. In many cases transport distance is long enough to bring crude temperature equal to surrounding water. This causes a steady-state condition. For MEG injection rates, around 20 ppm weight of total fluid may be expected. For LDHI injection rate, from 0.5% to 2% saturated water weight may be injected. It has to be noted that rates are defined on different scales. In-field LDHI injection rate may be 200 times less than MEG.

In addition, before start-up since piping on the platform is not pressurized, if the fluid from the reservoir reaches the platform (due to high-pressure reduction) it will freeze. To prevent this phenomenon, methanol may be injected in the well before allowing opening of down hole safety valve (SSSV). Injection rates of 120 L/h for gases may be anticipated. However, start-up operation and pressure equalization for each well may take a short time.

Demulsifier and Reverse Demulsifier

Emulsion is the phenomenon in which a material is distributed more or less uniformly in other material. This is different from solution. Emulsion can be separated by physical filtering, while to separate solutions actions like heating, vaporizing, and/or chemical reaction is needed. Demulsification is the reverse phenomenon. Demulsifiers accelerate coagulation and sedimentation. This is used mostly in separators. Demulsifier is injected at inlet piping of production and or test separators.

As explained, in a three-phase vessel, water, condensate, and gas are separated. This is achieved by simply reducing fluid momentum by impinging on baffles. Due to pressure reduction, some gas is separated. Then fluid is kept in the vessel for maybe 5 min. This will allow water, which is heavier than condensate, to settle. To accelerate this phenomenon demulsifiers are used.

Reverse demulsifier is injected upstream of the water treatment package to improve water condensate separation before disposing water to sea. One of the injection locations is at the outlet for produced water from production or test separators. For demulsifier and reverse demulsifier, injection rates of 20–40 ppmv may be expected. For low rate of produced water even 5 ppmv has also been used.

Antifoam
In the test separator, liquid level shall be clearly distinguished from gas to enable accurate measurement. Foam bubbles that may be created due to agitation in the crude do not allow accurate measurement. In addition, in the production separator they may trigger false level alarms or even shut down if detected at HH level. Antifoam is injected at inlet piping of both test and production separator to prevent this phenomenon. Based on crude composition, pressure and temperature injection ratios of 10–50 ppmv may be expected.

Corrosion Inhibitor
Corrosion inhibitor is injected to prevent piping corrosion. This may be injected downstream choke or to the infield or the export pipeline. Platform piping is normally selected from higher-quality material, therefore in a majority of cases corrosion inhibitor is only injected at the point pipeline is leaving the platform. If corrosion inhibitor is injected downstream choke and if sufficient volume is injected and under continuous monitoring, platform piping may be selected from carbon steel material. Since any damage to platform piping may have disastrous consequences, normally this is avoided. Injection to the export line is always done because costs of an export line from exotic material like Inconel are very high.

With some additive, MEG will act both as corrosion and hydrate inhibitor. Efficiency of each material in the crude depends on many factors and injection rate shall be carefully selected. Injection values as low as 5 ppmv have been used.

The chemical injection package consists of tanks, pumps, and control panel. Tank volume depends on the injection rate and shall be sufficient to provide required volume within supply boat return time plus some margin. In areas where seasonal storms are foreseen, larger volumes shall be selected. Tank shall be redundant to enable cleaning, repair without platform shutdown. In addition, suitable level gauges or switches to start/stop pumps or trigger alarm for refill shall be placed. Tank material shall be compatible with contained chemical.

These storage tanks are normally atmospheric but designed for a minimum pressure considering fluid static head plus blanketing gas pressure. Tank top and bottom plates shall be able to support concentrated loads during repair or transportation. Therefore in external faces stiffeners (angle sections) are common. Placing stiffeners inside tank is normally avoided. It may cause some inaccessible points in which slug and contaminant may settle and cause corrosion.

Pumps' discharge rate is normally not so high but their pressure is high. When injecting to the well, the pressure shall be equal to well flowing pressure and when injecting to the export line, it shall be equal to platform design pressure. In many cases displacement-type pumps have been used.

In addition to these two main items of the package there are several instruments. They may include pressure gauges and transmitters in the pump suction and discharge heads, level gauges and transmitters in the tanks, pressure reduction valves in the bypass lines to return excess chemical to the tanks, etc.

Package will have a local control panel to operate pumps in addition to transferring some alarms or signals like running, stop, general fault, etc. to central control system.

In some cases all chemicals may be combined in one skid. See Fig. 3.15.
1. First tank, which has larger volume, contains corrosion inhibitor.
2. Second tank is antifoam, which is injected in FWKO drums and test separator.

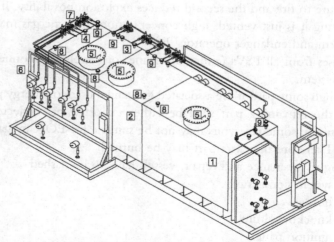

Figure 3.15 Chemical injection systems on one skid.

3. Third tank is for demulsifier, which is injected continuously in FWKO drums.
4. Fourth tank, which is smaller than all, is reverse demulsifier tank. It is only injected in the hydrocyclones.
5. Tanks' manholes are located at top. Only the smallest tank did not have enough width. Therefore manhole is placed at side.
6. Pressure indicators after injection pumps (operating pressure 125 bar) are all located in a line to facilitate site access.
7. PSVs of all tanks are also piped near each other.
8. Each tank has a vent, which is located near related manhole.
9. Filling nozzles are located at edge of each tank. This enables filling with hand pumps from side of the package.

3.1.10 Flare System

Gas separated from oil (associated gas) may be used in power generation or transferred via export line to onshore. It may also be injected to the reservoir to boost oil production. Nowadays both environmental protection regulations and economic considerations do not allow gas flaring. This means engineering design shall consider "no flaring" concept. However, this concept does not mean that flare does not need to be designed. Some flaring sources are listed here:

- In emergency conditions platform shall be depressurized, which means gas has to be either burned or vented. This is to minimize volume of containment and reduce its pressure. The first one reduces platform exposure to fire and the second reduces explosion possibility. If instead of flaring it is just vented, high concentrations of toxic gas may reach platform and endanger operators' life.
- Releases from all PSVs (either from lines or vessels) are connected to flare system.
- Although some part of the associated gas can be used for energy production, the remaining part shall be either exported or reinjected. Soil structure in some reservoirs may not be suitable for EOR by gas injection. Therefore the extra part may be burned.
- During maintenance and repair, vessels have to be flashed.
  Flare system consists of:
- Flare header
- Flare knockout drum
- Flare ignition panel

- Propane bottle rack
- Flare line and flare tip

Flare header is normally much larger than production header. Since releases may be from any point, high and low pressure lines may be separated. Flare knockout drum is a large vessel. The intention is to settle large oil droplets. They are later routed to closed drain drum. If they are directed to the flare tip large burning balls will drop from flare tip to the sea. Flare knockout drum is designed for very low temperatures.

Fig. 3.16 shows a flare knockout drum with fire coating. The side that is exposed to hydrocarbon or jet fire is coated with a suitable paint system. Other side has normal corrosion painting. Coaming is installed around vessel to collect spilled liquid.

Propane bottle rack and flare ignition panel are used to ensure that a pilot flame always will be present on the flare tip. Normally a two–inch line transfers propane to the tip. A certain volume of platform gas can be continuously directed to the flare tip. This is called normal flaring. Its volume is much less than emergency flaring, which directs platform design flow to the tip. It has to be ensured that there is always a flow of gas in the flare line to prevent air ingress to the line. Air presence in flare KO drum may blast the vessel. Flare tip is shown in Fig. 3.17.

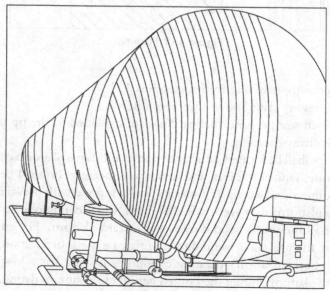

Figure 3.16 Flare knockout drum.

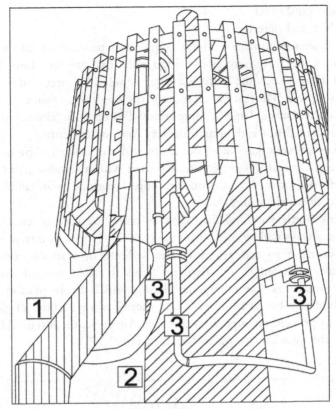

Figure 3.17 Flare tip.

1. Low-pressure flare line in this sketch is 12″.
2. High-pressure flare line in this sketch is 30″.
3. Two-inch size propane/fuel gas lines are extended to the tip to ensure pilot is always burning.

Flare tips shall be located sufficiently away and above platform elevations. Temperature, radiation levels, and toxic gas concentration shall be investigated by dispersion analysis. If dispersion study is done with prevailing wind direction, platform may seem safe. While in actual conditions flaring may happen at a time when wind direction is toward platform. For emergency conditions, the prevailing wind direction may be used. In addition, consultant may check platform operating conditions if wind has a low velocity toward the platform during the limited period of maximum flaring. For this duration, operators need not stay in the open area. However, equipment and instrument operating conditions shall not be exceeded. For normal flaring, wind direction shall be directed to the platform. A probable velocity shall be selected for this study.

If dispersion study requires flaring point to be long distances away and much higher above platform levels, another jacket may need to be installed. In some cases required distance is too much, which necessitates an intermediate platform to support connecting bridge and pipeline. Distances up to 180 m have been observed. Very high elevation may require a flare tower. Elevations more than 60 m above platform top level are common in large gas production projects. If distance is very small, a flare boom can be used. Its structure is very similar to vent boom shown in Fig. 3.26. For associated gas in satellite platforms maybe a flare boom is sufficient. But for gas platforms and associated gas in large oil production platforms a detached flare jacket/tower may be necessary.

To determine complex layout at start of the project a dispersion study shall be performed. Wind rose plays an essential role in layout selection. Total sum of the duration of any one of the hazardous conditions that may occur on the platform shall be minimized as much as possible. One of the main concerns is to ensure toxic, explosive gas is burned away from platform. Emergency conditions can happen any time. Designer shall select burning point (flare location) in a manner to have minimum wind blowing toward platform. Other considerations like existing platforms, pipelines, subsea cables, supply boat approach, etc. may also affect selected layout. In many cases flare is simply located in the direction of prevailing wind. For further explanation, see chapter "Disciplines Involved in Offshore Platforms Design," Section 2.7.1 "Complex/Platform Layout."

To preserve the environment, flares shall be smokeless. Smokeless definition is based on the Ringelmann chart and dates back to 1888 in Europe (Paris). Professor Ringelmann introduced this chart varying from 0 (all white) to 5 (all black) to give a measure of the smoke intensity.

3.1.11 Closed-drain System

Hazardous drains shall not be routed to sea. They shall be collected in a vessel and after separation its oil may be rerouted back to production line. The water may be directed to an open drain caisson. Gas can be routed to LP flare for burning. Normally it does not have a considerable volume to be used for other purposes. The duty of a closed-drain system is to collect all hazardous drains. It is composed of:

- A complicated network of drains from all headers, vessels, and major equipment

- Closed-drain drum
- Booster pumps
- Closed-drain skid main pumps

The network is completely separated from an open-drain system. If high concentration of H_2S is encountered, drain from each vessel is collected in a closed system via double block and bleed valves. These are normally closed valves. Only during maintenance does the operator open them to allow water/oil flow to the drum. In some cases, to avoid too many connections pipes from each elevation are connected to a common drain header and then connected to the drum.

Closed-drain drum is an atmospheric vessel located at the lowest possible elevation. Sometimes it is hung from below cellar deck elevation. This will enable gravity flow to it from all parts of the platform. The water from bottom of drum is drained to open-drain system. The vessel has two boots. Water being heavier than oil accumulates in the boots. Fig. 3.18 shows a closed drain drum hanging from below cellar deck.

1. Personal protection insulation is covered around it. Although this vessel is out of operators' normal route, since its operating temperature may become high, production platform can be installed.
2. Boots extend from bottom of vessel and collect water to be discharged to open-drain system.
3. In this figure closed-drain drum hangs below the cellar deck.

The oil inside drum is pumped up via booster pumps. Since closed-drain drum is operating at atmospheric pressure and ambient temperature, maximum separation has already taken place. Therefore even small head losses

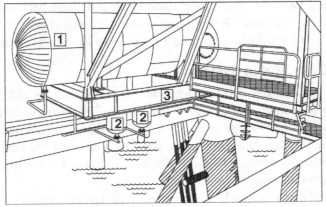

Figure 3.18 Closed-drain drum hung below cellar deck.

may lead to oil vaporizing at low pressures. To overcome this condition booster pumps shall have a very low required net positive suction head (NPSH). Their duty is only to lift oil from drum to the main pumps' inlet. Normally booster pumps are submerged inside boots. However, sometimes due to very low required NPSH they are placed at a deck below or at the lowest elevation of the drum to have positive pressure. Their flow rate is small.

Closed-drain main pumps usually are placed above closed-drain drum. They do not have NPSH problem because booster pumps have already provided sufficient head. They transfer oil to the export line. Therefore they shall have very high discharge pressure with low volume.

3.1.12 Fuel Gas Treatment

On a platform fuel gas may come from associated gas or directly from gas header. To be able to use it in a system, a minimum continuous flow shall be available. Adapting no flaring philosophy the gas flow rate used in normal flaring conditions is a reliable source. All or part of the discharged gas from separators and other vessels may be used for this purpose.

Gas from fuel gas package may be used for different purposes:

- Burning fuel gas for energy
- Purging flare vessel/line
- Blanketing vessels

In platforms where turbogenerators/turbocompressors are used, its main function will be to provide suitable gas for burning. In other cases it may be used for purging or blanketing. Purging is the term used for providing positive pressure in vessels like flare knockout drum and flare line to prevent air intrusion (which may cause explosion). A minimum flow is always maintained, which may also be used to keep pilot flare burning. Purging is normally used for flare line and has a continuous minimum flow. At the final stage, purging with fuel gas may be replaced by inert gas for operator entrance. Blanketing also provides a positive pressure in vessels like closed-drain drums which discharge to atmosphere via LP vent. This is again used to prevent air intrusion. Blanketing is normally used in atmospheric vessels without any gas flow.

The extent of treatment is defined based on fuel gas usage and downstream equipment requirements. Normally water removal is sufficient. For burning in turbogenerators, water content shall be low enough to provide sufficient energy. Water presence will not allow a clean burning. Nowadays H_2S removal is not a concern. Turbogenerator fuel systems can accept some levels of H_2S. If it is used for purging/blanketing, dispersion studies shall determine safe release distance so that H_2S content falls below permissible levels at the platform vicinity.

A fuel gas package may consist of scrubber, filter, and heater. Heater may be used to superheat discharge gas to ensure its temperature is well above dew point temperature. Temperature difference depends on the margin that the gas is expected to be cooled down during transfer to the consumption point. In most cases 10°C may be sufficient. This is useful for turbogenerators. In tropical areas and if gas is taken from outlet of degassing drum, high temperature difference may not be required. Therefore heater

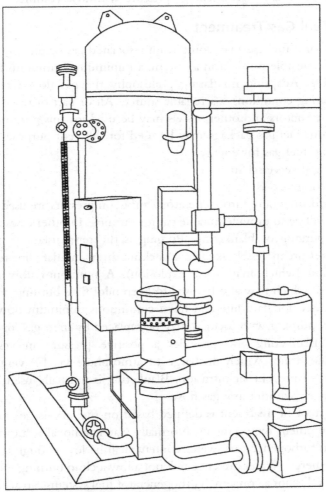

Figure 3.19 Fuel gas package.

may have very low power. In some cases a heat tracing facility may also be used. This is out-of-package scope of work and shall be provided by platform fabricator. Heat tracing is in fact heating the gas in the flow line to consuming point by an electrical wire wrapped around it.

Fig. 3.19 shows a fuel gas treatment package. This is a simple package and consists of:

- Scrubber vessel with level transmitter and drain
- Filter to absorb remaining vapor in gas
- Heater to increase gas temperature above dew point
- Control panel

As always, only filter and heater may have spare unit. Spare for vessels is not provided.

3.1.13 Compression System

Both gas lifting and gas injection are an EOR system. This means they are used to increase production from a low production or dead reservoir. Although equipment placed on the platform for these two systems have a lot in common, they are used for completely different conditions.

A compressed air system is used to produce sufficient volume of air with suitable pressure to enable pneumatic instruments to operate. Its operating pressure is relatively low (about 4–8 bars). A gas compression system is totally different. It may require very high pressures and may be used for:

- Export: in this case reservoir pressure may only be sufficient to bring gas to the platform level. This may include associated gas in oil reservoirs. Only one-stage compression may be used to increase pressure for the export. The intention is to overcome head loss during offshore transport to onshore tie-in point. In this case final pressure is much less than pressure needed for gas injection.
- Gas lift: After treatment gas pressure is increased to around reservoir flowing pressure. This EOR method is used in fields in which crude viscosity and specific gravity are very high and reservoir pressure has dropped to a value that cannot bring it to the surface. In other words crude oil is so viscous or reservoir pressure is so low that under normal conditions crude is not transported toward the surface. This means static head plus friction due to viscosity overcomes reservoir pressure. To continue/increase production, gas is injected below oil level. Gas injected in the reservoir will bubble inside crude oil and reduce its viscosity and specific gravity. With this new condition, reservoir pressure becomes sufficient to bring crude to surface.

- Gas injection: After treatment, gas pressure is increased to around reservoir shut-in pressure and back injected to reservoir. This will increase pressure on the crude to enable bringing it to surface. Gas injection is used to increase reservoir pressure and move oil from points far away to drill pipe. Therefore its release points shall be carefully selected. Otherwise it may push back oil instead of bringing it to surface. This study is under reservoir engineering specialty.

These two methods have something in common, which is gas treatment before pressurizing. Their difference is in volume, pressure of compressed gas, and injection well. In all cases drying the gas (dehydration) is one of the major requirements.

For oil in shallow reservoirs (for example, 500 m deep) static pressure of vertical column may be around 40 bar. This is simply calculated by multiplying weight of 500 m oil column by its specific gravity (in this case assumed to be 0.75). Providing this much pressure either in gas injection or gas lift scheme may not be sufficient to bring oil to the surface. Gas injection pressure must be enough to overcome head loss when injected in different locations of the reservoir in addition to head loss when existing crude moves from different locations in the reservoir toward drill pipe plus the static column pressure up to topside. For this condition a total pressure of 100 bar may be the minimum required value. For very deep (3000 m) reservoirs, this value may increase considerably and needs special compressors.

The point at which high-pressure gas is injected shall be carefully selected. It shall be at the edge of the reservoir boundary to ensure oil is pushed inside toward well. Otherwise oil will be pushed away to soil layers and can never be recovered. Selecting gas injection pressure and location is in reservoir engineering SOW.

For very high discharge pressures two- or even three-stage compressors may be used. In this case first the pressure of a certain gas volume is increased from P0 to P1. At this stage gas volume decreases while its temperature increases. Discharge gas from first stage is used as intake gas of second stage. Normal compressors can increase suction pressure three or four times. Table 3.3 shows sample capacity and suction/discharge pressures.

Table 3.3 Sample compressor stage capacities

Stage	Capacity MMSCFD	Suction pressure bara	Discharge pressure bara
First	79.8	14.3	50.5
Second	25.0	48.8	148.0

As per Charles and Gay-Lussac's law for a specific amount of *ideal gas*, numerical value of pressure multiplied by volume and divided by temperature is always constant. Increasing pressure with constant temperature reduces volume. Therefore discharge of two or three first-stage compressors can be directed to suction of a single second-stage compressor. This means that second- or third-stage compressors may have much smaller volumetric capacity with much higher pressure rating.

At each stage before compressor suction inlet, water vapor inside gas shall be removed to ensure moisture droplets do not harm compressor. This is done by passing gas through scrubber vessel and/or reducing gas temperature. Heat exchangers with any media can be used for temperature reduction. In some cases electrically driven air fans can be used. Electrical fans may have good efficiency if environmental conditions are cold. For tropical areas very large fan sizes with considerable electrical power may be needed.

Either moisture shall be removed to reduce vapor content below condensing ratio or temperature shall be increased to above dew point value. As explained before, dew point is defined as the temperature for any pressure at which gas cannot hold any more water vapor. Below this temperature vapor is condensed as water droplets. If gas is entered with this temperature to compressor suction until discharge point, droplets will not condense. Because with increasing pressure gas temperature passing from suction to discharge is increased, therefore water drop condensation cannot happen.

Generators' consumed power depends on pressure rating and gas volume. Normally part of suction gas may be consumed for power generation. For an estimate of the required power and gas consumption, refer to chapter "Disciplines Involved in Offshore Platforms Design."

Compressor package contains different items like filters, coolers, scrubbers, control panel, exhaust, power generation, etc. For a complete description, please refer to technical books.

3.1.14 Water Injection System

When pressure/yield of the reservoir decreases, one of the EOR methods to bring up crude oil to the surface is injecting water at the wells. This is achieved by taking seawater, treating it, and injecting it with high pressure to the reservoir. Water coming up with crude oil is also separated. It is then treated and used for injection. At the start of this process, seawater may compose a major part of injected water. Part of this injected water will again come up with oil to the surface. It will be separated from oil, treated, and reinjected to the well. This means that gradually the oily water portion share in injected water to the well increases. Therefore oily water treatment facilities shall be designed for the higher rate at end stages of reservoir life. At the same time seawater intake facilities shall also be designed for the high rate of seawater intake at start of injection. Therefore both packages shall have suitable flexibility to handle different volumes of water.

A reservoir engineering study shall identify required injection volumes at start/end of production in addition to crude water cut.

Original seawater treatment facilities' capacity is several times total production volume. Water cut is a portion of total production. Therefore seawater treatment capacity may undergo smaller decreases, but oily water treatment capacity will experience higher increases. Normally packages do not work with good efficiency if their design rate is much different from the operating rate. Efficiency of packages with large operating range is low. Therefore a solution may be to leave free space for additional packages in oily water treatment package for future extension.

Water injection volume depends on many factors like reservoir depth, crude viscosity, reservoir pressure, etc. However, at least twice to triple volume of oil production rate shall be considered.

Location, volume, and pressure of the injection water shall be carefully studied and selected by reservoir engineers. Here only some description of the equipment required for treatment and injection is given. Oily water treatment is described separately in Section 3.1.8.

The water injection system consists of:

- Seawater lift pumps will bring in required water. If seawater flow stops, platform production will shut down. Therefore based on platform shutdown consequences, suitable contingencies in providing seawater caissons and pump shall be installed. These pumps may require large volume with small heads. Their duty is to bring seawater from underwater to platform surface for treatment. Using one huge pump is not recommended. Its failure and platform shutdown may lead to loss of production. Instead, three caissons each with 50% capacity (total 3 × 50%) for final stage may be a better design. With gradual reduction in portion of injection water from seawater lift pumps, one of them may be removed from operating cycle.
- Particle removal filters prevent entrance of solids larger than a certain size into downstream piping. Caisson entrance is located well below

seawater surface and covered by steel bars to prevent fish from entering. However, it does not prevent entrance of particles less than a few mm size. Filtering may be done in several stages ranging from coarse to fine grades. Minimum size depends on many factors like well conditions. The major factor is to prevent clogging. In a project, 2-μm-size filters were selected.

- Pump strainer will stop small–size material entrance. Normally strainers are equipped with a differential pressure transmitter. This gauge measures pressure difference between two sides of a strainer. If it exceeds the preset value, an alarm will be triggered. This alarm will inform operator to shut down the pump and start backwash flow to clean strainer. Backwash lines should have been installed by vendor.
- Chemical injection system for treating injection water may be combined with process chemical injection skid. It has to be noted that some of the required chemicals are totally different from process system chemicals.
- Deaeration tower: The action will be to remove dissolved oxygen from seawater. Maximum allowable oxygen level shall be low enough to prevent any biochemical reaction or living organism reproduction.
- Chemical removal filters mainly remove sulfates from seawater. This is essential to prevent scaling in the well and reservoir.
- Injection pump skid: These are normally large-capacity high-pressure pumps. Pump installation, power demand, contingency, injection header, and lines shall be carefully arranged on each platform. Pump pressure may vary from well shut-in to above well-flowing pressure. At start of injection, high pressure is needed. Long distance and substantial water quantity has a large head loss. Injected water shall have sufficient pressure to push HC fluid toward well.

Seawater contains oxygen, debris, chemicals, and organic matter. All these substances are harmful to the reservoir environment below ground. Oxygen will flourish the organisms in the reservoir, debris will clog well, chemicals may induce scaling, and organic matter will multiply themselves and lead to further clogging. It seems salt is no problem other than piping corrosion. This is very good news. Otherwise installing desalters on offshore platform would require considerable space and electrical power demand. In addition, extracted salt disposal would become a severe problem. Imagine 25,000 ppm wt salt in 40,000 barrel per day makes approximately 160 ton salt on a daily basis. Water coming from reservoir may also have high salt content even several times seawater content. Dosage rate for chemical injections shall be calculated for each case.

- Biocide: Is injected to kill living organisms. If injected in vessels with a stilling time, may be batch injected. If injected at inlet or discharge of pumps with a continuous flow, shall be continuously injected.
- Oxygen scavenger: Free oxygen shall be removed as much as possible. Any reaction of a chemical element with oxygen may produce heat or start corrosion. To remove oxygen, chemicals are injected into the base of the deaerator tower. Maximum oxygen content allowed for injection to reservoir is 10 ppb.
- Filtering aid: Although last-stage filters can remove items with very small size, there may be smaller items. Chemicals to improve coagulation injected into the inlet of the fine-filter package will improve its efficiency.
- Antifoam: Foaming not only reduces efficiency but also by triggering false alarms or shutdowns may hinder production. Antifoam chemicals are injected into the inlet of deaerator tower.
- Scale inhibitor may be injected into the water outlet of the deaerator tower.
- Corrosion inhibitor: Piping material for water treatment facilities may be of exotic material that is corrosion resistant. Final injection point may be at wellhead of another platform. It will become too costly to extend exotic material downstream to treatment facilities. Corrosion inhibitor may be injected into the water injection pump suction.

For each chemical, storage volume for 2 weeks of maximum consumption rate is normally provided. However, this is not a universal practice. It depends on platform location, distance from land-based supply stations, weather conditions, available supply boat facilities, etc.

Seawater contains other dissolved materials. One of the most important ones is sulfates. It has two effects. First it increases corrosion in the platform piping. The second impact happens inside the reservoir. Two phenomena decompose sulfate and produce hydrogen sulfide (H_2S). At lower temperatures of 60—80°C, sulfate reducing bacteria decompose it. This process is called biological sulfate reduction (BSR). At higher temperatures (160—180°C), thermochemical sulfate reduction (TSR) decomposes sulfate. Both phenomena produce hydrogen sulfide. Compared to reservoir utilization life, both these phenomena are very slow.

Formation water contains some sulfate. In a project, 4100 ppmw was measured. A typical value of 2600 ppmw sulfate for seawater has been used.

Although this is much less than sulfate rate in formation water, if injected water volume is increased several times, total combined sulfate rate in the production volume will increase considerably.

Using nanofilters for separation of sulfate ions enables installing sulfate reduction modules on the platforms or floating storage units. In this way sulfate concentration can be reduced to 50 ppmw. Seawater passes nanofilters at the last stage after removing other particles. To reduce filter dimensions and increase efficiency, seawater may be pressurized. Each vendor has its own proprietary design for this package.

3.1.15 Pig Launching and Receiving Systems

The final product of each platform shall be directed to a pipeline. As said before, it is possible that the water or slug in the crude either totally or partially blocks the export line and consequently stops or reduces the crude oil or gas export. In this case pipeline has to be cleared from slug and water. To perform these activities some solid or intelligent pigs are launched. Launched pigs pass with the pressure of the crude or gas on their back and clear all the water or slug that has blocked the pipeline. Sometimes in addition to clearing the passage it is intended that the pigs obtain information about the pipeline situation. Special pigs having sensors with sufficient data storage capacity are launched through the pipeline. Their probes will save necessary information from pipeline length in memory that will be later recovered. They can store data about pipeline corrosion, clearance, etc. Intelligent pigs are bigger and require handling equipment. Some dimensional data are given in Fig. 3.21.

Ordinary pigs are normally sphere type. Cylindrical pigs are used for cleaning. A cleaning pig is shown in Fig. 3.20.
1. Painted steel cylinder connects two side portions.
2. End portions are made of rubber to have flexibility in passing bends.

Fig. 3.21 shows an intelligent pig. It consists of two segments that are connected with a hinged joint. This enables passing bends. At front cleaning pads and at end sensors are installed. Pig launcher consists of a sloped barrel with end closing door and controlling valves to direct the flow behind pig and bypass line. In some projects an automatic pig launcher has been used. This may keep several normal spherical pigs inside the barrel which may be launched by operator or from central control room (CCR). Automatic pig launching is not used with intelligent pigs. They are too heavy/big to be stored inside pig launcher barrel. In addition they are too costly to store for a long time, and launch may be once per

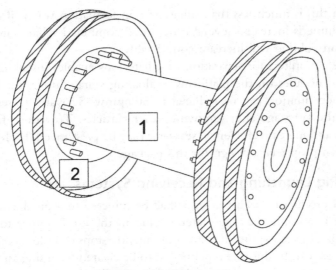

Figure 3.20 Ordinary pig.

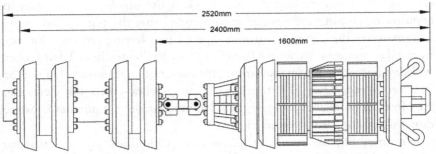

Figure 3.21 Two-segment 24″ intelligent pig, T = 2520 mm (ROSEN).

year. Pigging for cleaning purposes is more frequently done. Operators cannot remove heavy intelligent pig to place/launch a normal pig and then again replace it.

Pig launcher door has a special mechanism. This mechanism will not allow pressurizing launcher before door is closed and will not allow opening before inside pressure is completely vented. At the same time its opening and closure shall be rapid.

Different vendors may have different patents for this system. One of the commonly used systems is composed of a three-section lock. When the door is locked, these three segments are pushed to the barrel wall and door internal side. This pushing provides sealing against internal gas pressure. Therefore as long as the door is not locked (segments are not pushed to barrel wall/door) the pig launcher is not sealed. It is clear that when the system is not sealed any gas injection inside pig launcher to pressurize internal cavity and move pig will fail. Therefore the first goal is achieved. Before locking the door it is impossible to pressurize the pig launcher.

To open the door, first it has to be unlocked. For unlocking, the three segments shall move away from barrel wall toward inside. The point is that after the pig is launched, the inside of the barrel is still pressurized. This pressure exerts a force on the thickness of the three segments. This force prevents any movement. The first step is to vent it and release pressure in the back of the door. If this is not done, the door cannot be opened. This serves the second requirement. Opening the pig launcher door before its venting is not possible.

Another simple but very effective method is to use a set of keys. Each key is placed in a compartment that will only be opened when the previous stage key is inserted in a lock. These mechanical interlocks have been successfully installed in several platforms.

Since toxic and corrosive material is in contact with the internal surface of the pig launcher barrel, it is normally fabricated from cladded pipes. Fabrication from solid Inconel pipe will be very costly. As explained previously, the pig launcher enclosure is designed to enable pressurizing only after it has been fully closed and opening the door to remove pig is only possible after it has been completely depressurized.

In Fig. 3.22 a pig launcher with motorized valves and control panel is shown on the skid. The following items can be identified:

1. **Pig launcher door:** As explained before, it prevents pressurizing before complete closure and opening before depressurizing.
2. **Pig launcher barrel is sloped** and connects with a reducer to export line. This ensures liquid is not trapped inside barrel.
3. **First pressurizing valve and nozzle** are used for intelligent pigs. Since it is larger, gas shall be fed at the start of the barrel.
4. **Second pressurizing valve and nozzle** are used for normal pigging operations. Since in this case only the last sphere shall be launched, gas is injected between the two stoppers.
5. **Pressurizing is done by a branch** from the main gas line. Since normal pigging is automatic, main line and normal pig pressurizing valves are both motorized. Intelligent pigging is manual, therefore its valve is also manual. It is understood that this valve is large bore and works at high

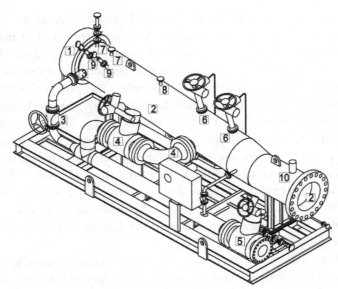

Figure 3.22 Pig launcher.

pressure. Therefore a gearbox mechanism or an electrical motor will open it. However, its open/close command is not coming from the integrated control and safety system (ICSS).

6. Two stoppers allow only one pig in ready position for launching. The second stopper prevents other pigs from sliding toward launching position.

7. One PSV and one vent nozzle are installed. Venting is done before any maintenance.

8. After venting, nitrogen is injected to ensure all toxic gas is released.

9. Pressure transmitter and gauge shows internal pressure condition to continue operation.

10. At entrance to export line, a signaler checks pig passage.

Automatic pig launcher requires a pig handling device. If multiple pigs are inserted, a flapper keeps other pigs from moving forward. This pig holder may work by pneumatic power. A simple calculation for the forces and required air volume is given in Section 3.3.6 "Motorized/Actuated Valves."

3.1.16 Electrical Submerged Pumps

In the beginning of this chapter and in Sections 3.1.13 and 3.1.14, some EOR methods were explained. The electrical submerged pump (ESP) is also an artificial lifting method. It is the first equipment in contact with hydrocarbon and as per the approach adopted in this chapter should have been explained before Xmas Trees. However, since the author had no previous experience with it and this pump is only used for bringing hydrocarbon to the surface, it is explained very briefly here.

An electrical pump is placed inside the drill pipe penetrating reservoir at an elevation above well pipe perforations. Pump, motor, seal, and inlet are placed inside the well as the tubing is run. ESPs can be placed in vertical, horizontal, or deviated wells. In spite of this ability, they shall be installed in a straight section of drill pipe for better efficiency.

They are manufactured very thin in the range of 4–6 inches OD. They consist of an electrically powered multistage pump. Its electrical motor is working with very high voltage rating (3–5 kV). Therefore a transformer is installed on the platform. Based on cable size, 1 km of cables may have large voltage drop.

These pumps are better suited to wells with low GOR. Sand will rapidly erode pump internals or block crude passage. It is said these pumps can handle crudes from a very low production of 70 BPD to as high as 64,000 BPD. They are widely used in North America and Russia. They are not used very much in the Middle East.

The author has no experience with these pumps. It is noted that ESPs are widely used as an efficient and cost-effective artificial lift method. In this edition of this book only their name and application is mentioned.

3.1.17 Export Meter

An export metering package may be used for oil, condensate, and gas measurement. It may fulfill two purposes.

- Internal usage: For example, when one company is responsible for both offshore platform and onshore plant. In this case production rate is already known via test separator and composition can be defined in the laboratory from samples taken offshore. Only the export quantity shall be monitored.

- External: Not only the quantity but also the quality of the crude is important. In this case it may need a package for energy content measurement. This will be the basis for unit cost and total payment evaluation.

If a metering system is used for export gas cost evaluation, a third party may be assigned as responsible for metering package sealing, maintenance, calibration, and operation.

A metering system in its simplest form is a flow measurement unit. Since the densities of fluid and gas are very different, different techniques may be used for density measurement. Different measuring tools using ultrasonic waves, magnetic field, turbine flow meters, and orifice meters are used.

Gas and oil density and viscosity differ greatly. Therefore thickness and shape of the boundary layer in contact with piping sidewall are very different. In addition, fluid velocity profile (inside a pipe) from side to center line varies greatly. Fluid turbulence after any bend, discharge flange, etc. affects velocity profile in the piping cross-section. A straight pipe run is needed to overcome turbulences. This length may be on the order of several times pipe diameter.

Some items may affect metering accuracy:

- Density: A metering system designed for a certain specific gravity like crude oil loses its accuracy if fluid density undergoes considerable change.
- Volume flow rate: Best accuracy is obtained at a specific flow rate. Increasing or decreasing flow rate (for example, during turndown conditions) to certain levels may impact metering accuracy.
- Velocity: Fluid velocity in piping section is one of the main measured parameters. If it exceeds certain values, some error may occur in its measurement.
- Impurities: Presence of solid material like sand, etc. even at small sizes may impact accuracy.

A majority of manufacturers claim to be able to cover a very wide range of these parameters. For each case technical catalogues and vendor track records must be carefully examined.

3.2 UTILITY SYSTEMS

Utility systems are those that do not change crude composition or do not increase its production rate but are necessary for platform operation. As explained before, other than safety packages, a universally accepted definition for categorizing systems and packages on a platform is not defined. In this book utility packages are classified under three categories.

The first category, utility systems, provides the driving force for equipment operation:

- Wellhead control panel and hydraulic power unit (WHCP/HPU) provide hydraulic force to operate main valves' actuators.

- Compressed air system provides pneumatic power to operate instruments.
- Diesel oil system provides fuel for power generation.
- Power generation system provides electrical power for all purposes.
- Electrical power storage facilities provide uninterrupted power supply.

The second category of systems is necessary for safe operation of the platform. It has to be noted that these are not safety equipment:

- Blanket/inert gas prevents possible air intrusion to spaces full of flammable gas.
- Atmospheric vent allows excess gas to be dispersed in a safe area.
- Open drain provides a stilling casing to remove the last portion of oil to be separated from water discharged to sea.
- Hypochlorite system prevents algae and living organisms to grow in the piping system.

The third category of utility systems is necessary for operators' work:

- Service seawater is used for housekeeping activities.
- Pedestal crane is used for transferring food, chemicals, spare parts, etc. from supply boat.
- Material handling system moves equipment, materials, and spares in the platform.

Packages in each system are described in brief.

3.2.1 Wellhead Control Panel and Hydraulic Power Pack Unit

The wellhead control panel and hydraulic power unit may also be categorized under "safety." Actuators of the Xmas Tree valves and emergency shutdown valves are all controlled by this package. It has to provide very high pressures at the discharge of the actuators to enable closing Xmas Tree valves within a very short time. In fact their actuators are kept open by fluid pressure. When the pressure is removed, they will close. This mechanism enables rapid closure and fail-close action.

Normally these valves are all fail-close type. Hydraulic pressure opens flow against a spring or mechanical device, which tends to close flow. Pneumatic drivers are not used because with very high pressures that have to be exerted, actuator sizes will become too large. Normally well bay area is a compact and dense area. Therefore using small-size equipment that can handle high pressures is very important and necessary.

For each well a board is dedicated that shows required information including pressure, temperature, flow, and position of the actuator. To reduce total package size, two or three actuators may share a common accumulator. Different Xmas Tree manufacturers may use configurations that are specific to their brand, but actuator driving force may be more or less similar. WHCP provides hydraulic pressure to the inlet of the actuator, and its internal mechanism provides required pressure at outlet of the actuator. Since an incompressible fluid pressure is uniform over area, using larger area at outlet generates much higher force than inlet. This force shall be sufficient for safe opening/closing of the valves.

Hydraulic oil passes filters/strainers before being charged to actuator inlet. To prevent inadvertent backflow, check valves are placed to prevent oil return without pressure release. For maintenance purposes isolation valves are located in each line of the hydraulic oil network.

WHCP/HPU consist of:
- Hydraulic oil storage
- Pumps and electrical motors
- Panels

Fig. 3.23 shows sketch of a WHCP package. Pumps are located inside the cabin. Gauges are installed at the wall to facilitate viewing without necessity to open the door. Hydraulic oil cylinders are connected by heavy-duty tubing. Each of the panels shown on the side of the package controls one Xmas Tree.

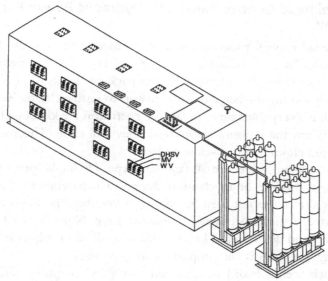

Figure 3.23 Wellhead control panel (WHCP).

Closing and opening of Xmas Tree valves shall be in sequence. Opening shall start from the lowest valve and proceed to top. At start-up or after any shutdown all valves (DHSV, SSV, and WV) are in close position. Opening sequence shall be DHSV, SSV, and WV. WHCP unit control panel (UCP) shall have an inherited logic to prevent opening of any valve if valves above it are not closed.

Well start-up shall be one by one. This is not only for the hydraulic power requirement but also is a safety criterion. The point is that when a problem occurs all flow shall be immediately directed to the flare. Therefore sufficient capacity shall be available. Closing shall be in reverse order and shall take much less time.

In gas platforms flow rate and normally operating pressure is larger than oil platforms. In spite of that, specific gravity is much less. In both platforms WHCP may operate at two or more pressure levels. DHSV operates at highest pressure level. This valve is designed for well shut-in pressure and is located well below seabed (80 m or more). The first pressure level is well shut-in pressure used for DHSV. In some designs SSV and WV may also be opened/closed with the same circuit or a second pressure level. SSV and WV are also designed for well shut-in pressure. But they are above sea level. Therefore their pressure head is at least reduced by the weight of the crude column compared to DHSV. In some designs a separate high-pressure hydraulic line may be used for SSV and WV.

The second (or third) pressure level is well flowing pressure used for choke, ESD, and other control valves. Choke is the first valve downstream MV designed for platform design pressure and is protected by a PSV. Choke actual pressure is higher than other shutdown valves in the platform. This pressure difference is equal to the head loss in platform piping.

Each pressure level may require a separate oil network.

3.2.2 Compressed Air System

Compressed air system provides suitable air for operation of instrument valves and air-driven equipment in addition to air required for platform housekeeping and maintenance activities. This system may include compressor, filter, heater, wet and dry air receiver vessels with required instrumentation, and control panel. Each item (other than vessels) may have a spare unit.

Air used by instruments/packages shall be free from oil and humidity. Oil will block the passages and humidity will start corrosion and induce malfunction of the instruments.

- In some cases oil free compressors are used.
- In other cases for cost purposes oil-lubricated compressors may be used with oil-separating filters.

Instruments are the main compressed air consumer. Therefore the main portion of compressed air shall become dry. The remaining part is saved in wet air receiver. To remove humidity:

- Air dryer filters are used.
- In some cases in addition to filters, air is superheated to ensure its temperature is considerably higher than dew point temperature.

It is important to note that super heating does not mean very high temperatures. Based on efficiency of dryer filter, superheating may only be a few degrees above ambient.

Dry compressed air is saved in a vessel. The vessel is equipped with suitable instruments. Whenever inside pressure drops below a specified value, compressor is turned on automatically and vessel pressure level is maintained.

Normal working pressures for actuators may be less than 6 bars. It depends on line pressure, valve size, and actuator opening mechanism. Air receiver vessel pressure will be higher to compensate for head losses in piping and tubing path and to enable saving sufficient air to actuate several valves simultaneously. Wet air vessel pressure shall be a little higher than dry air receiver. This is needed to compensate additional head loss in filter and connecting piping.

Each vessel is equipped with pressure transmitters. As per compressed air specification different pressure levels may trigger different activities, which may range from restarting main compressor to starting standby compressor. This is to ensure sufficient high pressure air becomes available within a short period. Pressure settings shall be decided based on air consumption.

Compressor capacity is estimated by number of air consumers and their required volume. To calculate air compressor capacity a document is generated to summarize all air consumption volume and necessary pressure. This is generated with suitable contingency at the start of the project and revised based on vendor data.

Air is mainly composed of nitrogen ($\sim$78%) and oxygen ($\sim$21%). The remaining 1% contains all other gases like CO_2, water vapor, etc. At any temperature air can contain a certain amount of water vapor in the gaseous state. The difference between two terms shall be well understood. First is the maximum amount of vapor and second is the available water vapor.

Saturation vapor pressure (P_{sat}) at atmospheric condition in Pascal can be calculated by the approximate formula where T is the temperature in degrees Celsius (Ref. [20]). Atmospheric pressure is about 100,000 Pa (accurately, standard atmospheric pressure is defined as 101,300 Pa).

$$P_{sat} = 610.5 \, \text{Exp}\left(\frac{T}{T + 237.3} \times 17.269\right)$$

$$RH = \frac{P}{P_{sat}}$$

This formula is for temperature above 0°C over water. For temperatures below it, constant parameters change but general format is the same. This formula can be rearranged to calculate dew point temperature at any saturation vapor pressure.

Relative humidity (RH) is the ratio of actual present water vapor to P_{sat} at the related temperature. For example, 100% relative humidity shows this body of air cannot absorb any more water vapor. RH is usually expressed as a percentage rather than as a fraction. The RH is a ratio. It does not define the water content of the air unless the temperature is given. BS5250 has a plot of this equation in different temperatures and RH.

The relationship between vapor pressure and weight is defined for any gas by the equation:

$$p = \frac{nRT}{V}$$

p is the pressure in Pa
V is the volume in cubic meters
T is the temperature in degrees Kelvin (degrees Celsius + 273.16)
n is the quantity of gas expressed in molar mass. It means weight divided by molecular mass. From basic chemistry, water includes one molecule oxygen, which is 16 g, plus two molecules hydrogen, which is 1 g. Therefore water molar mass is 0.018 kg.
R is the gas constant: 8.31 J/mol/m^3
Inserting these numbers in the previous formula for one cubic meter of air gives water weight in grams based on pressure in Pa and temperature in °C. P_{sat} or vapor pressure at any relative humidity can be calculated from the previous formula and inserted here:

$$w = \frac{2.166p}{T + 273.16}$$

Actual vapor pressure can be obtained by multiplying relative humidity to P_{sat}. Assuming an offshore location has 50% relative humidity at 50°C, using P_{sat} relationship the maximum water vapor pressure is calculated to be 12,328.6 Pa. Considering 50% relative humidity, actual water vapor pressure is 6164.3 Pa. One cubic meter of this air contains about 41 g water. This approach can give a rapid calculation method for the amount of water that shall be disposed for any volume of compressed air. A curve generated for the amount of water in 50% and 100% relative humidity per cubic meter of air is given in Fig. 3.24.

The curve shows at 25°C and 100% relative humidity each cubic meter of air (maybe 1200 g) has about 23 g of water. Since this is distilled water, its volume in CCs is almost equal to weight in grams.

Valve actuators shall be as small as possible to enable installation on a congested piping area. On the other hand they shall have suitable margin between their design and operating pressure. In addition air pressure due to any reason may reduce in the tubing. Valve actuators shall be able to fulfill their duty in this condition also. For example, required torque to open or close a valve may be calculated at a low value of 3—4 barg. But the actuator operating pressure may be at a higher value of 10 barg. At the same time its design pressure may be at compressor design pressure of 12 barg or higher.

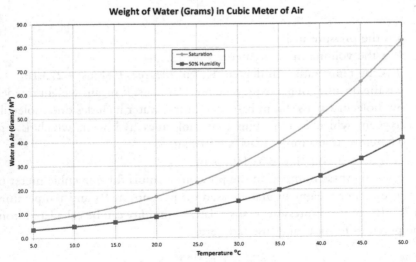

Figure 3.24 Water weight in air.

Compressor will work intermittently. It will pressurize wet and dry air receivers to a certain value. Since air is a compressible media, its pressure drops with gradual air consumption.

Fig. 3.25 shows sketch of a compressor package. This skid contains three main parts:

1. Main and standby compressors (each 100%) are located side by side. Gauges show their status and are installed toward access way to facilitate operators' view. Fresh air suction is located inside compressor skid. This way installing detector at air entrance is much easier. In addition vendor can connect it to package UCP more easily. By detecting combustible gas, compressor shall shut down. Compressor may have a cooling fan. Its discharge is normally directed to the top. This way blowing wind in operators' route will be avoided. Compressors may have prefilters to remove humidity. This way some energy is saved for compressing water vapor. At the same time its corrosive impact on package is reduced.

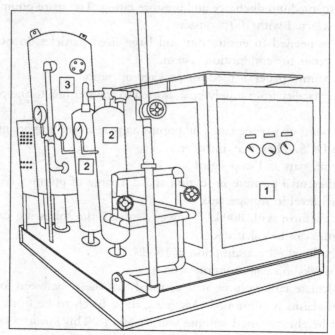

Figure 3.25 Compressed air package.

2. Postfilters are installed on the skid. This will facilitate piping from compressors to them. Each may be connected to a compressor or both may become online.
3. Receivers are normally installed out of package skid. First receiver is wet air and has larger capacity compared to dry air vessel. In some cases like the PID in chapter "Disciplines Involved in Offshore Platforms Design," air volume for nitrogen generation package is also taken from the compressed air system.

3.2.3 Diesel Oil System

A diesel oil package is needed to store diesel fuel for the power generator. Normally it may consist of:
- Storage tank: Its volume depends on supply boat arrival time. In locations where monsoon storms do not allow supply boat berthing, platform may be shut down to avoid hazards. In the Persian Gulf where maximum distance from platform to shore is 140 km, storage tank volume is determined for at least 2 weeks' power consumption.
- Feeding pump and spare: Normally these are low-capacity pumps with small-to-medium discharge and pressure rating. The spare pump is normally selected with 100% capacity.
- Filter: is needed to ensure dirt and large-sized particles do not enter diesel generator combustion system.
- Local control panel: is used for package operation.
- Feeding facility from supply boat: is used to transfer fuel to the platform storage.
- Instruments on storage tank and pumps transfer several important signals to the DCS. They may include:
  - Pump start and stop signal
  - Differential pressure in suction and discharge of pump
  - Fuel level in storage tank

For a platform with 800 kW power demand, the following capacities and volumes can be calculated:

Hourly diesel oil consumption = 190 L

Two weeks' oil consumption = 64 m^3

To calculate tank volume it is assumed that diesel between low- and high-level alarms is consumed. Assuming these levels to be from 0.10 to 0.90 tank height, required volume will be 80 m^3. This means a tank 5 m wide, 5 m length requires at least 3.2 m height to serve this purpose.

This tank shall have oil inlet and discharge nozzles. Vent to a safe location, drain, and level instrument nozzles are installed on the tank wall. Maintenance and access manhole is normally installed on top. This way leakage proof requirement becomes more relaxed. However, the access operator shall have sufficient head room between tank top plate and platform structural elements passing above the diesel tank. To limit sloshing impact, tanks may be divided into two or three compartments with holes for fuel passage at the bottom of bulkheads.

Diesel pump transfers fuel to a day tank. Although generator consumption may be very low, diesel transfer pump is designed to fill in day tank in a short time, say half an hour.

At ambient temperature diesel fuel does not have vapor tendency (high flash point). In addition, its flammability is low compared to other fuels. Therefore diesel tank may not require inert gas blanketing. In addition, other than water, which may be separated from diesel either in the tank or filters, it does not have severe corrosive property. Therefore carbon steel coated with some epoxy painting normally is used for tank material.

Diesel storage tank is normally classified as atmospheric tank. Its design pressure may be weight of water plus some pressure margin. This may be hydrostatic pressure of a column of water 4 or 5 m high (~ half barg). It is noted that walls are designed for low pressure. In a project hydrotesting a vessel to rapidly discharge water, 1 bar air pressure was applied. Vessel walls faced some plastic deformation.

Carbon steel plate with external stiffeners is sufficient for its structural strength. If angles are used as stiffener it is better to weld their edge to the steel plate and place their web away from the plate. This way the neutral axis of combined plate and stiffener section moves away from tank side plate and its structural strength increases. Care shall be exercised to check section modulus value instead of just the moment of inertia. This practice can be applied to all atmospheric tanks design and construction.

3.2.4 Power Generation System

Electrical power can be generated on the platform or supplied via subsea cables. Based on present technology, using solar energy panels is only feasible for minimum facility platforms where energy demand is very low.

For very big production platforms (especially with water/gas injection or lift facilities) or living quarters, electricity demand is so high that only gas turbine generators can be used. Associated gas may be used for this purpose.

Before using it as a power source, it has to be treated to match with generator manufacturer requirements. In many cases gas from the first-stage separator is suitable enough. Only water vapor shall be removed using a scrubber vessel. It will then pass a super heater and directly be fed to the generator.

In some cases gas is too sour. High H_2S, CO_2 content endangers GTG to rapid erosion. Therefore a sweetening package may be used to treat gas.

The major part of associated gas coming up with crude is methane gas. It has high energy content. As an example, assume crude from a field with 25,000 SBOPD produces 500 SCF/B. This field is able to produce 12.5 MMSCFD gas. Gas composition shows 40% to be CH_4. Gas gross heat value may be from 33 to 39 Mega Joule per cubic meter. Gas turbine generator efficiency in electrical power generation is assumed to be 30%. If for this gas content the net heat value is taken equal to 30 MJ/m^3, gas from this platform may generate about 15 MW energy. This is obtained by multiplying total gas volume by methane ratio by turbine efficiency multiplied by gas energy value for 1 s.

Total methane volume = 5 MMSCFD = 141,584 m^3/d

Total heat generated by this volume of CH_4 = 4,247,527 MJ/d

Generated electrical power = 14.7 MW

This is a relatively high-energy content, which may be used for a variety of purposes. Assume this field needs water injection. Normally to produce a certain amount of oil at least twice that volume shall be injected. As a minimum this field may need 50,000 barrels of water injection. Assume required pressure is 2000 psi ($\sim$138 Bar). Assuming pump efficiency to be 75%, required power for this injection is 1712 kW. This shows at least about 13 MW of power is left for other uses. Or the related gas can be transferred offshore for other domestic uses. The equivalent gas volume for 13 MW is about 11 MMSCFD.

Diesel consumption rate depends on several factors. Some of the major items include: manufacturer design, generator rating, motor RPM, ambient temperature, and pressure. As a general rule low power generators consume more fuel per unit generated power. Table 3.4 shows typical diesel fuel consumption and storage tank capacity for 2 weeks' consumption. Data are taken from manufacturer catalogues. There are minor differences in assuming fuel calorific value and density. Normally all manufacturers report consumption rate at 1 atm pressure. To prevent misunderstanding and avoid advertising in favor of a specific manufacturer in this purely technical book, I have removed the model number and manufacturer name. As an

Table 3.4 Typical diesel fuel consumption

Generator power (kW)	1800	400	353	272
Motor RPM	1500	1500	2600	3500
Diesel consumption (L/h)	469.8	103.7	97.9	75.0
Unit consumption (g/kW/h)	219	218	233	232
Diesel storage tank capacity for 2 weeks (m³)	185.7	41.0	38.7	29.6

average fuel volume from 0.25 to 0.3 L/h/kW can be considered. In calculating necessary storage tank volume, it is assumed 85% of its total volume is useful.

Increasing gas temperature by super heater is only to an extent to ensure its temperature is well above dew point. At this condition water vapor will not condense.

Diesel generators (DGs) are used for power demands about 1 or 2 MW.

A diesel generator system has four major parts. They are diesel engine, alternator, fuel system, and control system.

A diesel engine transfers chemical energy to mechanical energy and consists of the following parts. For a better description of each part's function, please refer to mechanical handbooks.

- Fuel pump/injection
- Combustion cylinders
- Gear box
- Lubricating system
- Cooling system
- Return fuel

A cooling system normally consists of closed-circuit distilled water that is either air or water cooled. A pressurized system can absorb heat above 100°C. Otherwise distilled water volume and circulation rate shall be sufficient to absorb excess heat. Circulating water shall be distilled demineralized water. Any impurity will settle inside cooling system tubing and either block water passage (reducing cooling efficiency) or induce erosion. Part of this water will be gradually lost as vapor. It has to be replaced by a suitable backup system.

If cooling system is water cooled, a heat exchanger can perform the job. Otherwise, a heavy-duty fan is needed to ensure sufficient cooling. It has to be noted that temperature change after contact with cooling media depends on contact surface/duration and original temperature difference. Air-cooled systems have better efficiency in winter than summer.

Combustion air is a small portion of cooling air. Very large fans may be needed to provide necessary air. In addition, air fans shall be redundant. A simpler solution may be to connect cooling air fan to motor shaft. Whenever diesel starts, its cooling fan starts. Since it is part of DG set, separate redundancy is not needed. This has a small cost saving, but more than that it has a simpler control system.

The alternator transfers mechanical energy to electrical energy and consists of following parts:

- Magnetic shaft
- Electrical coil system
- Automatic voltage regulator

Normally a diesel fuel system consisting of day tank, level transmitters, and closing valves, and is also supplied by DG set vendor. This is not a mandatory requirement. In fact this portion is relatively simple and ordinary manufactures can also fabricate it. However, in order to ensure proper operation of DG set, it is included in DG set vendor scope of work.

The control panel is very important. It gathers critical information like engine temperature, fuel pressure and volume, etc. from different portions and sends warning signals or even shutdown command.

Diesel generator packages have some miscellaneous items that are also important. For example, a drip pan to collect fuel, lubricating oils, and other drains and direct them to a hazardous drain system is very helpful in housekeeping of DG set room.

In offshore platforms, DG sets are installed over steel support. It is essential to prevent vibration transmission to supporting structure. This is done by installing several shock absorbers or vibration isolators in the skid. They can be inside the package or at its connection with steel supports.

A generator is normally sized such that the major portion (may be 80%) of its nominal power is consumed. Absorbing too low or too high load from a package will reduce its useful life. It has to be highlighted that a generator will only generate that much power that is absorbed by consumers. Power higher than generator capacity can be obtained for a short period. Normally its automatic voltage regulator allows even 110% load for 1 h duration. Larger loads may also be tolerated for a few minutes or seconds. Each manufacturer has a set point to shut down the generator if it is overloaded.

For higher power demand (on the order of several MWs) gas turbine generators (GTGs) may be used. Their efficiency is lower compared to DG set. One of the main reasons for switching to GTG is large size of required diesel fuel tank. For example, diesel required for 3 MW power within 2 weeks with average diesel consumption as specified in Table 3.4 is more than 237 m^3. It has to be noted that at best conditions only about 80–85% volume of a tank can be used. Therefore a suitable tank to generate 3 MW power will be on the order of 300 m^3. Providing this much diesel fuel in steel jacket platforms is obviously not easily feasible.

3.2.5 AC/DC UPS System

UPS is the abbreviation for uninterrupted power supply. Its function is to ensure that when electrical power generation (like DG set) on the platform fails, main safety features are able to continue a specific duration to ensure that the platform is brought to a safe and stable condition. This is done by storing electrical energy in a set of batteries. Battery sizing starts with careful calculation of essential consumers. When DG set is working, it continuously charges batteries. Related consumers are either fed from DC or AC circuits.

Since total shutdown of power generation system is an infrequent phenomenon, to prevent battery deterioration they are continuously in the power supply circuit. After any shutdown and after batteries are completely discharged and before any other start-up action, operators shall make sure batteries are fully charged. This is done with higher charging current and is known as rapid/boost charging.

Main alternating current (230 V) consumers may include process control, emergency shutdown system, and fire and gas systems, davit for lowering lifeboat, and telecommunication facilities. Radar sending alarm signals to approaching tugs shall work a specific time after platform shutdown.

Main direct current (110 V) consumers may include helideck lights, generator sets relays, and measuring and motor protection devices.

Electrolytes used in batteries may be acid or potassium hydroxide. Based on manufacturers' recommendations, electrolytes have to be recharged periodically. Electrolyte handling shall be carefully done to avoid burning hazards to operators. Normally battery room floor is covered with antiacid tiles.

Full (100%) sparing shall be provided for UPS system. Batteries are either stored in dedicated battery room or (if small) stored in a container in technical or switchgear room.

Battery room is a hazardous area. In addition to hazards due to electrolyte splashing, explosive gas is also emitted in the room. During normal operation and boost charging, water electrolysis emits hydrogen and oxygen in the battery room space. Emitted gas volume during boost charging is several times more than normal charging. Hydrogen concentration of about 4% may cause explosion. Forced ventilation is provided continuously to keep hydrogen concentration even lower than 1%. Hydrogen detectors are installed in the room to send alarm signals at early stages of hydrogen detection.

To calculate ventilation volume, two approaches may be used.

- Some standards specify a certain number of air changes per hour (ACH, 10 or 12) regardless of number of batteries.
- Ventilation rate (Q, m^3/h) is calculated based on number of battery cells (N) and their characteristic current (I, Ampere) equal to $Q = 0.05\ NI$.

Standards like EN50272 have specified ventilation of battery room (Ref. [36]). This does not mean air conditioning is mandatory. They only want to protect batteries against excessive heat. Different operators may have different approaches. They may require providing air conditioning for battery room. Battery manufacturers claim their batteries are able to work in a wide temperature range, say $-20°C$ to $+50°C$. However, useful battery life is reduced in higher temperature. Typical lifetime for nickel ($+$pole)/cadmium ($-$pole) batteries may reduce by 20% if battery room temperature is kept at 35°C instead of 25°C.

Purely from a cost point of view, accepting this lifetime reduction and replacing batteries after say 10 years may be more cost-effective. The point is that providing air conditioning and at the same time extracting air from a room requires a high electrical power demand. This needs a high CAPEX for the air conditioning package (which shall be explosion proof) plus high OPEX. If during normal operation ventilation rate can be reduced or during boost charging higher temperatures may be accepted, the air conditioning package capacity reduces considerably and makes this approach more attractive.

3.2.6 Blanket Gas/Inert Gas

Both fuel gas and inert gas can be used for blanketing. Inert gas is required for the last stage of purging before operators enter vessel. Air has oxygen. When in contact with flammable material, with sufficient heat it will cause flame or explosion. To prevent this phenomenon, in some spaces inert gas can be injected. Or gas derived from the platform may also be used. This action is called blanketing. The difference between blanketing and purging was explained in Section 3.1.12. As far as oxygen entry is prevented, fire and explosion hazards are stopped.

For example, a flare line contains flammable gas. If due to any reason the flare tip flame is extinguished, air may enter the line and progress backward to flare knockout drum. In this vessel if air, gas, and heat ratios approach dangerous levels, an explosion may occur. Therefore a minimum gas volume is continuously injected to prevent any backflow.

Diesel or other chemicals stored in atmospheric tanks may have the same problem. Gas is injected on top of the fluid with a small overpressure to ensure air intrusion is prevented. The injected gas may be fuel gas but instead of that inert gas can also be used.

Before operators' entrance to a process vessel all hydrocarbon shall be washed and drained. In addition, existing vapor gas shall be vented. To purge toxic or flammable vapor, inert gas is injected. Here fuel gas cannot be used.

If required volume is low, nitrogen bottles can be used. They are normally 200 bar cylinders, which are bottled onshore and transported to the platform. The bottle regulator nozzle is connected to vessel purging nozzle and dangerous vapors are vented from dedicated vent route.

If required volume is high or inert gas is required for continuous blanketing, then an inert gas package shall be installed. Nitrogen is freely available from air in large quantities. A supply of dry air will be taken from a compressed air system and nitrogen with more than 99% purity is generated. Nitrogen and oxygen separation is done using membrane separators.

3.2.7 Atmospheric Vent System

An atmospheric vent system in its broad sense cannot be categorized as equipment. However, since it has a collector piping and requires special design features, it is explained here. Two cases shall be distinguished:
- Small-scale venting from packages, vessels, instruments, etc.
- Venting from satellite platforms

The first group consists of several small diameter pipes, which are directed to a remote location in platform. Venting volume is very low. The second case is much larger. Installing a flare system may be very costly. Depending on the flared gas volume, distances more than 100 m from the platform may be required. Construction and installing a jacket and flare tower with necessary lines and system as explained in Section 3.1.10 is very

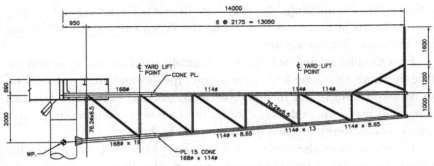

Figure 3.26 Vent boom.

costly. For satellite platforms that may be unmanned and have much smaller gas release, a boom may direct a release point to more than 10−15 m away from the platform. A truss-like structure may serve this purpose very well. At the end location a tip is left for venting. A careful dispersion study shall be performed to ensure vented gas will not have adverse impacts on operators' safety. Fig. 3.26 shows a vent boom. Truss structure extends 14 m away from platform. It may be triangular shaped. At top a walkway may be installed for operators' access to tip. Very small-sized tubulars shall be avoided to prevent vortex-shedding effects induced by wind action.

3.2.8 Open-drain System

An open-drain system is considered as a nonhazardous drain, which shall not be combined with a hazardous drain in a closed-drain system. Although it may contain oil, its ratio is very small and has no danger of gas release or gas trap. Open-drain system consists of a network of drains from different locations (mainly from housekeeping activities). The effluent water is finally collected in a caisson. Oil that is lighter than seawater will gradually float to the surface. A submerged open-drain pump will start by a command signal and transfer the oil collected at the top of the caisson to the tote tanks or supply boat collecting platform effluent.

Similar to the seawater service pump, an open-drain pump is also installed inside a caisson. In some cases caisson may be installed with topside. The service water caisson does not have much piping, but open-drain caisson has several pipes collecting drains from different locations. Therefore caisson internal tolerances are more important. Here a pump is installed at a higher elevation compared to service water and instead of water discharges oil.

In spite of service water, both discharge volume and pressure head are much less and a small 2″-size pump may be sufficient. The only problem may be the coordination between piping, mechanical, and structural groups. The first discipline installs internal piping, the second discipline procures pump, and the third discipline installs caisson and supports it by jacket members. Interface data on nozzle sizes, pump dimensions, and caisson arrangement shall be transferred between them.

3.2.9 Hypochlorite Injection System

Sodium hypochlorite may be injected both in sewage treatment system and in seawater service. The function and package complexity is different. In service water its duty is to inhibit marine growth and destroy biological organisms present in the caisson.

This section is limited to seawater service injection. The other one will be explained under sewage treatment topic. A hypochlorite injection package is very simple. It consists of a tank with a solenoid valve. Hypochlorite flow is with gravity. A solenoid valve is designed to allow intermittent discharge to caisson at specified time intervals. In addition, whenever the service water pump starts, the solenoid valve shall open and allow a much larger volume hypochlorite injection.

Tank shall be equipped with level instruments to inform operators for filling. Filling may be done with a hand pump.

3.2.10 Service Water System

The service water system is used for housekeeping activities. It may also be used as firefighting pump for small fire accidents. A fire water jockey pump provides minimum pressure inside fire ring before main fire pump is activated. The sea water pump main activity is to supply water to hoses for washing and housekeeping operations. In addition, it supplies water for firefighting foam system.

It takes water from a suitable depth below sea surface. This is to ensure oil and surface dirt is not taken in. A caisson is normally connected to the jacket. Based on seawater pump capacity, its internal diameter may vary. For a pump with 50 m^3/h capacity a 900 mm caisson is suitable. This caisson is supported in several elevations by jacket members. Several guides are installed inside the caisson to ensure pump shaft is located vertically. At the bottom some racks are welded to the caisson entrance. This prevents large fishes entering the caisson and getting trapped in the pump. In addition to this barrier, hypochlorite is injected to remove living creatures. A strainer is also placed at pump intake.

This pump has to discharge water to the highest platform level. This means it shall have at least 6 bar pressure. If used for extinguishing small fires, it shall have at least 16 bar operating pressure. Using 20% for design to operating condition its mechanical power may be as high as 26.7 kW. This means a minimum electrical power equal to 33.5 kW is needed.

In some projects jacket may be in the SOW of an EPCIC contractor other than topside. Seawater pump is designed by topside engineering. In this case either it may be given to jacket contractor as free issue equipment or caisson and pump are installed simultaneously with topside. Caisson guides shall be fabricated and installed by jacket contractor. First the caisson is lowered. Then pump is lifted stage by stage and lowered. Internal guides shall be preinstalled.

Pump LCP and hypochlorite system tank/pump may be installed near each other at sea deck level (drain deck). This is the first accessible level after the boat landing.

3.2.11 Pedestal Crane

Pedestal crane is the main equipment to transfer cargo loads from supply boat to the platform. Cargo may vary from food to spare parts, tote tanks, consumables, personnel via transfer basket, etc. Platform crane capacity and outreach depends on type of platform and its intended usage.

One of the main usages of the pedestal crane is at start-up. Lots of equipment, spares, and consumables shall be transferred to the platform. At this time platform generators/power system are not running. Therefore the crane shall have its own driving motor. Since considerable weight shall be lifted, using electrical power with battery is not feasible. Therefore cranes are mostly powered by diesel motors. In many platforms crane diesel fuel has been stored inside column supporting it. In other cases a small diesel tank may be constructed inside the crane base.

Battery or pneumatic system combined with mechanical driver can be used for diesel motor start-up. Starting system capacity shall be carefully designed for several starting runs. If it is pneumatic, a small vessel with sufficient pressure may provide required air. For electrical starters, a battery pack may be used. After starting the pedestal crane power-generating motor, the air vessel and battery pack shall be charged.

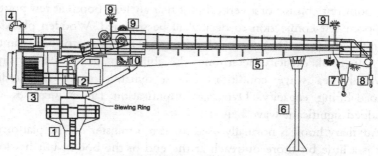

Figure 3.27 Pedestal.

Pedestal crane operation is for a limited duration during supply boat arrival. Therefore the diesel fuel tank need not be very big. Fig. 3.27 shows a typical pedestal crane. In this sketch the following items can be identified:

1. **Crane base** is installed on the main deck. A conical section connects larger base fabricated by vendor to pedestal fabricated in EPCIC contractor workshop. Slewing ring is located between top and bottom sections. Anti-spark ring connects electrical wires, which allows electrical power and instrument signals be transferred without interruption due to platform rotations.
2. **Operator cabin** is located exactly below boom to provide a better view. Operator seat is placed in side of the boom toward the platform. This allows a better view both on the supply boat when lifting a cargo and on the main deck laydown area when unloading the cargo.
3. Crane operation requires considerable electrical power. If this power is taken from platform generator, its sizing will be affected. Therefore a dedicated diesel generator is installed at diesel room behind operator's cabin. Sufficient noise and heat insulation shall be provided to enable operator to hear radio communications made by others (on supply boat/platform elevations or from CCR) and cabin heat, ventilation, and air conditioning (HVAC) be able to function properly.
4. **Wind sock** is located at end of the boom and gives a visual representation of wind direction and velocity. Wind sock material is flexible. They are calibrated such that circles around it mean a certain velocity limit. For example, in this sketch two circles are straight, which may mean 5-knot wind velocity.
5. **Boom** is a closed section box. It is hinged to crane. Luffing is done by hydraulic jack connecting boom to crane base. Its structural design impacts crane outreach.

6. Boom rest consists of a vertical post that anchors boom at rest position. Boom rest connection to deck shall be stiffened. Wooden planks are installed on top of boom rest column to prevent damage to boom painting.

7. Main hook capacity determines the ability to lift loads from supply vessel. Sea wave conditions add a dynamic factor that reduces load–lifting capacity. Dynamic amplification factor increases with added significant wave height.

8. Auxiliary hook is normally used for cargo transfer on the platform. It has a little bit more outreach at the end of the boom, but has lower lifting capacity.

9. Aeronautical obstruction light, operation light, and flood lights may be installed on the boom.

10. Inclination measure simply consists of a page with inscribed angles and a vertical indicator, which with boom movement shows its angle of rotation.

Below crane boom is an empty space that can be used to store tote tanks and other miscellaneous items like welding machines and consumables transported to platform for a temporary maintenance/repair operation. However, they shall not hinder cargo transfer to main deck and be protected against dropped objects. Operator cabin is always located to give a clear view of the supply boat lifting area and platform laydown area. That is why the cabin is always fabricated toward the platform and a little projected away from crane base. Pedestal crane is installed at the edge of the platform.

Cranes shall lift hatch covers above well bay area. Therefore their outreach shall include this part of the platform. In many cases if helideck is not extended outside platform deck via a cantilever, crane boon will pass over chopper landing area. Aviation regulations require a 210 degrees obstacle-free zone. To follow this regulation limit, switches may be introduced in the crane control system that automatically prevent passing over this area. In case transferring goods to a specific area requires violating this rule, the operator can temporarily override the limit switch.

Boom rest is a vertical post providing a seat with sidewalls to support crane boom during offshore transportation and at rest. Applied loads are small but it is several meters high and has to act as a column with only one end fixed and the other end free.

Pedestal crane has three movements. The first one is hoisting. This is lifting the load by turning the sling around lifting pulley. Normally this is powered hydraulically. To reduce hydraulic tensile force, the sling is turned several times around the pulley. In Fig. 3.27 four rounds have been used.

Sling tensile force is equal to the weight of cargo + lifting block + wire rope weight divided by number of slings. Using several rounds of sling reduces sling force but increases moving length. It has no specific impact on the total work that is needed to lift an object, but it reduces required power and sling size. For example, with four round slings when lifting a 20-ton object the sling safe working load shall be 5 tons.

The initial seconds of hoisting are critical. Supply boat moves up and down due to wave action. Tide action is very slow and does not have an impact on crane operation. Waves with one or 2 m height have a period around 2 or 3 s. When hoisting a load, tugboat may move downward or upward suddenly. Both movements impose a dynamic effect on hoisting system and crane boom. Cranes are designed to tolerate these movements by either a heave compensator or considering dynamic load factor in wire sling design, hoisting system, and boom design. Supply boats are not allowed to berth in rough weather conditions. Range 4 to 5 on the Beaufort scale is typically the maximum allowed condition. Therefore this heave compensating system may be designed for a significant wave height around 2 m. However, in emergency cases they may be allowed to come near the platform with full running engines and headed away from platform.

The second movement is luffing. This is changing crane boom angle in the vertical plane without turning the hoist. Fig. 3.28 shows luffing

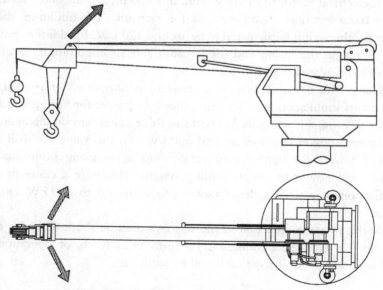

Figure 3.28 Luffing and slewing movements.

direction. This is done by injecting hydraulic oil to the cylindrical jack connected to the crane boom. This movement lifts both cargo and crane boom and therefore consumes the largest power.

The third movement is slewing. This is rotating the crane around its base in horizontal direction. To do this action, the crane motor has to overcome the friction factor on its base. Therefore slewing consumes the smallest energy of the three crane movements. Fig. 3.28 shows slewing direction.

A typical platform crane may lift around 20 tons in 10 to maximum 15 m lifting radius from its base. The lifting load at highest outreach (say 20 m) reduces considerably, maybe to 5 tons. Auxiliary hook capacity is normally less than pedestal crane lifting capacity at maximum outreach. Maximum lifting capacity is governed by its hydraulic system capacity. Maximum lifting outreach is governed by boom structural strength.

The main factor in crane power is hoisting speed. Assume a crane lifts a 20 ton load with a speed of 20 m/min to 30 m elevation. This is about one foot per second. Total work is:

$$Mgh = 20,000 \times 9.81 \times 30 = 5.9 \times 10^6 \, J$$

It takes less than 100 s to lift this load 30 m. The required power is then calculated to be ~ 60 kW. However, this is much less than the required driving system power. At least three efficiency factors shall be multiplied. They include wire rope turning around lifting block and pulley driving motor, electrical motor to rotating drum movement, and diesel to electrical power transformation. Assuming 80% for each one, the minimum diesel power for this action is calculated to be around 120 kW. In addition, power for lighting on the boom and inside cabin and air conditioning shall be separately calculated.

Most of the operators require a crane to perform two or even three movements simultaneously. Assuming the same power for luffing and half of it for slewing, if it is required to perform three operations simultaneously, the required power becomes around 300 kW. To this value we shall add required power for lights inside cabin, floodlights along boom, cabin HVAC, and power to charge starting system. Therefore a crane of this specification may need a diesel motor producing up to 400 kW engine power.

For this power range, a diesel motor may consume about 80 kg/h. This is about 100 L. Even considering 10 continuous hours of operation, a 1-cubic-meter diesel storage tank will be sufficient.

3.2.12 Material Handling System

Although this may seem to be part of miscellaneous items, no platform can operate without a proper material handling system. It may be difficult to give a universal definition for material handling equipment. In a production platform with a mechanical workshop there will be many more items. In a small wellhead platform there may be only a few. They may be categorized as:

- Transport trolley
- Fixed monorail
- Fixed winches
- Manual hoists
- Movable floor cranes
- Pulling tools (ratchet puller)

Transport trolleys are normally four wheel, hand driven. Their dimension is about 80 cm to 1 m width by more than 1.5 m length. Their height is less than 80 cm. They may carry up to a 3-ton load. It is interesting to note that a transport trolley in some cases may govern deck plate thickness design. Normally hard rubber wheels are used. Steel wheels will apply loads in a theoretical line to the plate. Stresses may become very large due to concentrated loads. However, if the same magnitude load is distributed over a finite area, maximum stresses will reduce considerably. After load is applied, the wheel contact point with steel plate takes a rectangular shape.

Plate design follows stresses caused by maximum internal forces due to uniformly applied load on a rectangular area (with specified dimensions) over a plated area. Secondary beams supporting the plate deck area determine supporting conditions. It is customary to be conservative in the assumptions. Excessive deflections may cause difficulties in trolley movement or deform deck plate plastically.

In determining internal forces, normally[2] tables from engineering literature are used and a separate finite element analysis is not performed.

Assume a four-wheel 3-ton trailer. Its weight is 290 kg and wheel footprint is 200 × 40 mm. Secondary beams in a deck area consist of IPE220 profiles spaced every 1000 mm center to center. A conservative design can assume the plate edge to be simply supported at the middle of

[2] Table 11.4 of "Roark's Formulas for Stress and Strain."

each protruding side of the flange. Therefore for a rectangular plate smaller edge dimensions will become $1000 - 110/2 = 945$ mm.

In topside design, maximum length of secondary beams is determined by allowable deflections. Assume in this case L/360 is selected. With local live load of 5 kN/m^2, 10-mm-thick deck plate, and weight of IPE220 the distributed load is calculated to be about 6.1 kg/cm. Assuming snipe end I beams (both ends simply supported), maximum deflection is calculated by $5wl^4/384EI$. This value shall be less than allowable deflection while stress criteria are also satisfied. Considering this limitation the maximum span of this I beam is less than 578 cm.

Using Table 11.4-1C formula (Ref. [32]): b1/b = 0.042 and a1/a = 0.035. Maximum stress ratio is produced for a/b = 2. Use the maximum value of 1.73 for stress ratio.

$$\sigma_{max} = \frac{\beta W}{t^2}$$

Two cases shall be considered. The first case is for normal loading in which applied load is distributed evenly between four wheels. In this case $W = (3000 + 290)/4 = 825$ kg and allowable stress is 0.6 yield stress. For S355 material minimum 8.2 mm and for S275 material 9.3 mm plate thickness are required.

The second case is for uneven loading on three wheels. In this case $W = 1100$ kg and allowable stress is increased by 33%. For S355 material minimum 8.3 mm and for S275 material 9.4 mm plate thickness are required. As can be seen in this design, a minimum 10-mm plate is required.

Although in the second case applied load is increased, allowable stress is also increased to cater for its lower probability of occurrence.

Wheel ball bearings may impose a rolling friction factor of around 1%. For a 3-ton load-bearing trolley about 30 kg pulling force is sufficient. Therefore only one man may be sufficient to pull or push transport trolley. This is true provided large deflection does not occur in the deck plating. Otherwise a factor corresponding to the slope shall be added. Besides at start of motion, larger pulling forces are required. Motor-driven trolleys are also available and may become more popular.

3.3 INSTRUMENTATION AND CONTROL SYSTEMS

With continuous progress in electronic devices almost all platforms are relying on different instruments to monitor and control all devices and operations. At each critical point gauges, switches, and transmitters measure flow velocity, discharge volume, pressure, temperature, and level. Based on the function of that instrument, gathered data are either displayed locally or transferred in the form of a signal to UCP or CCR to trigger an action. All unmanned platforms rely on an automatic control system. Even manned platforms follow the same approach. The main information of all systems and packages is displayed on graphical display units in front of the operator. Old systems as shown in the two pictures in chapter "Disciplines Involved in Offshore Platforms Design," Section 2.2 "Control" are replaced with new monitors showing full PID with all necessary signals. Numerical data may either change online or by clicking on the related icon. Signals can be divided into three categories.

Remote control is also possible. Necessary data are transferred via fiber-optic cable or satellite connection to a remote control room. This feature is especially utilized when several satellite platforms are staggered over a distance. All of them can be monitored and controlled via a CCR. When operators want to attend to a platform for maintenance, then with CCR permission control is transferred to the local ICSS unit stationed on the visited platform. This is necessary to ensure that system fault does not endanger operators' lives. Even if control from both stations is possible, at a specific time only one center can have control and the other one automatically changes to monitoring status. To transfer nearby satellites' data, in addition to fiber-optic cable and satellite transmission, radio link can also be used.

3.3.1 Process Control System

Signals related to normal operation of packages and equipment like start/stop of a pump/compressor, open/close status of a main valve, etc. are transferred to the process control system (PCS). It may also be called the distributed control system (DCS). Although the main CCR may be able to take action on PCS signals, for some of the packages they are handled via package UCP. In addition, packages may have an LCP that operators can use. LCP is located on package skid.

One operating station will be installed in the platform technical room. This will be used by operators during visits. All operating and control functions are normally done through the PCS including:
• Adjustment of set points
• Change of control mode
• Adjustment of flow parameters (choke opening, closing)

- Opening and closing of valves
- Start and stop of pumps and motors
- Monitoring alarm signals
- Reset some local shutdown levels

Selected signals from main systems are continuously controlled. They may include:

- ESD/fire and gas system (FGS), responsible for platform safe operation
- Power generation system, responsible for platform electrical power
- Compressed air system, responsible for instrument valve activation system
- WHCP and HPU responsible for Xmas Tree and emergency shutdown valve (ESDV) operation
- HVAC system, responsible for maintaining comfort conditions for operators and fault-free normal conditions for sensitive instruments, electronic and telecommunication items
- Chemical injection system responsible for preventing corrosion and keeping process phenomenon within normal controllable levels.

3.3.2 Emergency Shut Down System

Signals related to emergency conditions leading to partial or full platform shutdown are transferred to ESD. When the operating conditions approach the mechanical limits, ESD shall automatically drive the plant to safe condition. Only for some of local shutdowns is restart from remote control unit possible. For major shutdowns like platform total shutdown and well shutdown or fire (FGS function) operators shall attend platform and reset necessary instruments.

ESD/FGS signals and their cables' transfer route, cable specifications, and redundancy regulations have stricter safety conditions compared to PCS. To ensure instruments related to ESD can continuously send signals to ESD system, a line monitoring system shall be installed. This means that if the line is cut in any location a minimum electrical current, which is continuously passing through a preinstalled resistor, is cut. On the other hand, if a short circuit occurs, this electrical current will tend to pass through the minimum resistance point. Therefore a preinstalled resistor will not have any current. Both these conditions trigger an alarm for the operator to check the route.

Before entering the main cabinet, signals from different instruments are connected to one or two marshaling cabinets. Here they are categorized and correct signals are transferred to the intended terminal of control system PLC.

3.3.3 Fire and Gas System

Signals related to a safety hazard are transferred to the fire and gas system. FGS controls all signals related to fire and gas detectors. If fire, smoke, or toxic/combustible gas is detected at a location, an alarm or specific action will be triggered. Main sensors for this system include:

- Combustible gas detector (CGD): Methane is the major item. One carbon and four hydrogen atoms have a molecular weight equal to 16.04. This is much lighter than air (~ 29). Therefore after release from leak source it will be trapped by the roof of that elevation.

- Toxic gas detector (TGD): Hydrogen sulfide is the major item. One sulfur and two hydrogen atoms have a molecular weight equal to 34.08. This is much heavier than air. Therefore after release from leak source it will spread at the floor of that elevation.

- H_2 gas detector is only installed at battery room. During boost charging there is a possibility of hydrogen release. High H_2 concentration may explode.

- Heat detection is based on rate of rise detection. In addition they have a maximum limit, which is set high above ambient temperature. It has to be noted that steel structures in direct sunlight can reach very high temperatures. Black body temperature is normally 85—90°C. Heat detectors shall be installed away from these points.

- Flame detection basis is optical and detects ultraviolet (UV) and infrared (IR) spectral range of flame. In the platform there are many UV/IR sources including lightning, arc welding, solar radiation, light reflection, etc. Detector shall ignore these false signals and issue alarm only for flame spectral range.

- Smoke detection basis is optical and reacts to visible and invisible products of combustion.

- Fusible plugs are also another type of flame detection. They can be either metallic or glass. They are installed in a network that is pressurized by instrument air. For industrial environments, glass plugs may be used only with a protection cage. Metallic plugs melt at a certain temperature. Glass plugs contain a liquid inside them that will break the glass after reaching a certain temperature level. Unlimited air release and pressure drop will trigger fire alarm.

It has to be noted that combustible and toxic gas separation does not happen immediately after release. Degree of separation depends on several factors, including release pressure/velocity, leak point size, natural ventilation, distance from leak source, etc. Besides, separation is not an abrupt immediate phenomenon. The percentage of each portion gradually decreases. For this reason, CGDs are installed at higher elevation and TGDs at lower elevation.

Detectors located outdoor shall have suitable ingress protection (IP) degree. A normal selection is IP54 for indoor and IP65 for outdoor. IP54 means indoor instruments are protected against dust (limited ingress, no harmful deposit) and against water sprayed from all directions (limited ingress permitted). IP65 means outdoor instruments are totally protected against dust and against low pressure jets of water from all directions (limited ingress permitted).

3.3.4 Choke Valve

Choke valves are the first instrument placed after Xmas Tree discharge (downstream wing valve) in the flow line. It is not part of the Xmas Tree block. However, its function is directly related to the Xmas Tree. Pipe segment between choke and wing valve has connections for chemical injections. Choke valves are supported by topside piping. Xmas Trees are installed after jacket and before topside installation. Therefore this segment will be installed offshore to enable catering for Xmas Tree installation tolerances.

Choke valve duty is to reduce/regulate well pressure/flow fluctuations to platform design pressure and crude flow. Due to this function pressure after choke is reduced from reservoir flowing pressure to platform operating pressure. These two values may be very different. For a better explanation, please refer to Section 2.8. Therefore some chokes that shall handle very large pressure differentials may have a two–stage construction.

Pressure after choke valve (which is called platform operating pressure) has to be carefully selected. It shall be calculated based on onshore facilities requirement. The intention is to minimize pumping facilities and use well pressure as the main driving source for crude transfer. If platform operating pressure is selected to be too high, it may reduce downstream pumping costs in the onshore plant, but platform equipment and export pipeline shall be designed with a very high pressure rating. In addition to costs, this will increase explosion probability. Pressure drops in pipeline to onshore, in the platform piping system and equipment, and required pressure at onshore tie-in has to be estimated.

On the other hand, if a high-pressure reduction is selected, choke design may become very complicated. It may need to drop pressure in two stages. Platform design with too low operating pressure may cause unnecessary separation in the pipeline, which may block flow path.

Since during platform life reservoir pressure may drop considerably, therefore the choke shall be able to support a wide range of pressure reduction. Each choke is equipped with a PSV. Pressure safety valve is immediately connected to choke valve. Fluctuations in reservoir pressure within a range can be regulated by choke. Pressure fluctuations (surges) exceeding this limit (not confined by choke) will be controlled by PSV through venting to flare header. PSV function is to release excess crude pressure from set point to the flare line. This is to ensure pressure in piping/equipment downstream choke valve will not exceed platform intended operating pressure.

Due to its importance and high-pressure value, the choke actuator is controlled by the wellhead control panel. Its actuator requires considerably large hydraulic oil volume and pressure.

As a main difference, with on/off valves chokes may open crude passage from 0% to 100% in gradual steps. Of course its design shall be such that in normal operating condition less than 80% is open. To enable rapid closure of flow path, hydraulic power pushes a disk blocking flow passage to open hydrocarbon flow. Whenever HPU fails or hydraulic pressure is removed intentionally to shut down the well, this disk closes flow. Choke valve is a fail-close valve.

Choke sizing shall be carefully done. Fluid or gas velocity shall be limited. High fluid velocity with sand presence results in rapid erosion of the line. In addition high gas velocity while having erosion effect with flying solids induces vibration and noise in platform piping.

Choke valves change fluid direction sharply. This action induces a considerably large thrust force. It shall be supported by valve supporting system. Two different vendors may state different thrust force values for the same flow discharge and pressure change.

Choke body material shall be suitable for highly corrosive conditions of fluid passing the wellhead. In addition, all its electrical/instrument components shall comply with hazardous zone classification.

3.3.5 On/Off Valve

On/off valves are either open or close. They are not flow regulators. This action is different from choke valves, which (at least theoretically) are able to regulate flow from 0% to 100%. On/off valves are used for one of the actions described following:

- Emergency shutdown: shuts down platform in an emergency condition
- Shutdown: shuts down a train or a package
- Blowdown: allows full emptying of the vessel
- Isolation: enables isolating a part of piping or package from remaining platform connections

Based on size, flow type, and pressure, valve actuating system can be pneumatic or hydraulic. ESDVs are installed on the main line. Therefore their pressure is platform operating pressure and their flow is total flow. In addition, their closing time shall be rapid. These valves use a hydraulic actuator and are normally operated by the wellhead control panel. Other valves may use pneumatic actuators via compressed air system and controlled by PCS.

A specific case happens when flows of several satellites are directed to a manifold platform. After riser and before connecting to production manifold, an ESDV shall be installed. This is in addition to the ESDV installed in the satellite platform at the connection to the infield riser. Assume a problem happens in the satellite and shutdown command is issued by ESD. At this stage all inventory is directed to the flare. This means platform pressure becomes atmospheric. Since other satellites are still sending their crude to the common manifold, fluid will move toward lowest pressure. If an ESDV is not installed immediately before connection to the production manifold, flow will continue toward endangered satellite and either fire or a dangerous situation will continue. It seems a similar situation escalated the fire in the Piper Alpha platform.

3.3.6 Motorized/Actuated Valves

Based on line size, pressure, flow, and requirement of automatic or remote operation, these valves may have motors or actuators. Actuators work with a force transfer media like a compressed air system or hydraulic oil. Those working with air are called pneumatic and those working with oil are called hydraulic. Motors work with electrical power. For hand-operated valves the wheel may directly turn the stem or for additional force exert force via a gear box system.

Pneumatic actuators have a bottle or air storage that is connected to the compressed air system. The excitation signal removes the stopper, which starts opening or closing operation. Normally, in process lines valves are

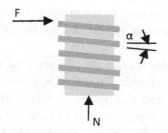

Figure 3.29 Thrusting shaft pitch angle.

fail–close type and in safety lines they are fail–open type. This means that in process lines a mechanical system like a spring closes the flow. Continuous actuation is required to keep it open. Compressed air inside a bottle may exert pressure over a disk, which pushes a piston. This piston may turn a valve stem via a gearbox system.

For large lines, big bottles may be required. A compressed air system may work in 5—7 barg pressure range. For contingency actuators, working pressure may be selected to be much less than this value, for example, 4—6 barg. Assume in a case that 100 kN thrust is needed. Thrust is pressure multiplied by area $N = P \times A$. For a circular disk with 5.0 barg pressure, disk area is calculated to be 2000 cm^2, which means disk diameter shall be more than 50 cm.

The pitch angle (Fig. 3.29) has an important impact on turning vertical thrust to horizontal force for rotating the shaft. A hypothetical case of changing pitch angle from 20 to 2 degrees shows 9.2 times reduction in horizontal force. Assuming pitch angle to be 20 degrees and using $F = N \sin\alpha \cos\alpha$, this 100-kN thrust may induce a horizontal force almost equal to 32 kN. The same case with 2 degrees pitch angle induces 3.5-kN horizontal force.

3.4 SAFETY SYSTEMS

Safety systems operation is essential for an offshore platform. Even if all process systems are available, unless all safety packages are installed and commissioned to be in proper working condition, operators are not allowed to start production. In this book equipment used for fire extinguishing, hazard mitigation, rescue, and evacuation are described under this topic. Fire/blast wall is not equipment, but when located between fire/blast source and equipment, it serves very efficiently to protect equipment located away from blast source. Therefore it is explained here.

Fire detection is covered in instrument discipline. Fire mitigation, extinguishing, and in the worst condition, rescue, is covered in safety discipline.

Fire is in fact a chemical reaction of a material with oxygen. In some materials like steel, this chemical reaction is a low heat, slow process. Its final result is rust and steel corrosion. In others like petrol, the result is excessive heat, and if sufficient amount of petrol is present there is explosion. To start a fire three elements are needed. They include fuel (any material that may burn), heat (to bring the material into a condition that enables its reaction with oxygen), and oxygen.

There are four fire classes (Ref. [37]):

- Class A fire is from wood, paper, and other materials that after burning produce ash.
- Class B fire is from liquid like petrol, diesel, paint, etc.
- Class C fire is from gases like methane, ethane, etc.
- Class D fire is from metals like Mn, Na, etc.

To extinguish a fire, one of the three components shall be removed. It means either heat must be reduced below reaction temperature with oxygen, or material that is able to burn is removed from location or oxygen contact with material is cut.

For class A fires, water is a very good solution. For other fire classes that induce very high temperatures, water may be the worst solution. High temperatures will decompose hydrogen and oxygen, increasing the flame. A famous example of this kind is cooking oil in the kitchen. If it burns, pouring water on the pot will increase fire and spread it to other parts. The best solution is to place pot door or drop a soaked towel on top of the pot to close off oxygen contact with fire.

For class B fires, foam is a better solution.

For an electrical fire in a confined space, CO_2 is better suited. Due to the adverse impact of CO_2 on the human respiratory system, using a CO_2 total flooding system requires installing certain safeguards, which are explained later.

For the same fire cases outdoors, dry powder and foam are better suited. One of the most simple and effective means of extinguishing small fires in onshore plants is a bucket of sand.

For class C fires (especially high-pressure gases), water, foam, and CO_2 are not effective. Gas velocity will push the fire-extinguishing material away from the affected region. Therefore the best solution is to cut gas flow by closing the flow line and directing remaining gas volume to the flare.

Adjacent equipment shall be protected from excessive heat to prevent fire from spreading.

In onshore tank farms the firefighters use water cannons to reduce temperature and prevent fire from spreading to adjacent tanks.

3.4.1 Firefighting Equipment

Firefighting equipment covers a wide range of items. In this section only pumps, piping, deluge valves, and sprinklers are described. Other items like CO_2, foam, and chemicals are covered under separate topics.

Installing a fire pump is mandatory for living quarters. For large production platforms that have a lot of expensive equipment, a fire pump is also provided. In design of unmanned or small satellite platforms, in some cases a burn-down philosophy is followed. For these platforms if a major fire happens, in the first step the flow of crude is shut down. All inventory is directed to the flare and the remaining crude within piping system and vessels is allowed to burn and finish. Service water package is normally used for housekeeping activities or single small-case fires. This means it is only used for localized fire cases. The main usage of water is to cool down adjacent equipment and vessels to prevent fire spreading.

The main items of this system include fire pump, fire ring, deluge valves, and sprinklers. Fire water demand calculations show necessary water volume. The fire water pump shall be able to provide the calculated amount of water with necessary pressure. Fire water is taken from seawater through a dedicated caisson. Caisson shall be installed such that it enables clean water intake. Racks at bottom of caisson shall prevent debris and fish entering the pump suction. Water volume demand and fire ring basic criteria for sizing are taken from NFPA 15. Some of the main requirements for fire water flow rate, velocity in piping, and pressure head follow:

- Fire water system pressure shall not be less than 175 psi (12.1 bar)
- Minimum inlet pressure shall not be less than:
  - 6 barg for fire hydrants
  - 7 barg for fire monitors
  - 6 barg for hose reels
  - 3.5 barg for deluge nozzles
- Maximum fluid velocity shall be less than:
  - 4 m/s in pipes feeding nozzles
  - 3 m/s in pipes feeding fire hydrants

- Fire water flow rate for process and utility area shall be calculated to provide at least:
  - Surface area of vessels 10.2 L/min/m^2
  - Projected area of skids and packages 20.4 L/min/m^2
  - Wellhead area 400 L/min/wellhead

Standards like the Norwegian Petroleum Standard (NORSOK) may have other requirements. For example, since the fire ring is operated only in emergencies and hopefully very seldom, therefore higher velocities in this piping are acceptable.

In a small platform, flow rate up to and even more than 600 m^3/h is common. Assume process area of one elevation has 22.0 m width and 33.0 m length. Well bay area in this elevation is 11.0 × 13.0 m and houses 12 active wells. Four package skids are installed with a total area equal to 60.0 m^2. This elevation includes two FWKO drum vessels each 2.4 m (dia) × 11.0 m (T/T) with semi-elliptical heads (long diameter 2.4 m and short diameter 1.2 m) plus two other vessels with 1.5 m diameter and 6.0 m T/T. Remaining area is covered with piping, manifolds, valves, and similar items. Required fire water discharge is:

Well bay area = 12 × 400 = 4800 L/min

FWKO Drum = 2 × (π × 2.4 × 11 + 2 × π × 1.2 × 0.6) × 10.2 = 1784.2 L/min

Vessels = 2 × (π × 1.6 × 6 + 2 × π × 0.75 × 0.375) × 10.2 = 612.8 L/min

Skid Packages = 60 × 20.4 = 1224.0 L/min

This level alone may require about 8421 L/min (505 m^3/h).

Considering 20 bar pressure for a fire water pump with a capacity of 600 m^3/h and 0.75 efficiency, its mechanical power is calculated to be 444 kW. This is a relatively large motor. Each fire water pump shall have its own dedicated diesel engine. Assuming another 0.8 for converting the pump electrical motor to mechanical power and another 0.8 for converting diesel engine power to electrical power, the gross diesel power for this pump shall be above 690 kW. This is a very big motor. It will consume about 140 kg of fuel within 1 h.

Fig. 3.30 shows sketch of a fire water pump. Note the important following items:

1. Power is generated by a diesel engine. It is dedicated only for fire water pump operation. Operators run it on a regular basis at specific intervals to ensure that it will operate when needed. Diesel fuel for several hours is provided in a specific tank.

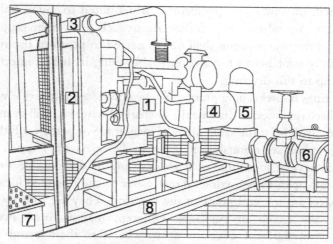

Figure 3.30 Fire water pump.

2. Diesel engine cooling is provided by a radiator.
3. Diesel engine exhaust shall be emitted in a safe area. If necessary, smoke and heat shall be released at a location sufficiently away from the area.
4. In normal motors mechanical power of diesel engine is changed to electrical power by an alternator. Here it is not needed and it can directly be connected to the pump shaft.
5. Fire water pump extracts water from a caisson.
6. Regulating valve connects pump to fire ring.
7. Diesel motor is started by a battery pack. Battery capacity shall be sufficient for a specific number of starts. In addition to it, another starter like a pneumatic exciter shall be provided.
8. Pump package is placed on a single skid. This facilitates its transportation, installation, and commissioning.

According to safety regulations, living quarters shall have another fire water pump with 100% capacity as backup. This backup pump shall be electrical. In this way only the living quarter shall have two fire water pumps with different running mechanisms. This means that even if any of the pumps or its power system fails, another one is available for operation.

To ensure immediate water spray on the intended location, piping connected to the fire water pump shall always be pressurized. This is achieved by a jockey pump. It provides a small quantity of water in sufficient pressure. Jockey pumps are normally electrical and are powered via platform switchboard.

As stated, in unmanned platforms it is permitted to install only a dry ring. In this case whenever a drilling rig or jack up is operating on this platform, its fire water pump shall be connected to platform fire ring. If a tug or supply boat berths for a considerable time, it shall connect its fire water pump to this dry ring.

In dry rings housekeeping pumps may be connected to provide water for small localized fire cases. In this case the discharged water rate will be much less but pressure shall be at the same order of magnitude or perhaps a little less.

Deluge valves are manually or automatically operated. They divide the fire ring to several locations, each serving one of the possible fire cases. They are fail-open valves, which direct water to the intended location.

Sprinklers are the end point of the fire ring. They are directly located above process vessels or packages. Different manufacturers may have sprinklers with different coverage areas. Therefore their layout shall be carefully checked to ensure the full intended area is covered.

Water is an incompressible fluid. In its ideal condition this means that if the jockey pump is stopped, after a single drop of water is discharged from the fire ring its pressure will drop immediately. Therefore after the jockey pump a check valve is installed. Then an accumulator is connected to the fire ring. Water enters this accumulator but is at the same time connected to an air compressor. Assume air volume is V_1 and it has 16 bar pressure. Due to water discharge air volume increases. Assuming it to be ideal gas, until V_2 becomes 33% more than V_1, fire ring pressure does not reduce to below 12 bar. This gives enough time to start jockey pump or main fire pump. Pressure reduction to initiate pump start is controlled by a pressure transmitter installed on the accumulator. Fig. 3.31 shows a simple PFD for this system.

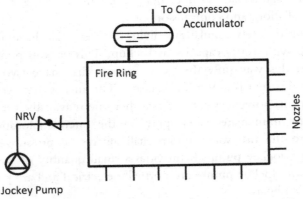

Figure 3.31 Simple fire ring process flow diagram (PFD).

3.4.2 Fire Wall/Fire Barrier

In previous sections it was explained that there are two active and passive types of fire protection. In active fire protection after a fire has started one of the extinguishing packages like fire water pump, foam, CO_2, dry chemical, etc. will start to extinguish fire. Water is normally used to cool down the adjacent packages to prevent fire spreading. Passive fire protection covers the package or area with insulation or a barrier to prevent spreading flames or heat or radiation. The main intentions are:

1. Prevent collapse of critical structural members.
2. Maintain the integrity of escape routes to the temporary refuge. For this purpose PFP shall not contribute to fire effects such as smoke, nor shall it emit toxic fumes whilst subjected to fire.
3. Limit the effects of jet fire and pool fire.
4. Maintain the integrity of the fire-rated prefabricated firewall panel connections at main and secondary steelwork locations.
5. Protect control modules and enclosures containing critical control equipment.

For main shutdown valves and packages, a fire jacket, and for utility areas, a fire wall, are constructed. Fire wall is a method of passive fire protection. Its main element is a steel plate stiffened with structural members. Hydrocarbon pool fire has different criteria than jet fire. A jet fire in addition to increasing heat and temperature exerts considerable pressure to the wall concentrated over a small area.

Both yield stress and modulus of elasticity of steel material reduce with increasing temperature. The American Petroleum Institute (API) and Centre Technique Industriel de la Construction Métallique (CTICM) have provided typical reduction factors for yield stress and modulus of elasticity to be used in structural design. These tables are found in chapter "Disciplines Involved in Offshore Platforms Design," Section 2.10.2.3.

Two types of fire including standard fire for typical material found in offices/houses and hydrocarbon fire are distinguished. Detailed temperature—time numerical data were given in chapter "Disciplines Involved in Offshore Platforms Design," Section 2.9.3. Fig. 3.32 presents them graphically. It is seen that hydrocarbon fire temperature within 3 min rises to a very high value of 887°C, while standard fire rises to 502°C within the same time span. The final temperature for the hydrocarbon fire approaches 1100°C and that of standard fire reaches 1150°C.

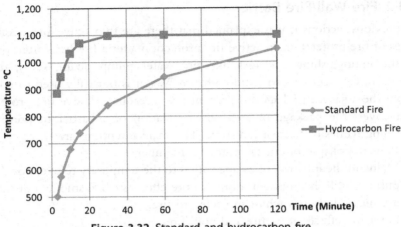

Figure 3.32 Standard and hydrocarbon fire.

In order to prevent fire from spreading to adjacent areas, several actions shall be done:

1. Prevent flame from access to adjacent area. This is done by providing a fix barrier like a steel plate in front of it. The structure shall be able to tolerate applied external loads. Therefore it shall be stiffened by rolled sections or other types of stiffeners.

2. Prevent heat increase over the steel plate. This is done by applying specific heat-resistant mixtures over steel plate. Intumescent material is very good in this regard. After being exposed to fire, they expand rapidly. With their very low temperature conductivity ratio, very little of fire temperature is absorbed by steel material. Steel has a very high conductivity ratio. In chapter "Disciplines Involved in Offshore Platforms Design," Section 2.1.2 a comparison was made between steel and rock wool material.

3. Prevent dramatic heat increase in the other side. This is again achieved by placing relatively thick insulating material. All material if subjected to heat flux for an unlimited time span will have the same temperature at both faces. The main benefit of using insulating material is the time gap provided before reaching to a certain temperature level. This time gap provides necessary time to initiate mitigation activities or in extreme conditions to abandon the platform.

The blast wall absorbs blast energy via deformation of structural elements. It keeps platform integrity and may be later replaced. Fire wall structural requirements have been discussed in chapter "Disciplines Involved in Offshore Platforms Design" and are not repeated here.

3.4.3 CO_2 Total Flooding System

CO_2 is an inert gas. It is much heavier than air. For a rule of thumb calculation, air is approximately assumed to be composed of 78% nitrogen and 21% oxygen. The ratio of other gases (including water vapor) sums up to be 1% and is about 0.5 in weight calculation. Nitrogen (N_2) molecular weight is 28 and oxygen (O_2) molecular weight is 32. Combining these two elements, air molecular weight is calculated approximately to be 29.0. This is much less than CO_2 molecular weight, which is 44. After spreading over the fire, CO_2 stays above it and extinguishes the fire by preventing oxygen from reaching it.

CO_2 total flooding system is effective in extinguishing several classes of fire. They include:

- Class A fires: Ordinary combustibles such as paper, wood, and textiles.
- Class B fires: Flammable liquid materials and engines utilizing gasoline and other flammable liquid fuels.
- Class C fires: Electrical hazards such as transformers, switches, circuit breakers, rotating and electronic equipment.
- Fires in hazardous solids.

To extinguish an electrical fire in an enclosed area like switch room, control room, and generator room, CO_2 is used. High concentration of CO_2 in breathing air causes health hazards. Table 3.5 shows percentages of CO_2 concentration in the air and their health hazards.

The CO_2 may be automatically or manually released. For full fire-fighting effect, all room connections with outside fresh air including doors, dampers, fans, etc. are closed. Therefore any personnel trapped in the room

Table 3.5 CO_2 concentration impact

Concentration of CO_2 in air (%)	Responses
5	Breathing becomes difficult for some individuals, inhaled for up to 1 h without serious aftereffects
10	Approaches threshold of unconsciousness in 30 min
12	Threshold of unconsciousness reached in 5 min
15	Exposure limit 1 min
20	Unconsciousness occurs in less than1 min
25	Lethal dose in a few minutes

face a death threat. To avoid this situation, before any CO_2 release a suitably located visual and audio alarm shall be initiated. CO_2 release will start after 30 s of alarm. If an automatic release system does not act properly, a manual push button is normally located outside exit door from each space, which can be operated by escaping personnel. It is located outside exit door to ensure operators start CO_2 release from a safe location.

To calculate required quantity of CO_2, first room volume shall be calculated. Removable equipment volume is normally neglected. For confined area below false floor, an occupancy ratio for cables, cable trays, and supports that are nonremovable may be considered.

CO_2 concentration is taken to be 50% equal to 1.33 kg/m³. Assume a technical room is 15 m long, 8 m wide, and 3.4 m height above a false floor and 60 cm below it. The volume is calculated to be 408 m³ above the false floor, assuming 15% occupancy below it, the total free volume is 469 m³. Based on CO_2 concentration, total required weight is calculated to be 624 kg. Using standard 100-pound (45.4-kg) cylinders, this room requires 14 cylinders.

As per standard, each area shall have 100% contingency. Therefore two sets are required.

3.4.4 Dry Chemical Extinguishing System

A dry chemical extinguishing system is intended for helideck firefighting. This skid contains three main sections:

- Storage tank: Contains necessary amount of extinguishing dry powder such as potassium or sodium bicarbonate based on manufacturer design.
- Activating cylinders and nozzles: Nitrogen inert gas is used to flow dry powder. When powder is fully dry, rapid flow of high-pressure nitrogen in front of storage tank nozzle induces a negative pressure. This negative pressure sucks dry powder out of the tank and spreads it over the fire.
- Hose reel shall extend from package skid to helideck area. Its nozzle shall be directed toward fire to spread dry powder on top of it.

In standby condition, operators shall check at least on a monthly basis storage tank sealing condition. In addition, powder shall be replaced periodically as per vendor recommendation.

At least two skids shall be placed in sides of helideck at a platform below it. This is to ensure that fire at any point of helideck can be extinguished. Platform elevation shall be such that nothing is more than 25 cm above helideck.

Normally each dry chemical package operation needs two operators. One of them shall unroll hose reel and move toward fire case. The other operator stays near storage tank and manually opens nitrogen bottle. After tank pressure reaches sufficient value, operator opens its nozzle to allow powder flow toward the hose reel. At this time the first operator shall open hose reel end nozzle to discharge trapped air. After that he can discharge powder over fire to extinguish it.

3.4.5 Lifeboat/Davit and Life Raft

Lifeboat is one of the items of rescue equipment. It consists of a totally enclosed package that has the ability to move away from the endangered platform. Some main features include:

A — Body/Arrangement

1. It shall be floatable with self-righting ability to enable it to float in an upright position after being lowered/dropped from platform. In fact this is not a mechanical system. It is achieved by proper design that ensures center of gravity of the fully loaded lifeboat is geometrically located below its center of buoyancy. Gravity is downward and buoyancy is upward. This induces a rotating torque that turns it upright even if it is dropped upside down to the water.
2. It shall have enough seats for the number of personnel. In addition, it shall at least be able to transfer one injured person on a stretcher. Stretcher can be placed in the middle of two side rows of seats. Individual seats shall have safety belts, which save personnel from sudden movements/jolts. In addition, seatbelts prevent personnel movement. This will ensure fully loaded center of gravity is in intended position as per design.
3. Lifeboat structure shall be light and at the same time robust to be able to resist rough sea conditions and heavy wave loads.
4. Material of construction shall be resistant to offshore environments. This includes humidity, temperature, salty waters, direct bright sunlight, UV radiation, etc.

B — Lowering System

5. Normally a lifeboat is lowered by a davit, which is electrically powered. In case of power failure during escape, it will be lowered by a mechanical system. This system is operated from inside lifeboat and consists of a

break and sliding system. In rare cases it may be dropped to the water from short heights. The lifeboat body shall have enough strength to take the impact due to dropping from a height. This is to cater for the emergency condition in which the davit is not working and the boat shall be dropped to the water.

C — Driving Engine

6. It shall have an engine to enable moving away from the hazardous area. Sea surface around the platform may be covered by oil spillage. This may trigger fire hazard if any spark from motor exhaust or high temperature comes in contact with volatile crude. Therefore the exhaust system shall be below the water surface.

7. Any diesel engine needs fuel and oxygen. When under water the oxygen shall be provided from an internal system. This means oxygen cylinders with sufficient capacity shall be provided to cover both human breathing and engine consumption requirements. This is for a specific period of time like 10 min. Considering lifeboat speed to be approximately 8 knots, this is sufficient to move about 2 km away from the platform.

   Assume a lifeboat has 26 passengers capacity. For comfortable working conditions (moderate energy consuming works — Ref. [21]) each operator needs 5—8 L/s. For emergency escape less comfortable conditions can be acceptable. Therefore consider 200 L per person per minute. Using these assumptions, minimum air consumption for operators will be 52 m^3. Typical engine air consumption for small motors (30—40 kW power) is 250—300 m^3/h. Manufacturers normally use water-cooled engines to avoid consuming too much air for cooling. Summing up the two values for 10 min moving below water surface, a minimum of 100 m^3 air is needed. If 160-bar pressurized bottles are used, total capacity equal to 625 L is needed. The compressed air system installed inside lifeboat shall be able to fulfill this requirement. Regulatory valves shall be robust enough to reduce 160 bar pressure to tolerable values for operators' breathing and motor combustion. This criterion is an important factor that shall be checked by buyer during factory acceptance test and by classification societies before issuing any approval document.

   In addition, the fuel system installed inside boat shall also have sufficient fuel capacity.

8. It shall have a cooling system for the engine, which is able to use seawater.

9. Since lifeboat shall be able to move through fire caused by hydrocarbon spread over sea surface, a cooling system for its top surface is needed. This cooling system shall take seawater from beneath the boat, which is free from hydrocarbons, and spread it on top of the boat. The main function is to keep boat surface temperature below a certain level. This is needed both from ignition point and also personnel comfort.

D — Personnel Comfort

10. It shall have an air system to allow personnel inside the boat to breathe normally.
11. It shall have robust seats accompanied with proper belts to prevent personnel from hitting each other during sudden movements.
12. Inside lifeboat suitable medicine is provided for each person to prevent vomiting and nausea.

E — Personnel Rescue

13. It shall have some survival equipment like water and solid long-life rations to cater for personnel needs.
14. It shall have first aid kits to treat injured personnel. This may range from basic antibiotics to antipain pills to bandages and antiseptic materials.
15. SOLAS requirements define the minimum items that shall be present onboard a unit.
16. It shall have some telecommunication equipment to allow personnel to inform their whereabouts to the rescue team. Other than flashlights, hand/rocket flares, smoke signalers, and VHF radio, two main types of telecommunication equipment are needed.
    • Search and Rescue Radar Transponder (SART): It receives radar signals of nearby boats and sends a signal at the same frequency to indicate position of lifeboat. This will enable pilot of the searching boat to find accurate location of the lifeboat or raft.
    • Emergency Position Indication Radio Beacon (EPIRB): When in contact with water, it automatically starts to send signals in a universal distress channel to satellite. Through satellite connection, the nearest boat or ship will find lifeboat/raft location.

Area near lifeboat is usually used as muster area. This area shall be sufficient for the number of personnel allocated to that lifeboat. Since lifeboat is overhanging from the platform side, an access platform shall be constructed. This will be from the sides for normal entrance and from the end for injured personnel moved by a stretcher.

A life raft does not have a motor. It is dropped from the platform by gravity and is opened after floating on sea surface. Some main features include:

1. It shall be self-inflatable with a neutral gas.
2. It shall be floatable and totally enclosed. However, the personnel inside life raft have the ability to open a window and paddle to move it.
3. It shall be self-righting. Therefore after being dropped from any height and floating to surface it shall turn to upright position. This enables it to be useable for rescuing personnel after being inflated. Personnel shall drop themselves from platform to the water and after life raft is inflated come overboard. Normally small life rafts do not have self-righting capability and shall be pulled upright using a strap connected to it for this purpose.

3.4.6 Safety Shower and Eyewash

In process plants crude may spill over operators. It may splash over any part of the body like clothing and face. This induces a hazard to personnel safety. Crude oil may have volatile components in hot temperatures. Exposure to sunlight and proximity to hot surfaces or flame may trigger serious safety hazards. The first action is to reduce spilled material as much as possible. This may be achieved by wiping it off or washing it. The second action is to reduce its temperature. This is achieved by pouring water on it.

For eyewash freshwater is used to avoid possibility of any chemical reaction and salty seawater being poured over scars/injuries. For clothing any water may be used. Safety shower and eyewash package may consist of the following items:

- Connection to freshwater system
- Standpipe having separate eyewash and shower nozzles. This shall be operated by a pedal to have free hands for wiping and other activities.
- Regulatory valves to change pressure on eyewash compared to shower.
- Regulatory valve to check water temperature is below burning point for human body.
- Drain system

Since eyes may not be opened under pressure, the eyewash shall have a sink to pour water into it. Then operator shall put his head inside it and open eyelids to enable washing external crudes. The sink has to be clean so that the possibility of infections is not added to the chemical hazards.

3.4.7 Oxygen Breathing Apparatus

Fixed and portable types of breathing apparatus are used. In brief, they contain an oxygen cylinder that is connected via a regulator to the mask. Personnel use them when working in an area that has a possibility of toxic gas release.

Portable breathing apparatus contain the following items:
- Harness and strap: This is used to carry the set easily while hands are free for operation. It is made of lightweight durable material.
- Oxygen cylinder: Container is made of lightweight durable material like aluminum or corrosion-resistant composite material. A normal-weight human may consume 250 mL/min pure oxygen. Based on activity, this value may increase much more. Pure oxygen cannot be injected to lungs. Normal air contains only 21% oxygen. Besides, only a quarter of available oxygen in breathing air can be consumed. British Standard (1991) "Code of Practice for Ventilation Principles and Designing for Natural Ventilation" [21] has defined 1.8 L/s as the fresh air requirement when a person is "seated quietly." For "very heavy work" the required fresh air volume may increase to 14 L/s. It is understood that heavier personnel working in hot environments may consume more. Cylinders may be pressurized up to 200 bars and more. An average of this pressure is used for cylinder volume calculation. Considering 11 L/s for fresh air requirement plus 20% for contingency and some for leakage and wastes, it is concluded that for 1 h operation at least one cylinder (10 L volume) is needed.
- Regulator and valve: To have maximum oxygen volume, cylinder pressure is very high. Direct connection to lungs threatens human life. A valve with regulator controls discharge pressure. Since during operation personnel need to concentrate on their job, location of this valve and regulator shall be such that solid objects may not impact it. Regulator breaking may seriously endanger personnel life.
- Alarm system: During operation personnel may not monitor elapsed time continuously. Therefore an alarm should be available to warn personnel. This alarm may be both visible and audible. It shall be triggered when at least 10% of consumable air is left. Of course this duration shall be enough to enable personnel to move from a toxic to safe area.
- Compressor package: At least one shall be provided in each platform to enable filling the empty bottles for reuse. It shall be powered by normal electrical outlets provided on the platform.

In many movies (especially war movies) film actors and actresses have been shown wearing a breathing mask. Therefore breathing air apparatus (BA) is a fairly familiar instrument and a photo is not shown.

3.5 ACCOMMODATION SYSTEMS

This categorization is not an accurate one. For example, HVAC is used for control rooms as well as accommodation. Its proper operation in a control room is even more critical. Platform control system may trigger shut down if technical room HVAC is shut down due to any malfunction for a long time. Some control systems may have a temperature limit. Exceeding that value may generate false signals, which may have adverse impact on platform operation and safety. Heat generated by working of several switch-gears in addition to environment temperature may be dangerous to some control and telecommunication equipment.

A potable water system may also be connected to safety eyewash equipment. Again it is emphasized that categories used in this chapter do not follow a universally accepted system. In addition, there are several other functions in an accommodation area that are not discussed in this book. Some examples may include:

- Storing meat, butter, etc. in freezer and cold room
- Preparing food in galley and serving it in messroom
- Washing and ironing clothes in laundry
- Recreation facilities like cinema, gymnasium, etc.
- Laboratory facilities
- Maintenance and repair workshops

The equipment, spaces, and personnel needed for these functions are not described in this book, but in fact their importance for the living quarter (LQ) is not less than, say, power generation!!

3.5.1 Heat Ventilation and Air Conditioning Systems

The HVAC package performs several functions including:

- Maintaining enclosed area temperature in a comfortable range against very high outside temperature variation from cold to hot environment
- Providing sufficient oxygen for breathing of operators based on their number and activity level
- Removing indoor humidity generated by human activities/added by outside fresh air and maintaining it at comfortable level

- Removing H_2, CO_2, smoke, and odors generated by human or equipment activity in an enclosed space to maintain healthy, nonannoying and nonexplosive concentrations
- Providing sufficient combustion/cooling air in enclosed spaces that contain fuel-consuming motors for power generation or other process activities.

This system is generally required for living quarters. Temperature, in CCR, electrical room, telecommunication room, generator room, battery room, etc. is also an important factor that shall be kept within acceptable limits to allow proper functioning of the equipment. Switchboards and control cabinets are a major source of heat release. Control systems and related instruments are very sensitive to high temperature and their malfunction may trigger unacceptable hazards to platform safe operation.

Any abnormal increase in control room temperature shall at least trigger an alarm. If crew cannot control temperature within limits, then they shall take the next steps for platform partial shutdown. Many instruments are able to function properly even to temperatures exceeding 40°C. Therefore in some area like generator rooms only adequate ventilation may be sufficient. In the generator room a large cooling air volume is necessary for diesel motor operation. Cooling this air volume to temperatures as set for CCR or cabins may require a dedicated power generator. Any LCP or instrument located at this room shall be able to tolerate a hot, humid environment.

In addition to temperature, the pressure of all rooms used by personnel shall also be carefully controlled. These areas may include cabins, offices, control room, telecom room, entertainment room, battery room, toilet, bath, meeting room, galley, etc. Two philosophies, providing positive and negative pressure compared to atmospheric, shall be followed.

- Cabins, offices, control room, generator room, etc. shall be protected against intrusion of gases, which may either be toxic or explosive. To achieve this, their pressure is slightly increased compared to the surrounding space.
- Toilets and baths may generate odors and bad smells. These spaces shall be ventilated to an open area. The negative pressure intakes fresh air from adjacent clean area and removes bad smell.
- Batteries during charging may generate hydrogen. If this space is not ventilated, hydrogen may reach exploding levels. Internal negative pressure intakes fresh air from safe area to reduce hydrogen concentration ratio.

Negative or positive differential pressure is very small about 30—50 pa. Atmospheric pressure is a little above 101 Kpa. This means a very small pressure difference is sufficient.

HVAC package may contain several items like louver, filter, fire damper, fan, condenser, evaporator, inlet/exhaust ducts, shut off damper, heater, TGD/CGD detectors, diffuser, and humidity/temperature/pressure transmitters.

Each room shall be completely isolated. Several ducts may connect an enclosed area to outside air. All doors and any windows installed in offshore platforms shall be gas tight. Air-handling units are also gas tight. In addition to them, we have four entrances:

- Fresh air duct brings outside air from a safe area to supply oxygen demand.
- Return air circulates part of inside room air to reduce cooling load. This is a major part (maybe 90%) of air circulated in air handling unit (AHU).
- Supply air duct brings in combined fresh and return air after being cooled by condensing unit.
- Relief duct regulates room internal pressure.

During any gas release fire dampers will be closed. Their control is done by platform FGS system. HVAC UCP or platform DCS may have no control on them. At any entrance to an enclosed area like fresh air, return air, supply air, or relief air duct a fire damper shall be installed. Their actuation system is pneumatically powered.

Internal ducts may have manual dampers to allow air flow adjustment by operators. They include a frame with several blades that are connected to a handle. Turning it will partially close or open the available area for air passage.

Air ducts may have humidity absorption filter to control humidity. Drain below it directs condensed water to outside.

Shutoff dampers are installed at entrance of AHUs. Operating and spare units may be turned off/on temporarily. During their shutdown, internal cooled air shall not be able to escape or leak outside. Shutoff dampers are gas tight and their construction is similar to fire dampers. However, similar to AHU these dampers are controlled by HVAC UCP or DCS.

Louvers at entrance of all outside ducts prevent entrance of solid objects. They consist of a metal frame with several bent plates. In front view it seems that the total duct area is closed, but in the section air passage is available.

All these items can be flange connected to each other. This allows easy assembling and removal for maintenance. Access area in front of each item shall be provided.

3.5.2 Potable (Fresh) Water System

Potable freshwater system may range from a large-scale package producing fresh clean water from seawater to a small watercooler using bottled water transported via supply vessels. The large complicated package is used in LQs. Freshwater package installed in satellite platforms may cover a wide design range. Minimum facility satellites only include steel structure housing Xmas Tree on each well connected to the combined manifold and transferring crude to the production platform via natural well pressure. If located at nearby distances to the main complex, they will be serviced by tugboats stationed in the vicinity. Short travel distance enables operators to return to main LQ after finishing routine daily job. Therefore no other equipment is installed on them.

Some other satellite platforms are bigger or installed at longer distances. They may either have a technical room and only enough space for refuge or a temporary shelter. In both of them operators are not residing permanently. They will visit either in scheduled routine inspections or specific task visits. Operators may stay on the platform for short or long terms. These two are different concepts.

In one concept residence on the platform is limited to daily visits, and if bad weather is encountered, one or two nights' stay is possible. In this case only a temporary refuge is designed. Freshwater for small, unmanned facilities may be transported by supply boat and stored in a tank. For 1-day visits to small minimum facility satellites, operators may take cold drinking water in bottles. It may include only a refrigerator for keeping food, a microwave for heating it, and a minimum of two beds and toilet. In this case the freshwater system consists of several water bottles brought from central LQ. Refuge air conditioning may be taken by connecting a damper to technical room space. For safety reasons residence of a single operator on a platform is not permitted. In many regions this design is not accepted. Minimum expected comfort level for operators after a full day's work is much higher than a refrigerator and a bed.

If a limited number of operators need to stay on the platform for several days, then a shelter may be provided. It may include bedroom, toilet, shower, kitchen, food storage, cloth/dish washing facilities, etc. Capacity of this equipment depends on the expected number of operators. In this case it shall have dedicated air conditioning, water heater, and freshwater system. Even in this case, a freshwater system may include a tank that is filled periodically by tugboat. Its sterilization, heating, and cooling shall be designed. In unmanned platforms only a potable tank is installed, which is periodically filled by supply boats. But the quality of water shall be kept properly.

Seawater has a lot of minerals. In reverse osmosis some of the minerals are also taken in. With vaporization (combined heating and partial vacuum) produced freshwater has almost no minerals, which is not good for drinking purposes, so some minerals need to be replaced, which is done by adding some tablets.

For large living quarters required freshwater volume may be too high for transportation and shall be generated offshore. As an example, using 250 LPCD criteria for a 60-man LQ, at least 15 m³/day freshwater is needed. To this volume other consumers like closed circuit cooling systems shall be added. This does not mean that freshwater generator capacity shall be limited to 15 m³/day. Based on operator requirements, it may be decided to generate required water volume within 1 or 2 h or in a long continuous operation within several operating hours. Therefore freshwater package capacity may be several times more than this value.

Freshwater is used for the following purposes:

1. Drinking is a small portion of freshwater (less than 3 L per person, per day)
2. Washing (cloth, shower, dishes, etc.)
3. Diluting sewage water
4. Eyewash
5. Part of the freshwater may be used for cooling water in a closed-circuit system. This part shall have no minerals because they would be deposited and clog the cooling system.

Two issues shall be noted. Freshwater used for cooling systems shall not contain any minerals. This may be achieved using a demineralization package. If freshwater is generated by reverse osmosis, demineralization is needed. If evaporation is used, then water is already demineralized. For a cooling system, demineralized water is okay. Freshwater used for drinking needs to have a certain amount of minerals. Therefore certain tablets shall be added to supply minerals needed by humans.

For eyewash, freshwater is used, but for a safety shower only filtered seawater may be used.

Total freshwater per person, per day may vary from 180 to 250 L. A freshwater system may consist of following sections:

1. Water is taken from a caisson suitably located to enable intake of water free from oil. Caisson opening shall be covered with a rack system to prevent fish entry.

2. In addition to this racked entrance, pump suction will be equipped with a strainer system. The package may also consist of several filters arranged from coarse to fine sizes. Since gradually filters will be clogged, a replacement or clearing system shall be provided. This depends on the vendor. But normally a backwash system using freshwater is the easiest and most practical. Starting a backwash system may be done periodically as per recommended duration in the maintenance manual or via automatic command issued by a differential pressure transmitter in the suction and discharge of the filters or manually by checking a differential pressure gauge.

3. The first step may be injecting hypochlorite or some other disinfecting agent in the suction water. The hypochlorite tank is atmospheric and to resist a high corrosion rate it may be fabricated from glass fiber—reinforced plastic (GRP). A small dosage pump may be used for this purpose.

4. Seawater contains a lot of minerals. The main item is salt. Freshwater may be obtained by reverse osmosis or heating/vacuuming. Inducing vacuum is used to reduce water vapor pressure. As a general rule, in standard atmospheric pressure water will boil at 100°C. In addition, it needs 2260 kJ/kg as vaporization latent heat. Providing this much energy increases freshwater production costs. Inducing vacuum reduces the required energy considerably.

5. Sterilization is usually done by UV light, which has a very effective sterilization effect.

6. The last section of this package may be storage and heating for required purposes.

3.5.3 Sewage Treatment System

The job of the sewage treatment system is to collect gray drains from kitchen, showers, and wash basins plus black drains from toilets and combine them with washing drains. These are all classified as nonhazardous drains that do not contain oil or gas contents. The package then mixes and reduces the particles to small sizes. The main intention is to perform several tasks before discharging to sea:

• Biochemical oxygen demand (BOD) of the system is reduced to a certain level (30—45 ppm). This will enable dumping to sea with no hazard to the environment. BOD is the amount of oxygen that microorganisms present in wastewater consume to decompose organic matter.

It is measured for a span of 5 days in dark environment for 20°C water temperature. This is a measure of water purity. Drinking water may have a BOD of 1–2 ppm.

- Reduce particle sizes to prevent piping clogging and at the same time enable their rapid disintegration after disposal to sea.
- Discharge water is diluted enough to reduce total suspended solids to acceptable levels (30–50 ppm). Increased suspended solids limit water transparency and may impact fish breathing if trapped in their gills.
- Disinfect the accumulation with sodium hypochlorite to enable its safe discharge to sea.

These are performed in several steps.

- Black and gray water are mixed in a tank to have a uniform mix.
- This mixture is disinfected with hypochlorite sodium to kill parasites.
- The solids are fragmented to small sizes.
- This solution is diluted enough to acceptable limits with freshwater.

In living quarter platforms that have a large volume of sewage material, to dilute it to an acceptable limit a large volume of freshwater is required. The main portion of solids may be filtered. This is either burned in an incinerator or kept in another tank to be transported onshore via supply boats. The remaining materials will be discharged via sewage caisson to sea. In many platforms this is a nutrition point for fishes. It is normal to have fish colonies gathering around these points.

Tanks can be made of mild carbon steel. Only hypochlorite tank shall be of suitable material such as plastic or GRP to prevent corrosion.

A macerator will disintegrate solids to a smaller size before being diluted with freshwater. Therefore it shall have a hardened surface to resist wearing of its internal surfaces.

Package capacity depends on the number of personnel on the platform. For unmanned platforms that are only visited by a limited number of crew during maintenance, a minimum capacity will be specified. For others, 100 to 150 L/day for each crew may be used.

3.5.4 Incinerator

In offshore living quarters a lot of domestic, operational, and maintenance wastes are produced. These wastes may vary from food remnants (rice, vegetables, meats, bones, fats, cooking oil, spoiled fruits, dairy products, etc.) to solid objects like metallic cans, broken glasses, plastic containers/bottles, papers, boxes, wooden pallets, etc. After some waste treatment fluid items can be discharged to sea. This treatment may be no more than diluting and adding biocide/antiseptic chemicals. Please refer to Section 3.5.3 "Sewage Treatment System." Transferring all solid parts to an onshore waste disposal system may not be economical. In addition to necessary space allocated to its temporary containers, the effort to store it and possibility of odors/fouling is always a nuisance in LQ. Some items are allowed for offshore burning.

The incinerator package has a burner room (combustion chamber) in which all items are packed. Then electric or fuel-induced flame is applied. To ensure full burning, continuous oxygen supply shall be provided. In the remaining ash only a limited amount of unburned materials are allowed.

The burner room shall have suitable thermal insulation (like fire-resistant bricks, rock wool insulation, etc.) to prevent its external wall temperature from exceeding allowable limits above ambient temperature. Burner room construction materials shall be suitable to resist high temperatures and corrosion in a marine environment. Its smoke exhaust shall be directed to a location away from the living space and far from inlet ducts. It is understood that an incinerator is not burning continuously. Therefore it may be possible to start burning when the wind is blowing away from the quarter direction.

All safeguards for hot surfaces and ignition sources shall be implemented for incinerators.

It has to be noted that gradually more stringent regulations on smoke intensity and type of allowable material for burning are introduced. In some cases an incinerator may be abandoned. For these conditions heavy duty mixers are used, which can combine all food solid materials to a liquid stage. Similar to grey/black waste, the combination can be treated for living organisms and then discharged to sea. Plastic/metal cans and other solid wastes plus wastepapers can be separately packed/compacted to small sizes for transfer to onshore and later recirculation.

3.5.5 Helifuel System

A helifuel system is used in living quarters. In other platforms, although there may be a helipad this package is not provided. It may also be called an aviation fuel system. Aviation fuel is not in continuous usage. If platform is located very far away in offshore waters, helicopters may need refueling. For nearshore platforms helicopter fuel tank capacity is sufficient for return flight.

For safety reasons this package is normally provided in at least two separate skids. It may also be accompanied by a starter unit.

Normally the first skid contains a tank for storing fuel, filter, pump, and control panel. Other than static items like the tank, other mechanical items like pump and filter may need 2 × 100% spare unit. The sparing philosophy may follow general platform sparing approach. However, for small power units like small discharge pumps a complete spare unit is normally provided. This skid is placed away from the helipad in a protected area. Due to high fuel volume that is kept in the tank, it is vulnerable to impacts.

The second skid may contain flow metering and hose reel with fuel nozzle. This skid may be placed in a location, which although protected and away from the helicopter obstacle-free zone, the operator can easily take the fueling nozzle and rapidly approach the helicopter. The path shall be free from obstacles and have no turning points.

Tanks are designed under static condition. However, a minimum design pressure is always set as per project requirement. Tank volume depends on the selected design helicopter. Helicopters used for transporting crew and able to carry 15 persons have relatively large fuel tanks. Their capacity may reach well over $1-1.5 \text{ m}^3$. Assuming supply boats bring new tote tanks in 2-week intervals and that within this duration two or three times refueling has happened, required tank capacity may reach to 5 cubic meters. Tank filling is done by tote tanks. For protection, in some platforms in which helipad is overhanging, the first skid (containing fuel tank) is placed directly below the helipad. Therefore pedestal crane has no or very limited access to it. Chain blocks connected to lifting eyes permanently installed above the tank or material handling trolleys may be used to transfer tote tanks from the location left by the pedestal crane to nearby fuel tank. Necessary safety precautions shall be undertaken both in transferring fuel tote tanks and during tank-filling operation. Small manual pumps can be used to transfer fuel to tank. All tanks shall have proper venting to a safe location. They shall have level gauges and drain/overflow nozzles.

The first skid shall have a suitable drip pan to collect all spilled fuel from the tank, pump, and filters during all operation and maintenance activities.

To enable helicopter filling within a short period, 5-min refueling duration is selected. By this assumption pump capacity is calculated to be above $12 \text{ m}^3/\text{h}$. Pump's electrical driving motor and all its junction boxes, control panel, and electrical connections shall all be explosion proof.

Filter unit is placed downstream of the pumps. For heavy fuels, at inlet and outlet of filters a differential pressure transmitter is placed. If volume of trapped solids increases, it may hinder free flow of liquid. It has to be considered that aviation fuel is normally a clean/light fuel. For this reason in many cases using pressure gauges at inlet and outlet will be sufficient. It is again repeated that in offshore platforms a helifuel system is not a frequently used package. Water separation in filters may be needed for a helifuel package. Particle sizes above 1 μm and water content above 15 ppm may be removed.

Filter cleaning can be done by backwashing with liquid or compressed air. During this operation it shall be separated from the system. Therefore a bypass route is needed in the package design. Since this is not a lengthy process it can be performed in periods during which platform crew are sure a helicopter will not land.

Control panel can also be local and only a few start/stop or general fault signals to be issued to main control system. Other signals or audio/visual alarms can be local. They may include tank fuel level (high and low), internal pressure, pump suction and discharge pressure, filter high-differential pressure and fuel flow to second skid.

Second skid (dispenser unit) may include metering system, hose reel, and fueling nozzle. This may also have its own control panel.

Helifuel construction package material shall be able to withstand offshore environment.

3.6 PACKAGE PROCUREMENT

Procuring a package is a complex task and requires cooperation between project control, procurement, yard, quality group, and engineering team. In the engineering team this task involves several disciplines. Not only each engineering discipline has to purchase several items but also they shall cooperate in purchasing other discipline packages. Procured items may range from simple bulk items to complicated process packages. Each group wants to get optimum conditions for his duties.

- Engineering team has to handle technical aspects and wants minimum deviation from project specifications.
- Commercial department wants to minimize payments.
- Fabrication yard needs packages on time to fit in fabrication schedule.
- Quality control (QC) group has to ensure good-quality package is installed.
- Planning team has to ensure general project plan is not deviated from.

Obtaining optimum conditions for all involved parties may not be feasible. Vendor normal production line may have some deviations from project requirements. Full compliance with project specifications or time-table may have additional cost/time impact. Vendors normally comply with minimum QC requirements as per code. A project's additional QC requirements may impose cost/time impacts. A final compromise may be reached between different parties based on the decisions that may be made by project management team.

In procuring a package the following main steps are noted:

- Engineering team finalizes data sheets, specifications, and other required documents that shall be included in material requisition (MR) for quotation. This step may need multiple correspondence/meetings with EPCIC contractor, certifying authority (CA), and client team members. Main contract documents, normal engineering practice in the region, minimum code requirements, etc. are the basis to close this step.
- Commercial department receives quotations from vendors. They may send invitation to vendors in client approved vendor list (AVL) or those that were involved in their previous projects or simply contact new manufacturers.
- Engineering team finalizes technical clarifications and issues technical bid evaluation (TBE). It is necessary to issue consolidated comments from all disciplines. To expedite this process, after receiving vendor responses unclear points can be discussed in a technical meeting with short-listed vendors. Using Internet communication facilitated video meetings with foreign vendors is possible at low cost and with short notice time.
- Commercial team uses TBE to finalize negotiations with approved vendors and prepare commercial bid evaluation. At this stage they have to finalize all scope of supply, cost, time, and delivery conditions. Services after package arrival in yard, responsibilities of each side to supply/replace missing or damaged items, commissioning and hookup technical services, unit rates for onshore or offshore services, etc. shall all become clear at this stage.
- Based on project manager decisions, engineering team issues MR for purchase and financial team (or commercial team) issues purchase order (PO) for selected vendor. This may not always be for the minimum cost or vendor with minimum technical deviations or minimum manufacturing duration. Each of these criteria may become the governing criteria at a certain stage. In normal conditions the project manager's decision is based on a general evaluation of pros and cons of acceptable quotations.

- After receiving PO or AP vendor starts generating vendor data to be reviewed by necessary groups. Parts of vendor data need to be reviewed by planning team, parts by commercial department, major parts by engineering team, maintenance parts by operators, etc. Each department needs to ensure intended manufacturing scheme matches their plan.
- After a certain group of engineering documents are approved, vendor starts procurement of necessary materials and manufacturing of the package. Some deviations may become necessary at this stage. For example, a material that was already approved may need longer supply time and does not match package fabrication schedule. Or detailed calculations may show a vessel size that was understood to fit in the package general arrangement is not adequate from a technical point of view. All these changes need the engineering team's careful review and adjustment in design.
- Before start of manufacturing, TPA, EPCIC contractor, and client may have a meeting with vendor (or visit its workshop) to ensure manufacturing and QA/QC procedures, necessary machinery, know-how, and expertise are available. This check may be much easier for reputable vendors that are already in client AVL. However, some degree of cross-check is necessary. For example, if a vessel with 170-mm-thick plate shall be fabricated, rolling machines with necessary capacity shall be available. Otherwise, vendor shall introduce capable subvendors. Outsourcing parts of the job may be necessary but this has cost/time impact. Unless the two vendors have long-term cooperation experience or they are subsections of the same plant, any outsourcing (whether small or large) is a stand-alone contract. Therefore all items necessary for a contract shall be followed. This takes time and has some cost impact.
- Inspection and test plan is one of the major documents (generated by vendor and commented by TPA, client, and EPCIC contractor) that defines who and at which stage of the manufacturing can and shall review or approve outcome product. Stage-by-stage and continuous inspection prevent defects being hidden by covering sections. In some conditions EPCIC contractor may assign some of its own responsibilities or approval authority to TPA. Normally EPCIC contractor representative may attend manufacturer workshop only at certain stages, but TPA representative may have daily presence.
- Before assembling a package, its different items shall be properly inspected. For example, a vessel may need hydrotesting or postweld

heat treatment (PWHT) before installing on the skid. Some activities shall be closed before others. For example, all weld tests shall be finished and repairs (if any) shall be performed/closed before placing vessel in PWHT furnace. Release certificate of each stage to enable start of next stage shall be carefully examined and archived.

EPCIC contractor shall consider contract with manufacturers as individual projects. Full project control, production expediting, QC measures, etc. shall be exercised. Manufacturing shops may be scattered all over the world. Therefore assigning resident representatives may not be feasible for all packages. For a contract to manufacture a ship, barge, platform, and similar items, assigning resident representatives to handle all engineering, financial, and executive issues is acceptable.

One of the most important stages before dispatching a package to EPCIC contractor yard is performing a factory acceptance test (FAT), which is normally witnessed by concerned parties. For complicated packages like turbogenerators, sweetening or dehydration packages, the presence of all concerned parties will help to find shortcomings. In some cases in which coordination may be time-consuming or costly or a package is small, TPA may be allowed to participate on behalf of EPCIC contractor. FAT allows checking different parts of the package, their operation and conformity. Before starting FAT, all release certificates shall be carefully evaluated. This ensures intermediate stages have followed necessary QC measures. FAT approval does not (and shall not) relieve manufacturer from its obligations. Some of the factors that necessitate continuation of vendor responsibility follow:

1. Some packages (like control system) have numerous signals. It is not possible or feasible within a few days to check each of the signals one by one. Only during loop check in commissioning period signals are tested one by one. If vendor has made a mistake in transferring a few of the signals, or the logic triggering necessary action is wrong, vendor shall rectify them at its own cost even after arrival at EPCIC contractor yard. This mistake may vary from transferring wrong signals, neglecting some of the signals, transferring signals with wrong power rating, to transferring them with the wrong consequence. In chapter "Disciplines Involved in Offshore Platforms Design" an example was given for wrongly connecting flowmeter cable instead of temperature. Control system for this platform will continuously shut it down because it will always detect high temperature in normal condition. The author has seen cases in which signals from level transmitter A started (or shut

down) pump B instead of pump A. It is understood that this problem was fabrication yard mistake and not related to the control system manufacturer.

2. Some shipped loose items may not correctly match project requirements. For example, a power cable size may be sufficient for transferring normal electrical current but may experience some raised temperature due to high starting current. This may increase its operating temperature, which may cause malfunction in hot, humid environment.

3. Some mistakes in QC may be revealed at a later stage after package delivery or even installation. It is a routine that QC only controls existence of QA procedures and checks their usage by manufacturing personnel. QC representative only inspects randomly selected items. For example, after ensuring electrode selection, storage and transport, welder qualification routines, welding procedure specifications are all in place and operative, QC may select 10% of welds for X-ray tests. As another example, assume there are 100 flanges in a package. QC representative may take the following steps:

   • Control that manufacturer has a procedure for procuring flanges.
   • Control that procurement is from manufacturers with established QA/QC system with valid ISO certificate.
   • Control that flange manufacturer QC department or a TPA has approved supplied flanges.
   • After these steps, QC representative may only select a few of the procured flanges for inspection.

   It is clear that implementing these steps reduces the possibility of damaged items. However, it has to be kept in mind that even after all these efforts, items with some defects may still appear in the final package. Replacing these items at any time, without additional cost to purchaser, shall be in vendor SOW.

4. In some packages transferring monitor and control signals between UCP and ICSS may not be as per consultant or project plan. Vendor may have an established design with already clear lines of communication. It may require considerable change in software or hardware to adapt it to project requirements. Since final guarantee is to be issued by manufacturer, in certain conditions vendor may decide to follow its own customary design to avoid further disturbance. This may not be easily clear at FAT. Only after commissioning when they want to connect actual signals may the missing/neglected items become clear. Therefore vendor responsibility shall extend to this duration and after.

Some vendors may perform these changes to save their reputation or when they expect to get future contracts. In spite of that, EPCIC contractor shall retain a portion of vendor payments after ensuring these required modifications are performed. Based on the previous discussion, while participation in FAT is necessary, it shall not be considered as replacement for vendor responsibilities. Both EPCIC contractor and vendor shall consider that any remedial action in EPCIC contractor yard is much more costly than in vendor manufacturing plant. Similarly, repairs offshore will be much more costly and time-consuming than in the yard. Only mobilizing a supply boat or helicopter to transfer people, equipment, and spare material to offshore platform may cost more than total repair or replacement activities of the whole package in the manufacturing shop.

3.7 BULK AND LONG-LEAD ITEMS

Offshore projects are awarded to EPCIC contractors. There are a large number of interface issues between E (Engineering), P (Procurement), C (Construction), I (Installation) and C (Commissioning) sectors. Clients try to avoid as much as possible these interface headaches. They assign all responsibility to contractors. Each team is responsible for its own specialist activities. The engineering team cannot procure equipment. Similarly the procurement team cannot fabricate topsides. The same applies for all these five main sectors. Technology, personnel, methods, and tools to perform these activities are different from each other. Even inside a team there are several specialist tasks.

Chapter "Disciplines Involved in Offshore Platforms Design" discussed engineering disciplines and described their main documents. For example, it was explained that in the mechanical discipline, an engineer experienced in rotary equipment may not be a static equipment expert. A similar issue applies to other teams. In the procurement team a negotiation/contract expert may not be so familiar with forwarding and custom clearance rules. In the fabrication yard a copper nickel pipe welder is not so good in stainless steel welding, let alone electrical cable installation.

The point is that only the combined efforts of all these very different activities may lead to a completed project. Any error, deficiency, and delay in any section impacts final outcome. Impacts may range from negligible to very large. For example, an engineering error may be rectifiable with a small reassignment (like reissuing a document) or it may cause long delays in final platform handover due to reordering a complex package or a large sum of shortage materials.

Each team has its own plan. Internal interfaces shall be carefully studied when issuing a plan. For example, detail design piping material takeoff (MTO) can be issued only after layout is determined, line sizing is done, and pipe stress analysis is finished. Compressed air system material requisition can be issued only after system specification and data sheets are ready. It has to be noted that data sheet defines compressor capacity, which is only identified after required air volume of consumers (at least main consumers) is known. This MR is issued by the mechanical discipline but its main consumers are instrument valves. This shows the importance of internal interfaces. The same relation applies to the interface between the five different sectors.

Each fabrication yard may have its own priorities and sequences. Factors like availability of experienced workforce, machinery/equipment, bulk items (exotic and normal), approved procedures, budget, etc. may affect project execution planning. For example, yard managers do not keep expert welders continuously if they do not have a project. Some heavy lifting cranes may be rented for short periods. Offshore fleet vessels are assigned only for a specific timetable. In some cases they shall be booked well in advance.

In addition to the above limitations, they shall consider project logical sequence. In an exaggerated term they cannot plan to erect a final level before first and intermediate levels are erected. The author is not familiar with yard work sequence, but a feasible schedule for constructing topside may be listed as follows:
1. Fabricate main joints of each elevation on temporary supports
2. Fabricate main and secondary structural members and install deck plates
3. Fabricate and install piping and equipment supports
4. Fabricate large-size piping spools
5. Install equipment and piping spools
6. Fabricate and install columns and braces
7. Fabricate and install next level
8. Fabricate and install remaining piping
9. Fabricate and install cable route, install electrical and instrument items, and lay cable.

There are many prerequisites to accomplish this schedule. For example:
• AFC main, secondary, and tertiary structural member drawings and materials are needed to perform steps 1 and 2.

- AFC piping arrangement drawings, models, and isometrics in addition to piping bulk material are needed before items 3, 4, and 5 are possible.

It is obvious that all items need not be ready simultaneously. However, when a portion of work starts, related prerequisites (documents, materials, workforce, and machinery) shall be ready. The first two items are provided by E and P team. Yard itself shall provide approved procedures, qualified workers, and machinery.

Successful and on-time completion of a project consists of successful and on-time completion of a series of tasks so that each one completes a portion of the job. Each team shall perform its share correctly. For example, to start and finish main structural members:

- Engineering team shall finish permanent and temporary analysis, issue-related drawings and MTO, and obtain client and TPA approval to send AFC drawings to yard.
- Procurement team shall receive quotations for structural materials based on MTO/requisition issued by engineering, obtain their technical approval, short list selected candidates, obtain management approval, allocate budget, issue PO, expedite vendor production, and transport material to yard.
- Construction group shall obtain TPA approval on fabrication procedures, perform necessary qualification tests, and mobilize qualified man power, equipment, and machinery.

It is clear that any delay or shortcomings in any part of this chain may delay the final product. It can be said that the main portion of procurement cost is for package equipment. However, the importance of bulk items is not less than packages.

After packages are installed on the topside, some work front to connect their utility including electrical power, control instruments, compressed air, fuel, etc. is opened. The most favorable schedule for the construction yard is to receive packages at a time when access from the top is still available. In this condition they can easily lift install it. If they have to install the package by skidding, some structural members or piping that may block the passage shall be removed and later reinstalled. This rework has cost/time impact for them.

But this may not be necessarily the best case from a procurement point of view. Based on the signed contract between procurement team and manufacturer, each package has a guarantee period. Duration depends on contract terms and conditions and package complexity. Durations like 24 months after delivery from manufacturer site to 12 months from installation on the platform are common. Assume total duration of a project is 36 months. If a package located on cellar deck is installed at month 20, then before offshore commissioning its guarantee may expire.

Bulk items on the other hand open a considerable work front. Depending on type of material, sizes and total volume, manufacturers' schedule, etc. bulk item procurement may be easier than equipment. Depending on bulk item diversity and type of material, manufacturer eagerness to obtain the contract varies. For example, procuring 1000 tons of carbon steel 24″-diameter, 25-mm-thick A106 grade A pipe is very attractive for manufacturers and is much easier than procuring 350 tons of CS, 150 tons of copper nickel, 200 tons of SS, and 300 tons of Inconel of different sizes in an order. For the second condition, the manufacturer may not be willing to assign its production line to small jobs. Maybe referring to a stockist is a better solution for the second case. The point is that after bulk material arrival to the yard, activities like spool fabrication, installation, and QC activities, which are essential for project completion and consume considerable man power, can proceed.

EPCIC project team obtains each sector's plan and combines them to a realistic project general plan. While each sector issues planning from start to end, project planning team may start from the end and move backward. Yard cost is much more than engineering. Therefore their needs and minimizing their cost have higher priority. Project team obtains "Ready On Site" schedule from yard and proceeds backward to define intermediate steps like: "custom clearance," "shipping," "delivery ex-works," "FAT," "manufacturing," "vendor engineering," "issuing PO and operative LC," "issuing TBE and MR for purchase," "issuing MR for requisition," etc. Only main intermediate steps are listed. For example, before issuing MR, related data sheet and specifications shall be issued in IFC, IFA, and later AFC state. Assigning necessary duration for each step with contingency may sum up to a total time much more than project contractual duration. Some steps may be shortened. For example, instead of shipping by sea routes, truck transport or airfreight may be used. Or manufacturer may be asked to assign overtime workers. It is understood that each expediting approach has its own costs.

Several trials may be necessary to finalize the project plan. Based on items that happen on the project critical path, changes may be implemented in other sectors' plan. For example, engineering team may be instructed to expedite issuing certain MRs earlier. Procurement team may be instructed to allocate available budget to certain packages. Construction team may be instructed to assign more man power (or work overtime) on specific job items, or change installation plan for certain equipment. This way the project team fits each sector plan in the general project plan. In addition to original trials, the project plan is continuously revised.

Almost in all projects manufacturing duration for certain complex equipment may become a governing factor. Long-lead items may be classified by either EPCIC contractor or client (end user). This term refers to the equipment (identified at the early stages of a project) to have a delivery time long enough to affect the overall project completion schedule. Typically, large compressors, turbine generators, control systems, ESD valves, pedestal cranes, etc. may need more than 1-year delivery time. Adding engineering and procurement formalities duration plus transport, these items may need half or more of total project duration.

In most cases capacity and size of main equipment (with some contingencies) is defined in front-end engineering design (FEED) stage. Either from previous experience or equipment definition, client (end user) has an understanding of items affecting total delivery. There is no universal, unique definition for long-lead item (LLI) identification. It depends on project total duration, budget, location, complexity of equipment, expected revenues, etc. If daily export value of a platform is more than US $1 million, then it is natural to pay US $5 million more to start production 1 month earlier.

A course of action taken by end user and EPCIC contractor for LLI may be different. For both of them, LLI packages will not follow the same supply chain as the other equipment. In normal procurement schedule by EPCIC contractor, LLI items may add well over 1 year to the overall project schedule. To avoid this delay, either of them may adapt specific measures.

End user may identify LLI at FEED stage. Then they start bidding process in parallel with EPCIC contract tenders. The engineering company performing FEED may support client in this process. Client selects LLI manufacturer before awarding EPCIC contract. They finish financial negotiations and issue "letter of intent". Selected EPCIC contractor is instructed to issue PO in favor of selected LLI manufacturer.

This process saves considerable time. However, since detail design is not completed, inevitably some changes may happen. The setback for this process is that LLI vendor has upper hand. They may claim large sums of money for any change or engineering team is forced to adapt detail design to vendor production. Adaptation may range from changes in structural support, platform layout, utility consumption, or interface with other packages. Impacts may only be weight, power demand, fuel capacity increase, etc. or only added cost. EPCIC shall study carefully to decide whether engineering team or vendor shall be instructed to adapt. This decision will be based on impact evaluation study.

A solution for these kinds of disadvantages is a detailed FEED for these packages. It is understood that basic design and FEED cannot identify details as in detail design. For example, assume in a project gas turbine generator is the selected LLI. Diesel generators normally do not happen in critical path. Generators' good performance is in a specific range of their nominal power. For example, if a generator designed for 2 MW is continuously working at less than half MW, it will require frequent maintenance and overhaul periods. Therefore before ordering, electrical power demand shall be determined as accurately as possible. Many of the consumers will be determined at later stages of detail design. Again it is understood that main consumers' size impacts generator selection. Using $3 \times 50\%$ scheme helps both in reducing costs and operating conditions.

LLI packages have their own UCP.

First of all they are complex packages. Therefore manufacturer shall be responsible for all operation and control issues. Vendor shall not be allowed to state my package malfunction is due to a wrong signal received from your control system. Second, platform control system is ordered at later stages of design. EPCIC contractor cannot wait that long for LLI PO.

If EPCIC decides to treat an item as LLI, they may instruct procurement process with IFC documents. At the same time they may ask vendors to comply with engineering changes that are clarified before vendor starts fabrication process.

In both cases the vendor can work in parallel with other engineering and construction activities to achieve an early delivery date and save time.

3.8 BROWN FIELD ENGINEERING

Brown field engineering covers a wide range of activities. It shall be distinguished from routine maintenance work. Although in the majority of maintenance work established and well-practiced procedures are followed, in some cases construction engineering (to some extent) may be required.

Brown field engineering may be required for several reasons. In all cases the engineering team shall consider offshore working has heavy costs.

Therefore as much as possible projects shall be fabricated in packages, allowing single-lift installation. Offshore work share shall be reduced as much as possible to a few weld lines at tie-in points and QC activities.

Brown field engineering may vary from small changes in a single platform to very large changes in all platforms of a complex or even adding new platforms. Expected increased revenue, water depth, proximity to onshore, new development costs, and other financial factors play a major role in decision-making. As a tentative list, from minor to major, we can state the following reasons that may necessitate brown field engineering:

- Field instruments (gauges, switches, or transmitters) may have lost their accuracy or may be damaged. Due to rapid technology changes an exact similar type may no longer be available in the market. Either power or transmitted signal specification may change. To adapt power cabling or UCP/platform control system to new signal types, some detail design may be needed. For this type of work some operators may have their own team or sign a man power supply contract with a consultant.
- Platform equipment like pumps, compressors, generators, etc. may not function properly. Although periodic overhaul improves their performance, after design life they may need to be replaced. To procure, install, and commission new equipment, considerable engineering studies in several disciplines may be required. This type of service can be done by operators' engineering team or by general man power supply or by small engineering services agreements.
- Updated safety or environmental regulations may impose some small-to large-scale changes. For example, cable type may change or new detectors may be required. These are very small and can be done by operators. Some middle-sized changes may also occur. For example, allowable discharged oil ratio may be reduced. It has actually reduced from 40 to 15 ppm in recent years. To cater to this change a new oily water package may be installed. Although each vendor may have its own patent, a credible solution may be to increase the number of hydrocyclones, recycle part of oily water, add absorbing filters, etc. Large-scale changes may also be anticipated. For example, allowable H_2S level in discharged water may be limited. In this case, H_2S stripper package may be required. This package may require considerable power that may not be available on the platform. In this case additional packages to supply necessary power shall also be installed. This may range from an extended deck to a new platform.

- Decommissioning of existing platforms after their design life is finished or damaged in unexpected storms needs specific engineering services. This is not considered as brown field engineering. However, if topside is totally removed and a new one is installed, it is categorized as brown field engineering. Damages due to storms, accidents, war, etc. fall in this category. In some cases substructure is saved and only topside is removed.
- Reservoir conditions may change. This may cover a very wide range of characteristics. Changes in GOR/water cut and presence of sand or heavy hydrocarbons are some of them. Changes in separator sizes may adjust volume dedicated to each water, oil, and gas compartment. Coalescer size, dehydration package capacity may change to remove added water. Oily water separator package or type may change to increase its efficiency. A new gas pipeline may be added to export gas. Increased gas volume may be dried and injected to boost oil production. Sand management system can be added to remove sand from production trains to prevent plugging or erosion. Solvents can be injected to prevent slug formation.
- Reservoir pressure may reduce or even stop production. EOR methods like gas/water injection or gas lift may be employed to bring production volume to profitable margins. Otherwise, platform may need to be decommissioned. Implementing EOR plans require a great amount of new equipment and a lot of electrical power for compressors, injection pumps, etc., which inevitably leads to new platforms. In addition, adding new equipment may require continuous presence of operators. This may justify constructing a new living quarter or increase capacity of existing LQ.
- New studies may show reservoir has more production potential than originally expected. New platforms may have to be constructed with tie-in to the old one. This may be a satellite platform linked by infield pipeline or a bridge linked platform, which includes new horizontally drilled wells.

As can be seen brown field engineering may vary in extent and complexity. Information about exiting platforms is vital in successful implementation of the project. This may be gathered in several ways.

If considerable engineering activity is expected, owner sends its team to investigate actual conditions. The outcome of this exercise shall clearly identify SOW for bid.

Data gathered by operators during production including composition, water and hydrocarbon contents, presence of sand, wax, asphalt, etc. shall all be reflected in bid documents.

Either owner or successful bidder may prepare a 3D model or update existing one. Laser-scanning technique is used extensively in brown field projects. This technique takes several photos from different angles of each point and combines them in a 3D model. Accuracy of this method is very high. Spools can be fabricated with several millimeters' accuracy.

As-built documentation is an inseparable part of documents needed to start engineering. Compared to detail design documents, many changes may be implemented in construction of platforms. These changes may include structural, electrical, instrument, piping, and other disciplines. Each one of them may affect final outcome. Size and location of structural elements, piping arrangement and sizes, electrical power distribution, and control system signals all may have changed. Brown field engineering shall be based on as-built documents.

The consultant responsible for detail design of a green field has many interfaces with EPCIC contractor. For brown field projects these interfaces are twice as much. Like containerizing goods for marine transport, brown field projects need close cooperation between all E, P, C, I, and C sectors to define green pieces/golden joints' location.

PART 2

Management, Budgeting and Scheduling of Offshore Platforms Design

CHAPTER 4

Balancing Between Client and Task Force Engineers

Concepts discussed in the second part of this book are very general. They are not solely intended for oil and gas consultants. All consultants in different engineering fields may use these ideas.

An engineering team is comprised of human individuals. With recent ease in workforce movement between countries, all sorts of people from different races, ethnic groups, religions, nationalities, political attitudes, etc. may be present in one company. Other than technical qualification of each member, the team must be able to work with each other in uniformity. A team of average-level engineers who respect each other and know how to properly manage interfaces will produce better results than a team of technically superior individuals who are working in isolated islands.

Once more I repeat that men and women alike can serve on an oil and gas engineering team. Throughout this book when referring to an engineering team, both he/she is used or reference to gender is removed. If anywhere only one gender is mentioned, it means both men and women, without any discrimination.

Within a family even brothers and sisters may quarrel. This is not a problem. Two colleagues may have heated disputes and arguments about a technical issue or even have widely separated sociopolitical attitudes. This is acceptable. The point is they shall be able to work with each other tomorrow. Another technical issue may need their close cooperation the next day. Part of the information required by a second engineer and not transferred properly (completely and on time) by the first one may be a determining factor in the second engineer's work.

No matter how many sophisticated rules and procedures are set in place, the first prerequisite for their effectiveness is that individuals want to follow them. Any engineer can always find a legal excuse for not fulfilling a requirement. Out of the office it is everybody's private life. Inside the office, professional ethics shall come first.

The responsibility for coordination between engineering disciplines rests with the engineering manager, and coordination between members of each

Practical Engineering Management of Offshore Oil and Gas Platforms
ISBN 978-0-12-809331-3
http://dx.doi.org/10.1016/B978-0-12-809331-3.00004-1
349

discipline rests with the related lead engineer (LE). Coordination includes at least:

- Transferring required data that impacts other disciplines or engineering documents
- On-time and complete commenting on interdisciplinary circulated documents
- Following an agreed time frame to allow other disciplines to progress for their portion of the job
- Transferring required data within each discipline

Responsibility of coordination/interfaces between the engineering team and client, contractor, construction yard, third party authority, marine warranty surveyor, etc. (if applicable) is with the engineering or project manager.

For many consulting projects direct meeting between two engineers from client and consultant is not allowed. In some projects it is allowed only if the LE has already been informed. It is a fact that in these meetings some unwanted/uncontrolled information may be transferred. This may lead to difficulties. The author believes that direct contact between client and engineering team members shall be allowed with LE prior information. However, the consultant is only responsible for documents issued via official transmittals.

A — Transferring Data Between Disciplines

Engineering work is a multidiscipline activity. A lot of data needs to be transferred between disciplines to enable a final acceptable outcome.

For example, the piping discipline is responsible to generate the platform 3D model. To do this they need information from other disciplines like:

- Structural-steel model including primary, secondary, and in certain cases tertiary members. The structural model may proceed in several steps. Normally main primary members are designed first, and later secondary and tertiary members are fixed. Many of the software programs allow data transfer without requiring to model from scratch. It is the consultant manager's responsibility to procure and install software that allows data transfer.
- Some structural model information (which at first glance may seem trivial) like star plates around leg joints, if not transferred to a 3D model may cause several clashes during yard fabrication. Their dimensions become clear during the miscellaneous design stage. This is long after

main and secondary members are designed. Until that time piping can't stop working. In the absence of accurate data they assume some clearances around critical joints.

- The architectural discipline defines accommodation, office, and technical areas. Different walls have different fire ratings. While wall location is defined by the architectural group, their rating is defined by the safety group. Piping requires exact wall, window/door dimensions, and arrangement. Photos have been circulated on the Internet via e-mails showing doors opening to nowhere or stairs with a closed wall in front. I don't know if they are real or prepared by Photoshop or similar software. One example is shown in Fig. 4.1, which I hope is the creation of a humorous mind and not an actual project.
- The mechanical discipline is responsible for the majority of package information. However, model generation starts from the beginning phases of the project. Even before receiving vendor quotations, the mechanical discipline shall provide preliminary information like package dimensions, arrangement, number/type/size of tie-ins, required utility, etc. from a previous database.

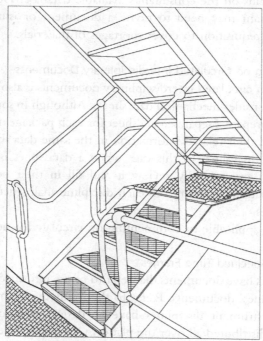

Figure 4.1 Stair leading to nowhere.

- Electrical and instrument disciplines shall provide information about number of switchboards, relative sizes, access requirements, and other information.

The above is only an example for a piping 3D model. More or less the same is applicable to others. For example, structural secondary members shall provide support at intended locations for the equipment and access platforms. The structural team can't leave empty space below equipment and declare that the job is finished. At the same time, until vendor design for a pressure vessel becomes available, the mechanical team can't provide accurate data for saddle location. Does this mean that the structural team shall not issue material take off (MTO) until receiving mechanical data? Absolutely not.

Philosophers can discuss for an eternity whether the egg or chicken was created first. But we engineers have to build platforms to provide oil and gas for human beings' energy source within a specific time span. We can't wait to find all the minute details and information first and then start fabrication. A structural team shall order steel material as one of the first requisitions in the project. This is all done by certain assumptions. Accuracy of the assumptions depends on the consultant's available database. However, even the best consultant may need to revise a document or issue a complementary MTO/requisition to cover shortages of materials.

B — Commenting on Circulated Interdisciplinary Documents

Commenting on circulated interdisciplinary documents is also an important requirement. Consider mechanical data sheets. Although in some cases each discipline may prepare a separate data sheet for each package, in many cases all disciplines fill their related information in the same data sheet issued by the mechanical discipline. In this case when a data sheet is issued other disciplines shall very carefully review it and fill in their portion of the required information. In doing this each discipline shall strictly follow the project deadlines.

The same is applicable to commenting on circulated vendor data.

C — Following Specified Time Frames

Many disciplines have documents that refer to or use information contained in other disciplines' documents. Refer to the following examples:

- Electrical/instrument disciplines have to define hazardous ratings of their items distributed all over the platform. In some projects hazardous area classification drawing is prepared by the safety discipline, which may not be so much used by safety itself. This doesn't mean they can plan to prepare this drawing at a later stage of the project. It shall be prepared very early.

- Mechanical discipline is responsible for the majority of material requisitions. Normally they refer to several documents generated by other disciplines like general criteria for electrical, control, and instrumentation inside packages, painting specifications, accelerations acting on individual equipment, and environmental data. Material requisitions shall be issued as soon as possible to enable the project team to obtain quotations for technical evaluation. Therefore all the so-called accessory documents shall be prepared within a proper time frame as per project requirement. Packages that have a long manufacturing time are called long lead items (LLIs) and shall be issued as early as possible.
- Almost all disciplines refer to piping and instrumentation diagrams (PIDs) to define their equipment capacity, rating, etc. It is notable that generating this document process relies on the information submitted by others. Therefore the process shall target to issue PIDs very early.

D — Transferring Required Data Inside Discipline

Many documents can only be prepared after others are to a certain level of completeness. They shall start with certain assumptions and upgrade information at each stage. For example:

- In permanent analysis of jacket design topside information is needed and vice versa. Due to high volume of work, two or more engineers will be assigned to each part. In jacket analysis, topside members may not be so accurately modeled and loaded and besides their stress evaluation will be skipped. The same applies in topside analysis for jacket members. However, both require other teams' modeling information. Therefore, at issued-for-comment stage both will proceed with a relatively coarse modeling of the other team. Later at issued for approval and approved for construction, models will be upgraded.
- In generator sizing, electrical discipline requires loads consumed by all motors. Therefore, electrical team may internally define generator size with some assumptions and upgrade it with more refined data on the load list document.
- In the piping discipline, the stress calculation group needs piping general arrangement (GA) and routing; however, the GA group may change the piping route based on new vendor and equipment layout information.

In all these cases it is necessary to upgrade the model with new information. The main point is that this information shall be provided with a predetermined plan to avoid excessive rework design and mistakes in procurement or construction.

E — Interface Manager

For small projects the engineering manager may take this role and for big projects they may assign a member of the engineering team. Other than documents that will be issued as per schedule (defined and agreed) in the master document register, there is much data that shall be distributed properly. Some data shall be supplied from the client side.

An important aspect of this coordination is that different sides may have some conflict of interest. For example, the construction yard may want lighter structures while the client wants heavier structures. The same is applicable to fewer nondestructive testing (NDT) or in the NDT requirements fewer radiographic tests, etc.

It is the engineering manager's responsibility to bring these diverse requests close to each other while following minimum standard requirements and ensuring quality of work.

For an engineering team it is essential to be a neutral judge and maintain professional honesty and dignity amid various requests.

4.1 CLIENT REQUESTS

In a majority of cases projects are given on a lump sum basis. Although at bid stage both sides (client and contractor) try to define project requirements/specifications as accurately as possible, it is impossible to define everything. Therefore during the engineering phase and later in the construction stage many disputes will arise. Managers from all sides shall understand that in many cases they shall compromise on certain items. This will prevent project stoppage and help it proceed faster. Profits of each single day of production ahead of schedule will far exceed many of the disputed costs. Change requests/orders from the contractor's side shall be reviewed keeping this issue in mind.

Clients normally require the best of the best at the beginning of the project. Unfortunately later during the production stage the operation team may neglect many items. These requests apply to both individual team member qualifications and equipment specifications.

4.1.1 Team Member Qualifications

Normally the client retains the right to accept or reject some LEs. To do this, in the engineering contract a clause is placed to review LEs' CVs and then interview them in a kick off meeting (KOM) or place certain minimum qualifications on their academic background or experience. This is a legitimate right. Unfortunately, the author has encountered several cases in which the rejection was based solely on the nationality of the respective engineer and not his or her knowledge of the job in hand.

Some clients may have some predefined requirements. For example, 8 years of experience in offshore design for an LE. Client engineers shall consider the project from a more general point of view. That is why engineers who have served several years in design, construction, or operation perform better as client team members. Previous experience in these fields enables him or her to better evaluate engineering team members in the interview or review their documents. Engineers who directly join a client team after graduation from university can't help client interests. As a client team member they have many contractual rights that may be misused if not accompanied with actual experience. If evaluation of an LE is based only on individuals' work experience duration and list of projects in his or her resume, the outcome may not be a perfect evaluation.

4.1.2 Project Specifications

An unexperienced client team member may have a wrong impression from project minimum specifications. This may range from material property to equipment capacity, design pressure, temperature, flow, etc. He may insist on using a certain material (alloy) that may only increase cost without affecting quality or due to longer production duration delay project completion date. One common request is to use exotic material in the flow line and providing more automation in manned platforms.

Client team members shall understand that a single issue doesn't have a single solution. As the saying goes, "there are many ways to skin a cat."

Nobody starts a project with an intention to deliver it later than contractual deadlines. In spite of this good intention, in many cases delay becomes inevitable. Therefore project managers from both sides gather evidence to prove the other side is responsible for delay. It is a fact that in a majority of cases a single side can't be blamed with full responsibility for all the delay.

As a general conclusion both sides shall understand that leaving an issue open will harm everybody. A decision shall be taken on each issue to enable both sides to proceed with the project. A Persian saying is, "A mad man may drop a rock in a well which requires forty wise men to take it out." This means an unreasonable request by an inexperienced engineer may take considerable time and effort from all sides to rectify the problem. Agreement/compromise in the long run benefits all sides. This is a matter of give and take. Normally client shall accept to have the biggest share of give. Unfortunately bureaucratic regulations on the client side may be the biggest obstacle in achieving this goal.

4.2 ENGINEERING TEAM REQUIREMENTS

Although in most consulting firms engineering members are already working in the company and are simply assigned to different projects, in many cases it is necessary to recruit new engineers. This may be due to relocation/quitting of present engineers, receiving new projects more than the present capacity of consultant, or the necessity of providing a specialized study that is not available in the company.

Recruiting is normally done by the human resources (HR) department using announcements on the consultant's Website or by recruiting agencies. In all cases engineering managers or LEs shall actively participate in the recruiting process. After all, a new member will be added to their team. They must be able to tolerate each other at least 8 h a day. For HR if someone has a university degree and has had certain years of experience and asks for a low salary that is enough. But this is not the case for an engineering manager. He may be willing to pay more to someone with less work experience but more productivity. Later in this chapter will explain more about the salary system.

4.2.1 Minimum Qualifications

An individual shall have certain qualifications to be nominated for a position. Some of the major items may be listed as below:
- Academic degree in related field is the absolute minimum requirement for any candidate in engineering fields. However this is not enough. Academic career never teaches step-by-step procedures for design of offshore platforms. Bachelor of Science (BSc) is the minimum. Having higher degrees like Master of Science (MSc) and PhD is a privilege.

- Only through experience gained in several projects can an engineer master design process. Normally an engineer with up to 3 years' experience may be classified as junior and above 6 years as senior. To become lead, more experience up to 8 or 9 years may be required. To become engineering manager, greater experience up to 15 years may be required. It is a fact that these are not sacred numbers that shall always be followed exactly. Highly motivated and knowledgeable individuals may surpass these criteria. Higher academic degrees may require less experience.
- One can gain sufficient design knowledge only by combining project experience with continuous study of major topics. This may be either by official education or by individual study. In the first case certificates are required and in the second case he or she shall prove acquired knowledge in the technical interview.
- Nowadays software plays a major role in engineering offices. Each engineer shall be familiar with the software that is used in that discipline in addition to general purpose software. It is worth highlighting and emphasizing that mastery of the software is not enough to be a good engineer. One has to continuously improve his basic knowledge of engineering understanding.
- Another important factor is the ability to transfer ideas. A very brilliant idea may be useless unless it is correctly described to other team members and properly implemented in design.

The engineering manager or LE shall consider all the above factors when recruiting a new comer. A grading system will be explained later in this chapter, which may be useful.

4.2.2 Matching to Construction/Installation Capabilities

The engineering team shall closely adapt platform design to construction and installation team capabilities. A basic knowledge of yard fabrication methods/standards and installation procedures is essential for design team members.

For example, the structural team shall understand construction yard capabilities and limitations. If a yard has only 200-ton crawler cranes, designing decks that may be divided to substructures of 300 ton weight will not be efficient. Of course it will not be impossible to fabricate this deck. The construction yard may temporarily rent two or more cranes with higher capacity or perform multiple crane lifts. But this will increase either the cost or fabrication time.

Adapting to construction capabilities doesn't mean that for a small yard, platform size shall be reduced. Platform general dimensions are governed by project requirements. Therefore the contractor shall provide suitable equipment and machinery.

If a yard has main skids that are installed at 20-m spans, then it is possible to alter topside design to use this span. The author has used a loadout support frame shown in chapter "Disciplines Involved in Offshore Platforms Design" for transporting at least four topsides. Internal spans were changed to adapt LSF dimensions.

Piping discipline is another example of construction yard capabilities. Design of long spool pieces or very congested lines on the same support or route will create difficulties both in construction and later in operation.

Routing cable trays in inaccessible areas will not be a good practice. Yard electricians must be able to bind cables on the tray, pass cables from glands, and connect them to the termination points.

4.3 DISCIPLINES/ENGINEERS MOBILIZATION AND REQUESTS

Consultant power depends on its workforce and database. Database is accumulated through the continuous data generation in different projects by engaged engineers. Therefore engineers are the main factor for any consultant. Their recruitment, evaluation, and knowledge improvement shall be based on established criteria. Different individuals for a team are like components of a machine. Each engineer shall be fit for assigned functions in the team.

It is a fact that consultants will try to spend money on those engineers who have greater loyalty. This is considered as a positive factor. If someone is working one year in this company and the next year in another, this will be considered negative. Some companies may adopt the "easy come, easy go" policy, others may not. The author has heard that many Japanese companies act like a family. An employee after entering a company may continue until retirement. In this company he may pass through all stages of work from a junior worker to a senior managing stage. Due to fluctuations in workforce demand and economic uprisings/slowdowns this may seem optimistic and somehow nonrealistic in many cases.

The author is of the opinion that each engineer shall gradually move through all steps in a team and therefore shall produce all documents of that discipline that fall in his or her technical understanding.

For example, in structural discipline as an extreme condition a new engineer shall start from preparing drawings and MTO. Then they proceed to miscellaneous analysis and later to temporary analysis and finally permanent analysis and preparing design basis. He or she shall also be able to prepare material specifications for steel or rubber or wooden materials.

This is to his or her benefit to know how different documents in a discipline are interrelated and how each one has its own impact on the final product. However, at the same time putting overqualified individuals at lower ranks for a long duration may be demoralizing and subsequently may lead to less productivity and even departure from a team. A better solution may be to assign complicated analysis and simpler calculations like MTO, etc. to the same engineer.

4.3.1 Engineering Recruitment

Engineering recruitment shall be done through well-established procedures. Advertising for recruitment either in the newspapers or through employment agencies shall be carefully supervised. Review of the CVs, short-listing candidates, performing interviews or giving exams shall be carefully conducted. Normally after graduation and with several years' experience engineers are reluctant to participate in the exams. But it can be explained to them that the opportunities are the same for all. In these exams (if conducted) a major portion shall concentrate on general engineering understanding and not on very specific and advanced engineering analysis. As much as possible calculations shall focus on basic concepts.

If only one or two engineers are required, exams may seem too time-consuming. Interviews will be more time effective. Both of them shall follow specific guidelines. The intention is to provide a subjective means of evaluation. To recruit a large number of engineers from the same location exams may be a better option. When recruiting from overseas, exams may not be possible. In that case one shall rely on reviewing CVs and telephone interviews.

For short-listing of a large number of applicants, several criteria may be used like:
- Academic degree and graduating university. BSc from a highly credited university will have higher grade than MSc of an unknown university.
- Outstanding technical performances like papers, publications, innovative designs, etc. will be considered as privilege.
- Ability of using special technical software will be a positive factor.

- Years of experience and field of expertise. For offshore steel-type structures, an engineer with 6 years of experience in onshore concrete design will get a lower grade than an engineer with only 2 years of offshore steel-structure design experience.

4.3.2 Salary

Salary shall be carefully calculated and paid. Normally it is a private issue and shall not be disclosed to other colleagues in the company. But more or less people will talk to each other and at least the order of magnitude of each one's salary will be known to everybody. Therefore it shall be calculated on a predefined and clear procedure to avoid adding any uneasiness or resentment or feeling of unjust behavior among colleagues. Each company shall define a clear grading system for personnel evaluation and salary adjustment.

Several factors shall be considered and a weight is given to each one. Based on company conditions/practices the weights in each section or the topics may vary. Nowadays engineering is becoming a worldwide business. Competent individuals move fairly easy from one consultancy to another. Based on the country of residence and international job market, each consultancy shall prepare and annually update its salary payment system.

The following factors and their weight have been used by the author in some projects. However, it has to be emphasized that each case has to be treated individually. Only the general policy has to be determined by these figures. Always a considerable value has to be left to direct manager decision and to the individual productivity. It is noted that all items under one topic may not have the same value. For example, under technical qualifications, technical knowledge that is specifically related to the project in hand has much more weight than the general engineering knowledge. Similar to it, work experience in offshore consultancy has more weight than general consultancy work or working in the fabricating yard. But similar to what was explained for client team members, having construction yard experience will be a positive factor for recruiting.

Several approaches may be used. One of them is to define an average salary for each class of employees. Then define the unit payment and distribute the factors and points in such a way that a normal engineer may obtain 100 points to get this salary. In this case engineers with higher qualifications may get points above 100 and therefore higher salaries.

For example, average salary for junior engineer rank may be considered as US$2000. Anyone with a BSc degree from a credited university and 2 years' experience in this field and good technical knowledge about the subject may be given 100 points. If someone with an MSc degree applies for the same position, he may be assigned 120 points. Therefore, his salary may be US$2400.

In an engineering team the following ranks may be defined:
- Office clerk
- Drafters and designers
- Engineers
- Senior Engineers
- Specialists
- Lead Engineers
- Project Managers
- Engineering Manager

The second approach is to use the same rating system for all jobs. However, the unit payment for each job shall be different. In this case the highest qualification may receive 100 points and normal qualifications may get fewer than 100 points. However, the unit payment per each point is such that they receive the intended salary.

In defining unit payment, a stepwise system similar to tax payment may also be used. In this case each point range may have different and stepwise increasing unit payments. This stepwise payment can be used both between disciplines and inside each discipline. For example, consultants may define some qualifications for junior, senior, lead, and specialist engineers. Maybe one with longer work experience at the senior engineering level gets more points than someone at the lead engineering level. However, LEs get higher unit rates. Therefore the total payment will be higher. Specialists shall always be treated specifically and on consultant-need basis.

For example, an average applicant intended for an engineer position may get around 50 points, and another average applicant intended for senior position may get 60 points. The first one's grade will be multiplied by 40 and the second one by 50. Therefore an engineer with 50 points will receive US$2000 and a senior engineer with 60 points will receive US $3000. Assuming someone intended for LE position obtains 60 points, his grades will be multiplied by 60 and he will receive US$3600.

The author has only used the second approach. However, the general understanding is that the herein-mentioned approaches and some others are possible and adapting a stepwise increase in unit rate payment (similar to a tax system) will enhance the evaluation/payment system. As noted on several occasions, each consultancy may have its own grading system that may have proved efficient for many years.

One rating system is proposed following. Once more it is emphasized that the system shall be adapted to each company, location, and market situation. Oil prices are very fluctuating. Consequently, oil projects may increase in one year and be reduced considerably the next year. Therefore, the same company at the same location after some years may use a new rating system.

A — Technical Qualifications: (25%)
1. Academic degree (PhD, MSc, BSc, Technical diploma) (10%)
2. General technical knowledge (3%) (poor, average, good, excellent)
3. Specific technical knowledge (6%) (poor, average, good, excellent)
4. Technical publications (3%)
5. Training courses (3%)

B — Work Experience: (25%)
6. Work experience in offshore structures consultancy (8%)
7. Work experience in offshore structures as EPCIC contractor, operator, etc. (3%)
8. Work experience in other fields (2%)
9. Proficiency in writing technical documents/communicating in client language (5%) (poor, average, good, excellent)
10. Technical or administrative staff (7%)

C — Work Ability (20%)
11. Ability to manage a team (5%) (small, medium, large)
12. Ability to work in project environment under stress and with preset deadlines and with a team (5%) (poor, average, good, excellent)
13. Continuation with one consultant or company (5%)
14. Ability to write and communicate in other languages (2%) (poor, average, good)
15. Ability to convince client in technical meetings (3%) (poor, average, good, excellent)

D — Managerial Impact (30%)
16. Immediate manager evaluation of efficiency and productivity (20%) (poor, average, good, excellent)
17. Company need (10%) (average, high)

Company management shall try as much as possible to provide clear definitions of each item to avoid unreasonable grading. Definition of grading shall not be too severe. For example, in one case the author has encountered an evaluation system in which only a PhD holder with 30 years' experience and having passed several specialist training courses, mastering at least three languages, and having published several papers with very high productivity could get 100 points. Normal BSc degree engineers with no published papers and 6–8 years of work experience and good productivity could only get about 50 points. It has to be emphasized that using this system is also possible (if suitable unit rates are applied) but it is not recommended. As a general rule, "a bad law is better than no law." This means that adopting a grading system and improving it during projects is better than having no system at all.

In the previous list the author has proposed some percentages for each item. It is emphasized that this proposal shall be reviewed cautiously and reevaluated on a continuous basis.

The major setback of this evaluation procedure is that certain items are measured subjectively. Exact guidelines to distinguish between poor, average, good, and excellent are not available. Someone categorized as technically "good" by the first interviewer may be categorized as "excellent" by the second interviewer. Even the same interviewer with some questions may evaluate someone as poor, but with other questions the same engineer may be rated as good.

Another setback is that some criteria can only be evaluated during actual work. For example, team workability or ability to convince clients in a meeting is only found during actual project execution. This means that the evaluation shall be repeated periodically. Interviewers may assign equal values to all newcomers.

The author has performed written examinations with the same questions for all to have a fair/uniform evaluation basis. But many engineers (especially those with longer experience) are reluctant to take part in examinations. Besides, on occasions when recruiting is from overseas, written examinations for engineers are very difficult if not impossible. Interviews are more convenient for everybody. But the results must be carefully recorded. Recording interview results has two advantages:

- First, it keeps an auditable record. Therefore, when revisiting employee salary her/his performance and improvements can be evaluated readily.
- Second, the interviewer himself will be cautious in her/his judgments and grading. Hence better accuracy can be expected.

Normally an interviewer is the manager of the same discipline. Therefore, bias or different grading may not have very drastic impact because all are treated uniformly. A considerable margin is left for an employee's immediate manager.

After the engineer has joined the consultancy the grading system shall be continuously reevaluated. It is again emphasized that a uniform evaluation shall be applied to all employees of the same rank. It is understood that discipline managers are human beings. They may not have the same feelings or evaluations of the same issues at different times.

It is advisable to define a min, max limit for each position in the consultant team. All proposals in that margin can proceed. Any proposal far above or below shall be revisited by someone else before becoming effective.

It is most important to avoid any sort of preferential bias based on sex, race, color, nationality, etc. It is even acceptable if a company says that they are paying less compared to other companies. But they must pay with equal standards for all. Unprejudiced evaluation is not only a human rights duty but it directly impacts economics. The first minute an employee discovers his colleague is getting more with similar qualifications and productivity, feelings of resentment then impact productivity.

4.3.3 Working Conditions

Engineering offices like all others shall be tidy and comfortable. They do not need to be luxurious, but chairs, desks, computers, and the office environment shall be pleasant and serve the engineers' needs. With rapid progress in computers, what seems luxurious this year will be normal next year. Chairs and desks don't evolve that much, but the minimum requirement is that the chairs be adjustable in height and alignment. The area allocated to each engineer shall be sufficient to put his or her personal belongings like books, references, and project work documents in addition to having flexible space for possible discussion with other colleagues.

The author has not had the chance to work in a Japanese company but has heard that in order to maximize office space utilization they prefer to arrange desks in rows with no partitions and the manager sits in front to see all in one view. Although the author has always praised other merits of Japanese companies like continuous working and progressing through all stages of an organizational chart, he disagrees with this approach. Each person shall be given a personal area and a little freedom. Of course different cultures may dictate different approaches, but in my opinion engineers don't like a military environment. Remember the George Orwell novel *1984*. I don't like to be under continuous surveillance.

For a manager to boost productivity it is not required to see all the monitors at all times to make sure personnel are not surfing the Internet or not working. When a manager goes to a fellow engineer's desk for some discussion without prior notice it is better to approach the desk from a direction that if the employee is not working he can find enough time to conceal the monitor and save face. Even if once or twice the employee is caught red handed, the situation shall be treated casually with a joke. Of course if this is repeated continuously, then the manager shall take necessary action.

No one does effective work during all 8 h in an office. This is not important. Of course I don't mean if someone is idle for 4 h then we say okay, no problem. Companies also sometimes have less work and sometimes more work. When there is less of a workload, the manager can tolerate greater idle time for personnel, and when there is a high workload all personnel shall fulfill their direct manager's expectations in the assigned time schedule. This attitude shall be a two-sided relationship.

The point is no one deliberately tries to steal from companies. Especially, with the junior engineers, I have encountered some of them working overtime without claiming, or studying/preparing, documents on the subject in their hand at home. It is manager's duty to see these sacrifices and to appraise/reward them. The reward may not always be financial.

It is manager's duty to set deadlines for each individual employee and check that they are fulfilled. Within that deadline and the contingencies noted some freedom can be accepted.

4.3.4 Required Data

Engineers shall use a complete database. This may include Goby documents from previous/similar projects to photos/reports of site visits, international and company standards, documents enclosed with the main contract, etc. Let us admit something openly—no engineering company starts from the discovery of fire. Similar documents from previous projects are used as goby and further developed to comply with specific project requirements. This doesn't mean copying.

In addition there are some site/project specific data that shall be provided by clients. Project managers shall receive a list of minimum required data from respective LEs. This list shall be cleared at the KOM. Due to bureaucratic constraints clients want to save themselves from decision-making/responsibility. Therefore, if something is included in the contract they will adhere to that and will divert the rest to consultant responsibility. Even for the information that shall be gathered from site visits they don't take responsibility.

Some portion of required information may be provided by the client but a major portion falls under consultant obligation. This part may be called "consultant know-how." Each engineer joining a company adds to this know-how and takes some with himself. Although in all engineering contracts it is written that the documents generated in each contract are the specific property of that client and can't be used for other purposes, inevitably engineers will use the generated knowledge in future projects.

It is said that knowledge is the impact left in mind after reading everything and forgetting everything. Even if engineers don't take hard or soft copies of the generated documents, their engineering understanding evolves after doing several projects and their knowledge increases. This is called individual experience and is quite legitimate.

4.3.5 Work Discipline

A consultant company is far from an army unit but it should have work discipline. Engineers should attend to work at the office on a regular basis. Office hours are determined by HR units and have to be followed by all personnel. Each engineer shall fully recognize the team that he or she is working in and be responsible to his or her immediate superior. Office regulations may even extend to individual dress codes. They shall be followed. Although it seems nowadays the general trend is to accept a larger degree of individual freedom in all offices. Dress codes shall not violate individual beliefs and common practice regulations.

Since there are many documents to be issued in each discipline and many of them are the prerequisite of others, each individual shall be responsible for documents entrusted to him. This ensures a successful team performance. If anyone says five documents have been assigned to me and I can generate them in any order, this may impact others' performance. As in other societies, the team member shall understand that his or her freedom ends where the other one's freedom starts.

The author has not had the chance to visit Google's offices but has seen e-mails showing Google staff freedom in office hours. It seems they have great freedom on start and finish hours for their workday. They can even play and entertain themselves during office hours. Maybe in the future other consultant offices will move in this direction, but currently fixed-time office hours with half to 1 h floating time are much preferred by all.

4.3.6 Human Factors

Always remember that employees are human. They are not slaves or robots. They don't work to benefit consultant shareholders. They work to increase their own knowledge, to improve their financial position, to upgrade their position in the company, to improve their living conditions, and to enjoy their life with their family. It is consultant manager's responsibility to find ways that these goals are in line with increasing consultant profits.

In the present economic conditions each company is known by the value of its shares in the market. Company managers are shareholders' representatives to increase their benefit. But shareholders must remember that employees are not slaves with the sole intention of increasing shareholders' money.

Many years ago my colleagues had prepared an animation for Oromieh Lake bridge project. Considering special conditions in that location (low water depth, very high salt content, extra weak seabed to more than 40 m, etc.) they came up with many innovations and novel ideas. As an example, they managed to float heavy concrete boxes to be transported from the construction site to the installation site in shallow water or load out them from the yard to the sea. These large caissons were used as bases for deep pile driving and bridge support.

To present the idea to the company general assembly they prepared an animation showing some of these novel solutions. After it was presented, one of the shareholders stood up addressing all audience and said "...you are showing these cartoons to us...." To degrade the value of the presented animation he intentionally used the word *cartoon* instead of *animation* to transfer meaning of *child play*. Although I was not directly involved in that project, even today I remember the nasty feeling his statements caused in my mind. I was aware that some of my colleagues had even worked on that subject at their homes and on weekends. Some of the proposed solutions (with company available equipment) were brilliant, but the stupid shareholder couldn't understand how much money this animation was bringing to his pocket. In addition he was rude enough to shout loudly his ignorance.

I know that the end goal of any financial institution is to make profit. In Persian we have a proverb that says, "No cat catches a mouse for god's sake." Very few companies working with strategic/political motivations can be exempted from this rule. However you don't need to tell it to the employee every day. I am not your slave. I work for my family and my own wellbeing. If in this course and out of my work you get some money I don't care, so much to your benefit. Company shareholders/managers shall never directly refer to profit as the sole purpose of the consultant. What will you feel if a manager walks in the hall with a whip in hand and continuously repeats "work for my money"?

The employees spend at least 8 h per day in the office. Considering lunch time and overtime this will even become longer. If you deduct the sleeping time, this is about a half-effective lifetime. If the office environment is like a prison, good engineers will simply find a new location and leave. That is why I am very much against installing closed circuit cameras in the office. It only reminds you of Big Brother in *1984*. Regardless of market condition and earnings, and regardless of position, everybody shall be treated respectfully.

Every manager shall have friendly relationships with all employees. It is a fact that many of the employees are much better educated than some of the managers. Managers' office shall always be open to employees. Other than commercial discussions and private disciplining talks with an employee, which may need a private environment, there is nothing private for an engineering manager. Closing the office door doesn't show the importance of the occupant. It only produces feelings of isolation. Employees will simply tell themselves "he is not from us." I don't mean to disguise your actual beliefs. After several days anyone can tell a genuine behavior from a fake action.

I agree that based on acceptable norms, a friendly behavior in the office and outside of the office may be different. I have heard that if in Japan you sit on the edge of your boss's table it will be considered a rude behavior. The same action in America may not be so rude. However, in America your boss may fire you without even bothering to give an excuse. This is the basis for what is called "at-will employment." This means an employer can fire an employee without any reason at any time without previous warning. Again I have heard that in Japan no employer allows him or herself to fire an employee without proper explanation and apologies.

4.3.7 Consultant Expectations

Any company is a business unit. It is normal if shareholders are looking for more profit from each project. In addition they expect to receive more projects in future and to expand the business. But if all newly established companies were to grow continuously, then the world economy would explode. Companies shall be established and closed. Out of several newly established companies only a few may succeed and grow. Of course they will also experience ups and downs, some of which are out of the consultant manager's power. What happens may depend on the world's economic and political situations, the price of oil, and many other factors.

A consultancy expects its employees to adhere to the following principles:

- Perform their assigned tasks professionally (in my career I used to say, with "boy scout's honor").
- Produce documents efficiently.
- Reduce mistakes and carelessness in document preparation. Anybody doing a job may occasionally make some mistakes. Managers shall accept this and try to establish procedures to reduce these mistakes as much as possible. However, it is the employee's responsibility to ensure that the same mistake is not repeated.
- Save documents generated in the company.
- Use consultant facilities like computers, printers, etc. carefully and avoid damaging them.
- Follow a project's schedule as closely as possible and, if a delay occurs, follow the catch up plans carefully.

I never intended to preach in this book. But I have learned from one of my friends and teachers who first taught me how to use offshore structural software that even business rules are, in the final analysis, governed by moral values.

CHAPTER 5

Handling Design Documentation

Documents and drawings are products of an engineering consultant, similar to cars that are products of an automobile manufacturer. Therefore, quality assurance (QA) and quality control (QC) measures shall be implemented in their production. Chapter "Balancing Between Client and Task Force Engineers" described some of the QA measures, like engineering team member qualifications from both the client and consultant point of view, working conditions such as salary, office space, human relations, etc. These are QA measures to ensure that the engineering team has the potential to generate documents with necessary quality. This Chapter describes further QA measures to ensure generated documents have necessary harmony and conformity with each other. In addition to that, QC measures to ensure issued documents match project requirements are described. Different from the car manufacturer is that we normally don't do destructive tests and that documents are not mass produced.

Strict regulations shall govern document production and distribution. The intention is to ensure that documents will satisfy project requirements, will follow standard regulations, are as per normal engineering practice, don't have ambiguities/internal contradictions, are constructible, and after installation and commissioning will deliver intended hydrocarbon volume. These procedures shall govern QA and QC steps:

QA steps:
- Document technical quality
- Document completeness
- Conformity to other documents
- Checking steps

QC steps:
- Numbering system
- Typing format
- Document distribution
- Archiving

Bullet points 1—4 are covered in the responsible discipline. Bullet points 5 and 6 are covered in the document numbering procedure and may change from project to project based on client requirements. Bullet points

Practical Engineering Management of Offshore Oil and Gas Platforms
ISBN 978-0-12-809331-3
http://dx.doi.org/10.1016/B978-0-12-809331-3.00005-3

7 and 8 also depend on project requirements but at the same time are affected by consultant practice and are taken care of in the document control center (DCC).

The author tried to present material in this chapter in a flow chart format but it showed no specific advantage. Text format may seem a little boring. It has to be emphasized that list numbering used in different sections doesn't mean priority or less importance. It doesn't mean if a step is not finished don't proceed to the next. Each item can be individually checked or performed.

5.1 MASTER DOCUMENT REGISTER

Document preparation, issue, and progress check is controlled by a tool called: master document register (MDR). Preparing, updating and continuous monitoring of the MDR is a very valuable tool in achieving the previous goals.

Preparing the MDR doesn't require sophisticated software or procedures. In its simplest form it may be prepared by software-generating worksheets. Automated systems have been developed that are very helpful. But even a simple table may serve the purpose. MDR is the main list including necessary information for each and every document/drawing that is (or shall be) prepared by the engineering group. Throughout the engineering project's duration MDR is a live document and will only be closed in the final dossier report.

Some large consultants have developed engineering document management systems that can handle any document from start of production in an engineering discipline to end of approval by client. This system has the capability of assigning different access levels for each document to different users. It has the ability to prepare all necessary reports like time spent on this document by a specific user, progress stage, final status, prerequisite and trailing activities, etc. This wide range of information is necessary for the project team as well as other departments like financial, business development, human resources, etc. In this book I will emphasize only the simple and basic method of using a worksheet data base.

A sample MDR format is attached in Table 5.1. Only a portion of one page is given. As a minimum it shall include following information:
- Document number and title
- Document type, class, and weight in discipline and project

Table 5.1 Sample page of master document register

No.	Document Number	Document Title	Doc. type	Class	Doc WF %	Planning Category	IFC/IFI/IFB Date	IFC/IFI/IFB TR No	Endorse by CA	IFA/IFP Date	IFA/IFP TR No	AFC Date	AFC TR No	Client Response R.	Client Response Letter No.	Client Response Date	Client Response St.	Delay	Prog. %
1	GEN-ST-DW-4566	Decks Structural Design Loading Diagram	DWG	2	0.068%	Planned	12-Apr-08			24-May-08		21-Jun-08						0	
						Forecasted	13-Sep-08			25-Oct-08		22-Nov-08						0	0.0%
						Actual												0	
2	GEN-ST-DW-4567	Structural Perspective View	DWG	3	0.274%	Planned	4-May-08			-		-						0	
						Forecasted	5-Oct-08			-		-						0	0.0%
						Actual												0	
3	GEN-ST-DW-4568	Cellar Deck Primary Framing Plan (El. +11500)	DWG	1	0.365%	Planned	20-Feb-08			9-Apr-08		7-May-08		IFC	LT-1052	10-May-08	C	7	
						Forecasted	20-Feb-08			9-Apr-08		7-May-08						0	60.0%
						Actual	19-Apr-08	TR-1252	√									0	
4	GEN-ST-DW-4569	Cellar Deck Secondary Framing Plan (El. +11500)	DWG	2	0.274%	Planned	29-Jul-08			9-Sep-08		7-Oct-08		IFC	LT-1139	26-May-08	C	9	
						Forecasted	29-Jul-08			9-Sep-08		7-Oct-08						0	60.0%
						Actual	3-May-08	TR-1298	√									0	
5	GEN-ST-DW-4570	Lower Deck Primary Framing Plan (El. +16500)	DWG	1	0.365%	Planned	20-Feb-08			9-Apr-08		7-May-08		IFC	LT-2393	10-May-08	C	7	
						Forecasted	20-Feb-08			9-Apr-08		7-May-08						0	60.0%
						Actual	19-Apr-08	TR-1252	√									0	
6	GEN-ST-DW-4571	Lower Deck Secondary Framing Plan (El. +16500)	DWG	2	0.274%	Planned	29-Jul-08			9-Sep-08		7-Oct-08		IFC	LT-1559	26-May-08	c	9	
						Forecasted	29-Jul-08			9-Sep-08		7-Oct-08						0	60.0%
						Actual	3-May-08	TR-1298	√									0	
7	GEN-ST-DW-4572	Main Deck Primary Framing Plan (El. +25500)	DWG	1	0.365%	Planned	20-Feb-08			9-Apr-08		7-May-08		IFC	LT-2309	10-May-08	C	7	
						Forecasted	20-Feb-08			9-Apr-08		7-May-08						0	60.0%
						Actual	19-Apr-08	TR-1252	√									0	

- Planned, forecast, and actual issue dates in each stage like: issued for comment (IFC), issued for approval (IFA), and approved for construction (AFC)
- Summary of client response history
- Delays and achieved progress

As per each contract and consultant normal operating procedures, document register configuration and format may change. These items are described in more detail in the following section.

5.1.1 Document Register Contents

Contents of the document register may vary based on consultant internal procedures and/or project-specific requirements. Besides, a consultant may prepare one register for its internal usage and one for inclusion in the weekly reports and issuing to the client. The degree of information included in each of them may vary based on consultant policy.

Each MDR may include one or two worksheets or be presented in different files with internal links to ensure uniform data are presented in all files. This is easily possible with defining necessary access for different departments. Since different parts of this information may be used for different purposes in the same organization, they may require stand-alone files. This will enable each department to process input information and extract necessary data or reports for management decisions. The major usage is by project team members. MDR shall include:

- Number of documents in each discipline and each category
- Document number and title
- Document type and class
- Discipline weight in engineering team
- Document weight in discipline and in total list
- Planned/actual start date
- Allocated man-hours (only for internal use)
- Actual consumed man-hours (only for internal use)
- Responsible engineer (only for internal use)
- Prerequisite documents (only for internal use)
- Planned/actual interdisciplinary check (IDC) date
- Planned/actual issued for comment date
- Planned/actual issued for approval date
- Planned/actual approved for construction date
- Latest issue information and last revision number
- Tracing transmittals number and date in each stage

- Client/engineering, procurement, construction, installation, and commissioning (EPCIC) contractor and third party authority (TPA) commenting/approval status
- Achieved progress

Information presented in MDR may be used by several parties including discipline engineers, lead engineers (LEs), project manager (PM), and even client. This is in addition to other departments within the consultancy like financial, human resources, etc.

For example, the financial department may want to know at a certain stage how much progress has actually been approved by the client. This means the expected cash in value. Subtracting actual received invoices from approved progress or invoices will give them an idea of how much further money for a specific period can be expected from this project.

Similarly, human resources may want to know if the assigned man power to this project matches requirements. Do they need to recruit more engineers? Can they transfer some of the team members to other task forces?

The business development department may want to know if the estimates made in this bid match with actual personnel productivity. What unforeseen factors have arisen during project execution? What contingencies were extra or too low and need reevaluation?

These are some of the questions that departments other than project team members may need. There are many more questions and usages of MDR for other departments.

Some information shall not be disclosed to members out of the engineering team. For example, the engineering team may have a different document issue plan than the one issued to the client. The internal plan is normally stricter. This is only for internal usage. Issuing it to client will increase team responsibility and may be misused. In this case different files can be generated with interlinks to ensure confidential information is not disclosed to all while conformity is preserved.

Each engineer shall know at least the following:
- What documents are assigned to him/her?
- What timetable shall be followed in generating documents?
- What are the allocated man-hours?
- What prerequisite documents should have been prepared?
- What next-stage activities depend on his document preparation?
- What is the latest status of his documents?

5.1.2 Number of Documents in Each Discipline

Some may think this is trivial information, but it is not. It has to be highlighted that client PM and consultant managing director (MD) don't have and don't put in much time to review the registers in detail to be informed of the actual project progress. In fact, what they may need is only a one-page table showing the number of documents in each discipline, those issued as IFC, IFA, and AFC and the number of approved documents in each stage. This will give them a first-glance impression of the project progress (engineering status) and that is a major part of what they need.

For the lead engineers and PM this information (although useful) may not be sufficient. They need to go into further detailing. Introducing hyperlinks in the MDR list will help them easily find client comments, consultant response(s) and whether they are closed or still need further clarification. It is emphasized that even for lead engineers and PMs this table provides an important first-glance status of the project progress. A sample report of this type is given in Table 5.2.

The progress calculation in this table is based on assuming 60% progress for IFC, 80% for IFA, and 100% for AFC issue. These figures may vary based on contractual requirements. To simplify calculations, the total number of documents issued in each category will be multiplied by the difference of the related stage with the next. For the previous case it is 60% for IFC, 20% for IFA, and 20% for AFC. Weights are multiplied by the total number of documents in that stage and then simply summed up. It is understood that some contracts may assign different progress values for document issue stages.

In one summation approach discipline weight is taken as equal to the number of its documents divided by total documents. For example, process weight may be taken equal to 82 divided by 1182, which is 6.94%. Since out of 82 documents a total of 74 have been issued in IFC, 68 in IFA, and 55 in AFC, the progress weight in the discipline is equal to $(74 \times 0.6 + 68 \times 0.2 + 55 \times 0.2)/82 = 84.1\%$. This means that the process inside itself has about 84% progress.

This approach may not be so accurate. It doesn't take into account physical weight of each document and discipline. Man-hours spent for documents vary considerably. To prepare some documents expert engineers shall use sophisticated software to generate necessary data/information. The better approach may be to calculate weighted man-hours based on consumed man-hours of each category of the engineering team.

Table 5.2 Sample number of documents

Discipline	Total	Started	IDC	IFC		IFA		AFC		Progress
				Issued	Comment	Issued	Comment	Issued	Comment	
General	10	8	—	8	8	8	6	4	1	72%
Process	82	80	78	74	70	68	60	55	40	84%
Safety	45	43	40	38	30	28	20	18	5	71%
Mechanical	110	90	85	80	50	65	35	35	15	62%
HVAC	25	17	14	12	5	9	1	0	0	36%
Piping	125	100	90	80	40	60	20	15	5	50%
Electrical	90	75	70	60	35	43	22	25	8	55%
Instrument/Control	180	100	90	75	52	58	34	45	15	36%
Telecommunication	25	18	14	12	6	10	2	5	0	41%
Structural	450	250	—	240	150	180	60	110	40	45%
Architectural	40	25	—	20	11	14	3	8	3	41%
Sum	1182	806	481	699	457	543	263	320	132	50%

This will inform the lead engineers and PM how many senior and junior engineers or modeling/drafting or DCC staff may be needed.

To calculate project progress, individual discipline progress is multiplied by discipline weight and the total results summed up. Assuming all documents have the same weight, other disciplines' weight and total progress will be as per Table 5.3.

This is not necessarily a correct approach but is used here for easy illustration and calculation.

Documents' weights/importance are not equal. They are calculated based on the required man-hours in each specialist category to complete them. This is important both in payment from client and in checking necessary man power. Differences between invoicing progress and physical progress (actual weight) are defined in this section and Section 5.1.11.

Project progress is obtained by multiplying each discipline's achieved progress to its weight in total engineering. In some projects clients may require to assign a weight for management. This way they want to ensure receiving required reports to the end of the project. In other projects weight may be assigned to documents like correspondence/numbering/progress measurement procedures and weekly/monthly progress reports. In other projects these documents may be issued once (even before kickoff meeting; KOM) with no weight factor.

In addition they may assign a specific weight factor for vendor data review/implementation or issuing final dossier documents. This way they

Table 5.3 Progress in each discipline and its weight and contribution to project progress

Discipline	Total	Progress	Discipline weight	Discipline progress
General	10	72%	0.85	0.61
Process	82	84%	6.94	5.84
Safety	45	71%	3.81	2.71
Mechanical	110	62%	9.31	5.75
HVAC	25	36%	2.12	0.76
Piping	125	50%	10.58	5.33
Electrical	90	55%	7.61	4.20
Instrument/Control	180	36%	15.23	5.55
Telecommunication	25	41%	2.12	0.86
Structural	450	45%	38.07	17.09
Architectural	40	41%	3.38	1.39
Sum	1182	0.50	100.00	50.08

can assign a fixed amount of project lumped sum cost for these activities. Consultants shall accept this only if a mechanism is envisaged to ensure necessary vendor data (which are issued by EPCIC contractor and consultant has no authority on their delivery) arrive on time. If vendor data schedule is not certain and client insists on assigning a certain portion of project to them, then suitable compensation measures for delay in submitting vendor documents to the consultant shall be foreseen. For each case, planning discipline and PM shall adopt the report format to project requirements.

Actual discipline weight is the number of total man-hours in different categories multiplied by respective unit rates. The same will apply to document weight. Therefore, the actual invoicing progress calculation for Table 5.3 may vary. Of course the difference will almost always be less than 10%. With that approach, columns 3, 4, and 5 will have different values.

As an example, assume the lead engineer unit rate is 60 $/h, senior 45 $/h, engineer 30 $/h, and drafting is 15 $/h. If the design basis document requires 30 h lead engineer and 60 h senior engineer, and 40 h engineer time it will weight $5700. Assume a drawing will take only 5 h LE, 10 h SE, 40 h EN, and 75 h drafting time. This weight is $3075. Both documents consume 130 h. But design basis weights 1.85 times more.

In addition as explained later and since design basis is issued at an early stage of the project, engineering manager (EM) shall consider its weight at least 20–25% more than actual calculated weight, which gives it 2.2–2.4 times more weight than this drawing if it is issued at a later stage. This is to ensure that small progress at the start of project (first nearly flat section of S-curve) can yield suitable progress for invoicing to improve consultant cash flow. Some EPCIC contractors understand this logic and accept it. Others may think pushing consultants as hard as possible will help them have a better grip on the project. The previous information shall be specifically hidden from this type of client.

The two issues of actual weight and importance shall be clearly distinguished. Document weight calculation was previously explained. It is a quantitative method. Document importance depends on the observer. EPCIC contractor, consultant EM, and project engineers have different evaluations of document importance.

- For EPCIC contractor a document is important if it can be directly used for procurement of packages, purchase of bulk items, or construction. For them, material takeoff (MTO), material requisition (MR), or fabrication drawings are the most important ones.

- For consultant executive manager documents having highest weight (largest progress/invoice) are the most important ones.
- For engineering team members documents that open the road to generation of other documents are the most important. In this way design basis and philosophies are the most important documents.

As explained herein, the EM can artificially increase weight of documents generated at the start of the project. However, their weight may not be sufficient to compensate for total costs. Therefore, he may assign a group of drafting and junior engineers to develop and issue typical detail documents that are more or less constant and applicable to different projects. These two policies and receiving advance payment help consultant cash flow at project start.

5.1.3 Document Number and Title

Each project will have its own "Document Numbering Procedure." Since the client may have several projects in different locations and want to identify them easily (or simply because they want to adapt to their archiving system), the numbering system is normally determined by the client. However, in some cases the consultant may be free to use his own system. Some consultants insist on having their own numbering system printed on the documents even if the client has given a specific system. This causes some additional work and may sometimes lead to mistakes.

Document number may consist of several characters or digits. It may show oil or gas field location, project location, platform designation (if there are several platforms), discipline name that issues document, document type, a serial number for its identification, and revision number. A combination of alphanumeric characters may be used. In some cases it may be as long as 16 characters or more.

The main issue is that each document shall be identified by a unique designation. In some cases the client wants a combination of total characters and numbers to be unique and in some cases they may want only the serial numerical digits to be unique. All are possible without any problem. Only the search engine and archiving software shall be able to identify separate documents easily.

Document title is usually left to consultant selection. They are determined by discipline engineers' experience and are fairly constant. Title shall reflect what contents can be found inside a document. Data sheets and specifications have a fairly constant name. Only the package name shall appear after document type.

5.1.4 Document Type and Class

Document type only shows what is expected to be seen in the text, whether it is a design basis, calculation sheet/note, analysis report, study report, philosophy, block diagram (BD), single line drawing, heat and material balance, instrumentation, material takeoff, list, drawing, specification, data sheet, main/secondary/tertiary framing drawing, equipment layout/general arrangement (GA) drawings, cable route, escape routes, material requisition, technical bid evaluation, etc. Each discipline may have its own categories of document types, but inside a project within different disciplines it is better to use uniform abbreviations for the same type documents. For example, every discipline has a design basis. Different names may be used like basis of design, design basis, or design premise. It is better in a document register for a specific project to make it uniform. However, this is not a mandatory requirement.

A list of different document types in process discipline and their proposed abbreviations follow:

- Basis of Design, Process-sizing Criteria (DB)
- Platform Operation Philosophy (PH)
- Mass, Heat and Utility Balances (report and calculation files) (HMB)
- Process Flow Diagrams (PFD)
- Utility Flow Diagrams (UFD)
- Process and Instrumentation Diagrams (PID)
- Process Data Sheets for Equipment, Instruments, Systems, Packages, Vessels (DS)
- Process Design Calculations (CAL)
- Relief System Calculations, including Flare Network Hydraulic and Blowdown Calculations (CAL)
- Cause-and-Effect Charts, Safe Chart (CH)
- Chemicals Characteristics and Consumption List (LI)
- Process Data Sheets for Control Valves, Safety Valves, BDVs (DS)
- Emergency Shut Down Philosophy (PH)
- Hydraulic Calculations (CAL)
- Main Process Line Sizing (CAL)
- Main Utility Line Sizing (CAL)
- Vendor/Subcontractor Data/Drawings/Calculation Notes Review (VD)

Some documents may be general documents that although generated during an engineering project may not be listed in the MDR. They are not directly related to project technical subjects. These documents may be listed under a general topic. Documents like MDR format, Document Numbering Procedure, Project Correspondence Procedure, Project Quality Plan, Progress Measurement Procedure, etc. are listed under this category. Some clients don't want to pay individually for these documents because they are not directly used for procurement or fabrication purposes. Therefore they include some of them under project management discipline and calculate the progress based on total project progress. Others may be categorized under different discipline titles.

The reason is these documents can/will be prepared even before actual project start. Therefore, client shall pay progress for them while they can't be directly used for procurement or fabrication. Therefore, the client prefers not to pay for them unless other "useful" documents are issued. By "useful" (from client point of view) I mean documents that are directly used for procurement and or construction of the project.

Document class shows its approval rank. Normally three ranks are identified.

- Approval Class
- Review Class
- Information Class

Document class is selected based on its importance for Client. For example, PIDs are classed under "Approval." They define equipment capacity, rating, operating/design temperature and pressure. In addition they provide information on piping material, line classes, change in piping/equipment specification, sectionalization, and control and monitoring instruments on the package or on the piping. Although some of this information like equipment power consumption is transferred from other disciplines and documents, PID is a document that is more than others referred to and therefore plays a central role. For this reason clients always categorize it as "Approval Class." They want to implement their comments on this document before taking any action toward procurement.

The difference between document class and issue status shall be clearly understood. The definition explained next may not be universal but has been used in many projects.

Normally, for documents/drawings under "Approval Class" the consultant/EPCIC contractor shall wait to receive client comment/approval on the issued document at a certain stage and then proceed to the next stage. For those under "Review Class" they may proceed after certain duration has elapsed even if client comment is not received. For "Information Class" they may proceed without waiting for client comments. The normal time limit for commenting is 14 calendar days. In spite of this definition, in almost all contracts the client keeps the right to comment any time on any document. If it was an engineering mistake then the consultant shall rectify it at his own cost at any stage of the job and even during the guarantee period. In some cases client comment may only be a personal preference and not standard minimum requirement or contractual obligation. In spite of this, almost always the consultant will find it very difficult (if not impossible) not to implement client comment (issued long after contractual period) and convince them to keep the issued revision.

In almost all contracts clients put in a sentence that stipulates "client approval on any document doesn't relieve consultant of its responsibility." In essence this only refers to engineering mistakes. However, in a majority of cases clients misuse it to enforce implementing their comments in the latest issued document.

The other importance of document class is in defining invoicing progress milestones. Normally the first issue of approval category is IFC, the second IFA, and the third AFC. For review category this is IFR and AFC. For the information category the first issue is considered to be IFI. Normal weight for IFC is 60%, IFA/IFR is 80%, and AFC/IFI is 100%. Again, I repeat that these weights may change as per contract.

5.1.5 Document Weight in Discipline

Each lead engineer shall prepare his/her own estimate of the man-hours required for complete preparation of a document. These man-hours shall include man-hours consumed by discipline manager, lead engineer, senior/junior engineers, designer, drafting, and specialist study. The man-hours shall include all required time for the following stages:

- Preparation, distribution, and review of the hard copies and required goby documents for all relevant personnel in the discipline
- Preparing the draft analysis model/study report, etc.
- Review of prepared draft model/report by a senior engineer
- Performing analysis and preparing the report
- Discipline internal check/review by checker
- Revisiting model/report for touch-ups as per discipline internal check
- IDC/review
- Revisiting model/report for comments received from other disciplines
- Issuing in IFC stage

- Interface with client in responding to comments
- Revisiting model/report to implement received comments by client, TPA, and others.
- Review by the checker
- Issuing IFA stage document
- Interface/meeting with client for second or other rounds of comments (if any)
- Interface with other disciplines for hazard identification (HAZID), hazard and operability (HAZOP), model review, and vendor data impact, whichever is/are applicable.
- Issuing documents in AFC stage
- Design review and clash check man-hours
- Possible reissue(s) in AFC stage
- Final issue in final dossier stage

It is evident that lead engineers don't have and can't calculate separate assigned man-hours for each document and each stage listed herein. They will only provide a general estimate, which based on previous experience may cover above time-consuming factors. Individual time sheets are a valuable source to define relatively accurate man-hours for different documents and disciplines.

At the start of the project each lead engineer distributes the required documents for performing discipline assignments between the task force engineers. This single point responsibility (SPR) concept shall be strictly adhered to Ref. [38]. Each document shall be assigned to one person. He/she may receive help from others in the same discipline or outside. But he/she shall be responsible to circulate for internal comments as necessary, implement them, and issue as per project time schedule.

Normally the time sheets in engineering offices is not so detailed to enable department managers assign a specific figure for each of these activities. Even if time sheets were so detailed the engineers couldn't accurately differentiate man-hours consumed for each of these stages. Therefore usefulness of such detailed time sheets is under question. Of course its existence is better than not. However, drastic change in consultant well-established procedures may not be so yielding at least for the first few projects. Normally man-hours consumed for each document are only tentatively known. In some consultancies engineers only fill the time sheet separating projects, not the documents. It is the planning department that based on the number of involved engineers and the weekly plans and achievements issued by lead engineers calculates the general consumed man-hours.

It is not possible to give an accurate figure for each of the items listed previously. However, the lead engineer shall keep titles of applicable man-hour consumption stages in mind, not to underestimate the required time for each document. The department manager or lead engineer estimates on each document a number for man-hours consumed and then multiplies a contingency factor to account for other interfaces.

As explained, some documents are more important for client/EPCIC contractor. For example, data sheets and specifications are used to get a vendor quotation for procurement. For long-lead items this may be a necessity to overcome some of the project schedule bottlenecks. Structural MTO and drawings are used in construction yard activities. This is the first yard activity. Sometimes clients define milestones for issuing these documents. This is a two-sided blade. The advantage is that issuing them at an early stage helps the project time schedule. The disadvantage is that issuing them earlier than the time that all necessary data are available introduces an uncertainty/inaccuracy factor on their adequacy. EPCIC contractor and consultant shall be aware of involved risk. Even if issued MTOs may change in the final dossier it is the consultant's obligation to keep changes within acceptable margins so that additional procurement is not required or will not adversely impact the project schedule.

The PM has to add some margins for contingencies and any unprecedented events. The lead engineer may apply contingency factors to each single document, the department manager may apply contingency on total discipline summation, and the PM shall apply contingency on the total project summation.

Company overhead factors shall also be added by EM. Detailed descriptions of factors affecting overhead factor are defined in chapter "Proposal Preparation."

5.1.6 Document Stage Action Dates

Dates of document action stages are needed for two reasons:
- Planning department to check progress as per plan and highlight delays to lead engineers
- Invoicing team to calculate achieved progress

For planning group date of starting, issued for IDC, issued for comment, issued for approval, and approved for construction dates are important.

For the invoicing group only contractual milestones that lead to payment are important. In some contracts the client accepts stage-wise payment for start and IDC and in some they don't accept. Physical progress of each document is different from its invoicing progress. It is normally agreed that in IFC stage almost 60% of the job is completed. The remaining 40% may be divided between the other two stages. However, each contract shall be reviewed separately. In some projects client agrees to calculate progress for AFC equal to 100% and then deduct some certain value and release it after issuance of final dossier. In some projects client may only deduct the good performance guarantee.

PM and lead discipline engineers are very anxious about accomplishments of dates as per plan. Some documents may be delayed in the project course at any one of these stages. This may be due to several factors like:

- The estimated man-hours were less than actually required.
- The engineer assigned to this job is not competent enough.
- Client introduces certain queries/comments that need further investigation and additional studies.
- Some uncontrollable events happen such as the assigned engineer resigns and it takes some time for the newly assigned engineer to become familiar with the job.

In some cases two plans may be prepared. One plan is issued for the client and the other one is kept for internal usage. The internal plan always has earlier dates for each document. From another point of view these may be considered as early and late plans for the same project. The late plan will be issued to the client and the early plan kept for internal usage and progress calculation.

In defining the late plan overtime shall not be considered. In this manner possible delays can be covered with several days' overtime.

Normally 5—10% behind planned progress (delay) may be acceptable. This may be the difference between early and late plans. If it increases beyond this limit, lead engineers and PM shall prepare a catch-up plan. Therefore every week the progress report shall be carefully reviewed by them. Client also may issue warning letters and ask for catch-up plan.

Lead engineer shall only think of how to remedy delayed time and whether it affects next-stage documents. PM shall not only think of remedial actions but also shall see if the delay is caused due to client faults. In that case he has to prepare proper claims. Backup documentation to support claims is very important and is prepared from the planning team report.

Normally clients are very reluctant to accept they are the cause of a delay. But to be honest, a majority of delays are due to client action. The client enjoys a very powerful position without so much responsibility. In a majority of cases even after proving they have caused delay, the consultant has to undertake to rectify the status with its own man power and expect no compensation. The client only issues responses to enable consultants to finish the job.

If a client team is fully mobilized at early stages of the project and has competent engineers then they may contribute to the project progress. However, if their team comes late and insists on commenting on all documents even if they were issued much earlier this will have a disastrous effect on the schedule. Meticulous engineers from both sides may cause problems to project progress. In a majority of cases, instead of closing documents with a focus on major items, they request so many unnecessary details, reports and insist on their own understanding that the only way for the consultant lead engineer to get rid of them is to implement comments without further arguments.

5.1.7 Allocated Source, Budget, and Achieved Progress

As mentioned previously, invoicing progress may be different from physical progress. Other than PM and lead engineers normally no one is interested in physical progress. However, there are certain norms that are widely accepted. For example, when a document is issued in IFC stage, 60% progress is assigned to it. At this milestone invoicing and physical progress become the same. It means if at a certain date all documents are issued simultaneously at IFC stage and no other activity is taking place, then invoicing and physical progress are both equal to 60%. But between these stages, although the project team may have achieved progress (greater physical progress), it is not calculated by the client (less invoicing progress).

This time lag between achieving invoicing progress, issuing related invoice, approval by client, and actual payment is a buffer safety zone for the client, who is always sure they have received more work than paid for. At the same time this delay has a negative impact on consultant cash flow. To be fair, I have to admit part of this is compensated for with advance payment.

There is one important point here that has to be emphasized. In each project there is a gap between the reported invoicing progress by the project team and the accepted progress by the client. This is if the client is fair. A majority of clients don't even bother to tell consultants what is the acceptable progress and why some parts have been crossed. Only to withhold payments (with a contractual reason) to have more power in possible disagreements with consultants do they systematically deduct some progress rates. This is in addition to their contractual right of paying approved invoices with certain delay. In fact in every contract the client keeps a few days (up to 2 weeks) for progress approval and then some additional time (may be up to 1 month) for payment. The only safeguard that the consultant has is the advance payment (AP). Before instructing AP payment, client receives a bank guarantee for it. Normally bank guarantee conditions are very strict and client can confiscate them any time without necessity of a court verdict.

Unfortunately in none of the projects that the author was involved were either of these two dates followed. Delays of up to several months have been seen in actual payment compared to actual progress. In addition, some of the EPCIC contractors are stingy toward consultants and generous to vendors. Therefore it is a good practice (and in many cases necessary mitigation measure) to manually/artificially add weight of documents that are issued at early stages of the project. This provides a small help to consultant cash flow.

As an example, design basis and philosophies are issued at early stages (maybe the first one-fourth span) of each project. The author recommends increasing their weight at least by 20—25%. This has several impacts.

- Normally project progress curve is a flat S shape at beginning. This means in a specific time duration at early and late stages of the job, less progress is achieved compared to the same time span at the middle of the project schedule. In the middle of the schedule the progress rate is higher. Increasing weight of design basis and philosophies helps to increase the early slope of the S-curve. Seeing the higher weight discipline LE is encouraged to issue them at early stages of the job to get more progress, while keeping a suitable man-hour margin in the physical progress.
- At the early stages of the project client/EPCIC contractor time is less occupied. Therefore, they can comment on these major technical documents. This means consultant can upgrade them to AFC and use them in related documents as prerequisite.

- These documents are affecting other documents in the related discipline. Therefore receiving comments at early stages of the project and responding to them helps to get them finalized and open the path for other related documents like specifications, data sheets, analysis, etc.

The next-stage documents after design basis and philosophies are specifications, data sheets, and analysis reports. These are the basis for material requisitions and fabrication drawings. It is recommended to increase the weight of these documents as well. The rate of increase and selection of documents is to be done by PM. Lead engineers shall keep away from this practice. It may leave a negative impact on their evaluation of the project and lead to erroneous measurements of project actual progress. LE shall use actual physical progress.

Although PM shall always keep a close eye on invoicing progress, LEs are normally concerned only with physical progress. A certain number of engineers and man-hours are allocated to each LE. He/she has to finish the project with that budget. Almost in all projects actual consumed man-hours will exceed first allocated man-hours. PM and discipline head of department will always keep some reserve man-hours and release permit to use it after careful investigation of the situation.

5.1.8 Document Status

Each document's status shall be clearly known. In some projects construction activities may continue parallel to engineering. Therefore upgrading a document to AFC stage shall be carefully monitored. In some contracts clients don't care for the actual status of the document. With each new issue the revision number and consequently the status will raise one stage. However, in some other projects unless the consultant is certain of document quality and status they will reissue it at present status with a higher revision number. By quality only the engineering work is not meant. Receiving and implementing all concerned parties' comments, implementing interface impacts, reviews and implementing vendor data impact all have an effect on quality.

Upgrading document stage to AFC after ensuring all parties' comments have been implemented and vendor data impact has been foreseen helps to prevent rework in the construction yard. Almost in all projects EPCIC contractor can't wait for as-built vendor data to start fabrication.

From the payment point of view the second approach is not to the consultant's benefit. Besides due to high costs that EPCIC contractor may have in standby condition, they also need to start as soon as possible with available information. In the course of project progress some rework or additional purchases may be beneficial in total project cost savings. The unfortunate point is that although a majority of EPCIC contractors know the advantage of these small reworks they may misuse it against consultants.

Document status includes its stage and commenting status. Several parties may be involved. Client, EPCIC contractor, TPA, MWS, etc. all may be involved on a particular document. Although normally in all contracts it is written that consolidated comments of all concerned parties are to be issued simultaneously, seldom (if ever) does this materialize. Therefore consultants shall keep track of their comments that have been taken care and closed. This delay list is a helpful tool against client claims for document delay, although it may never lead to additional payment for consultant rework.

5.1.9 Who Shall Prepare the Document Register?

The document register shall be prepared by lead engineers of the related disciplines. In its preparation he or she may use their own personal experience or use the consultant data base prepared from employees' worksheets or get advice from other senior engineers in the department.

Lead engineer shall assign each document to a specific engineer. He or she shall strictly follow the single-point responsibility (SPR) rule. Each document shall only have one assigned engineer. The assignee may use other sources' information in IDC stage. In the IDC matrix it shall become clear which documents are to be distributed to whom and when. The responsible engineer shall check the received IDC responses to ensure received information is complete.

5.1.10 Who Shall Update the Document Register?

The document register is continuously monitored by the planning department. Each week they issue a report to the PM highlighting documents that are behind schedule. Some of the delays may be attributed to the client. However there are some delays that are due to underestimating man-hours required for preparing a document or some other reasons that at that time couldn't be properly managed to avoid delay.

Based on the mutual agreement between lead engineer and PM the plan for the document register can be updated. It is important to note updating is not a weekly job. The revised plan shall be used as the basis for progress calculation. Each time the first plan remains unchanged only the revised plan or forecast changes. The actual progress is calculated based on the actual document issue dates.

Internally both early and late progress shall be calculated. If there is no contractual requirement, it is strictly recommended to report only the late plan to the client.

5.1.11 Progress Status Evaluation

Two types of progress are defined. They are invoicing and physical progress. The first one is important for clients and is used as the basis for payments. The second one is important for lead engineers and EMs.

During bid stage EM receives (from LEs) estimated man–hours in each discipline distributed to lead, senior, junior, etc. categories. Each one has a specific unit rate. Multiplying unit rates to man–hours and adding contingencies each discipline weight in the project proposal is obtained. At bid stage a detailed MDR is not prepared. Instead shares of different document groups are determined. For example, design basis and philosophies get 10%, specifications get 20%, data sheets get 25%, calculations get 15%, main drawings get 15%, and others get the remaining 15%. Based on agreements at KOM a detailed MDR will be presented with weights assigned to each document. Since MDR is a live document, during project progress it is possible to add or delete some documents. Normally it is preferred that additions or deletions don't impact previous invoicing progress calculations/payments. Therefore, either the weight will be deducted from unissued documents or zero weight will be assigned to them.

There is some time delay between issuing a document (ie, obtaining actual progress) to a client for review of progress appraisal to actual payment. If everything goes smoothly a document that is issued during a month (even on the first day) will be invoiced at the end of the month. Assuming documents are uniformly generated throughout the month (on average) its progress is claimed with 2 weeks' delay. Client review will also take at least 2 weeks. After this stage consultants are allowed to invoice. Normal routine in client financial department and bank money transfer will take at least 1 month. This means that even if everything proceeds smoothly, actual payment will be received with at least a 2-month delay.

To overcome this delay and reduce its impact, two mechanisms may be foreseen. The first is AP. Clients accept to pay 10—20% of project value against an acceptable bank guarantee. With each progress invoice client deducts pro rata portion. For example, in invoice 10 if progress is 65% they will deduct 65% of AP value from invoice payment.

AP payment alone is not enough to cover large staff payments at the start of the project. EM shall consider higher weight for documents that will be issued at an early stage of the project. For example, if design basis and philosophies that will be issued at an early stage of the project are calculated to have 15% weight, they will be assigned 18—21% weight. Weight of specifications and data sheets are also increased but similar reductions will be made on remaining documents like drawings, MTOs, etc.

Physical progress is not needed by clients. But it is very essential for LE and EM. This may be calculated only considering man-hours without multiplying by unit rates. At each stage LE knows how many man-hours have been consumed and how much budget remains. It is to the consultant's benefit if compared to invoicing progress the physical progress is less. This shall not be misinterpreted. It doesn't mean fewer documents are issued than those accepted by the client. It means documents are issued with better efficiency than originally planned. This way the project team is able to invoice 100% when there are still some budget man-hours left from the original estimate. Physical progress is intended for internal use and shall not be disclosed to client team members. They will think the consultant is cheating them to invoice higher while their physical progress is less. This is not cheating. The difference is in higher unit rates for lead and senior engineers, which are above average project unit rate.

5.1.12 Catch-up Plans

Catch-up plans are needed to cater for delays incurred on the project schedule. In very rare cases in which client accepts their responsibility or when they instruct a change request, a catch-up plan may only include an extension of the engineering plan. In this case the same task force may be able to achieve this goal. In many cases discipline engineers shall accelerate document issuance by either working overtime, adding a number of involved personnel, replacing some members of the present task force with more qualified engineers, etc.

As a rule of thumb, PM/lead engineer shall monitor each discipline's progress on a weekly basis and initiate catch-up plan if delays become more than 10%.

At early and middle stages of project, PM and LEs will check progress on a weekly basis. At end stages PM may check progress in biweekly steps, but LEs shall continuously check it in weekly steps.

5.2 INTERDISCIPLINARY CHECKS/DISCIPLINE INTERFACES

The main item that distinguishes a team from individuals is the meetings and interfaces that they have with each other. This covers both technical and human interfaces, which should be extended to engineering and client team members. In some cases client engineers may find themselves in a situation separated from others. This isolated feeling is not positive. The author thinks very positively of human interfaces; however, this section is dealing with preplanned technical interfaces.

It will be an exceptional job to create feelings of a unified team between client/consultant team members. The author has in few cases succeeded in achieving this goal. This way in addition to getting salary, engineers enjoy working on this project. Two engineers working as good team members add to each other's capabilities and cover weak points. They may produce 2.5 unit results. Two engineers working as bad team members decrease each other's productivity. They may produce 1.5 unit results consuming the same man-hours.

5.2.1 Identifying IDC Checks (IDC Matrix)

For multidisciplinary projects it is essential to have comments from all involved disciplines. At the start of each project and after preparing each discipline MDR, a general meeting between lead engineers has to be performed by PM (EM). In the meeting each LE will mark documents from other disciplines that he or she wants to review/comment on and at the same time mark documents from his/her discipline on which they want other disciplines to comment. The extent of commenting may vary from checking or inputting a few interface items to completing sections assigned to that discipline. IDC matrix shall be discussed mutually to achieve a general agreement.

An example is given for package data sheets. In some projects each discipline like mechanical, process, electrical, etc. prepares a separate data sheet for each package. In order to issue that package MR, all data sheets have to receive comments from client and either implement or respond to them and upgrade to a suitable level to be included in MR. Based on the author's experience this is very difficult to manage if data sheets are issued separately by each discipline.

Consider a pump that may require process, electrical, piping, mechanical, instrument, and structural information to be completely defined. Some of the information to be defined by each discipline includes:

- Process shall determine design, operating pressure/temperature, discharge rate, suction, and discharge line sizes.
- Electrical shall determine power consumption including voltage/current intensity, cable sizes, feeders, and protection rating.
- Instrument shall determine controlling signals like start, stop, running, and fault detection, whether they are transferred to integrated control and safety system or unit control panel.
- Piping shall determine material of construction and pipe specifications.
- Mechanical shall determine pump type and other mechanical specifications like continuous or intermittent operation.
- Structural shall determine supporting conditions and accelerations experienced in different installation, transportation, and operation stages.

These are at least six different disciplines. Some of the information comprises only a few words or figures. Some others may need extensive descriptions. Minimum document status to be included in MR is IFA. If each discipline issues necessary information separately, this makes six documents to be issued at least twice. Client comments shall be received, reviewed, and either implemented or responded to at both stages. The solution is to issue only one data sheet by the discipline that is responsible to procure that package. Rotary equipment is normally included in mechanical discipline responsibility. During IDC related sections in a consolidated data sheet will be filled in by each discipline. Ultimately before issuing with MR only one document needs to be issued twice. This is more manageable.

In addition to this case, many other documents shall be reviewed and interface data filled in or comments affecting them to be reviewed and agreed to by both originator and the related discipline. IDC check requirement for same type of document in different disciplines is not constant. For example, process design basis shall at least be reviewed by mechanical, electrical, instrument, and piping, while structural design basis may only need to be reviewed internally. Packages' specifications prepared by mechanical discipline shall be reviewed by others like process, electrical, and instrument. Table 5.4 shows a typical IDC matrix.

This table shows seven process documents that are part of their philosophies, calculations, reports, and specifications. As is clear, the certifying authority (CA) has requested to review all except one. Process itself is the discipline responsible for generating these documents. Safety has requested to review three, mechanical five, piping five, electrical one, instrument four, and structure has not requested to review any document. This table can be agreed on at the first stage of the project and if needed revisited during project progress. As one of the QC steps in issuing a document, DCC will use this file to check if all IDC steps have been completed before issuing that document.

5.2.2 Dedicated Time/Priority

Normally all disciplines face man power shortages. When commenting on IDC documents each discipline has their own shortcomings. Therefore it is natural to give priority to their related discipline documents. This may be acceptable for 1 or 2 days. But more than that will have cascading delay impact. Therefore in the IDC transmittal each discipline shall mark the return date. Before return date an alarm date may be specified. This will help DCC to check document status. In case response is delayed, the related discipline shall be immediately notified to assign an engineer urgently to perform and release the job.

Similar to the delay list on issuing documents DCC shall prepare delay list in commenting on IDC documents. To avoid cascading delay impact, PM shall closely check this list and expedite required documents.

5.2.3 Weekly Meetings

Weekly meeting is an essential tool in project coordination. In addition to general meetings, private discussions with each discipline shall also be conducted. The following issues may be discussed:
- Each discipline's delay and mitigation measures to prevent further delay/remedial action
- Delayed IDC documents and related catch-up plan
- Client commenting status and its impact on project progress
- Cascading impact of delays (from all sides) on remaining documents
- Information circulation/interface between disciplines
- General issues that all disciplines shall follow

Table 5.4 Typical IDC matrix page

No	Doc number	Document title	Doc type	Class	Discipline							
					CA	PR	SA	ME	PI	EL	IN	ST
Process												
Philosophy, design basis, calculation, specifications												
1	NOR-GEN-PR-DB-3900	Process design basis	DB	1	y		y	y	y	—	y	—
2	NOR-GEN-PR-RO-3901	General philosophy for isolation, shutdown and draining	RPT	1	y		y	—	y	—	y	—
3	NOR-GEN-PR-CS-3902	Process and utility line sizing calculation	CAL	2	y		—	—	y	—		—
4	NOR-GEN-PR-SP-3903	Water treatment system—duty specification	SPC	2	—		—	y	y	—	y	—
5	NOR-GEN-PR-SP-3904	HP/LP flare TIP process specification	SPC	2	y		—	y	—	—	—	—
6	NOR-GEN-PR-PR-3905	Emergency shutdown and depressurization philosophy	PHL	2	y		y	y	y	y	y	—
7	NOR-GEN-PR-RO-3906	Flare, blowdown and depressurization calculations and report	RPT	2	y		—	y	y	—	—	—

Meeting schedule shall not be rigid. Otherwise it may waste time. Some issues that shall be observed in planning each meeting may include:

- At early stages of the project weekly meetings may be necessary. Later the schedule may need to change. Biweekly and even monthly meetings are recommended at later stages of the project.
- The first stage of each meeting may be dedicated to a review of agreements in the previous meeting and completion status of each agreement.
- Meeting agenda shall be distributed in advance to allow participants to gather required response.
- Ad hoc meetings may be arranged when necessary. Normally they may cover only one or two subjects.
- It is advisable to group meeting agenda in several subtopics. The first will be "general" to cover issues all or majority of disciplines shall know. Issues related only to certain disciplines may be discussed individually or the related discipline may be allowed to leave the meeting after closing their issues.

5.3 DOCUMENT CONTROL CENTER

The document control center is the main control center for document handling/distribution in each engineering team. Documents and information flow to and from engineering team is directed by this group. Normally DCC staff are from secretarial and clerk levels. Very clear and established procedures shall be defined to avoid mistakes in document flow. This team shall perform a number of tasks. The following list shall be adapted to each consultant and project requirements. Some of these items may be transferred to the planning department. In a majority of cases for a specific project, planning and project control team can be stationed within DCC staff as part of the task force assigned to the project. Some of the tasks listed following may not be required as per contract terms.

DCC tasks may be listed in three categories: administration and QC; document handling; and document control.

5.3.1 Administrative and Quality Control

Tasks under the administration and QC category include activities needed to ensure documents are issued as per approved project format. If consultant has a dedicated department to cover all administrative tasks, they may be handled elsewhere. However, in a majority of projects DCC team may undertake all secretarial work and coordination needed by the project. The major items are listed herein:

1. Review of document's format to follow project quality system: Normally this activity may be only a glance through the document. At the start of each project document's/drawing's format will be prepared and distributed to all disciplines. This includes cover page, tabulation page, table of contents, introductory paragraphs' text, and other pages' logo and format. Format may extend from page margins to font size, line spacing, indents, definition of headings, etc. In some projects client may dictate its own preference, and in others consultant may use their usual format. After distributing approved formats, all engineering team members shall follow the same procedure.

2. Coordination with client for their possible trips to reserve hotels, book tickets, transportation, meals, etc.: This is necessary in case client and consultant are at different locations. Normally client team members (especially high-ranking managers) expect a specific level of courtesy. Regardless of cost or consumed time they expect the consultant to treat them with respect as VIP. This is neither illegitimate nor wrong.

3. Coordination for engineering team members for travel and meetings with client, EPCIC contractor, yard fabricator, vendors, etc.

4. Coordination with engineering subcontractors (if any) for general meetings like design review, HAZOP, etc. to book tickets, transportation, reserve hotels, meals, etc. For example HAZOP chairman shall be out of the engineering team. It is not cost-effective to recruit specialist study engineers on a permanent basis. Normally their services are hired for a specific task. Chairman or specialist may be stationed for a short duration in consultant office. Cost-wise it is efficient to devote their entire time for the assigned task. Even if their contract is on lump sum basis, it will save on the total cost to help them in these trivial matters with an office staff.

5.3.2 Document Handling

Project document handling is the main task for the DCC. This task covers a wide range of activities. Following, the list numbering continues from the previous section.

5. Establishing project filing system (including hard and soft copies) and continuously updating it: Consultant tendency is toward paperless office. Well-established office network with clearly defined access

routes is a real necessity. Client access to information shall be as per contract. For more safety it is even possible to physically separate networks. This may require additional effort in document uploading and increased costs in hardware installation. Software for document management is available. The author, however, has no experience with them. All their capabilities can be incorporated in a well-defined office system. Main capabilities of these document management systems include:

- They make it possible to assign different access levels to individual documents per user.
- They prepare a report of working time on each document by connected users. This is used in project control and may later be used in estimating man-hours required for each document during bidding.
- In addition they enable document issuing after release permit by predefined authorized user is granted.
- They can assign schedule to each document, which helps in planning and project control.
- They prevent excessive printing and help to move toward paperless office.

Hard copy archiving requires a lot of space. After each new revision old ones shall be transferred to a discard archive and kept only for the sake of history. Of course all correspondence including comments, responses, minutes of meetings (MOMs), actions shall be kept in a traceable manner. Any time in a project a newcomer may open a long time ago disputed and closed issue. Especially if he is from client team he expects to receive a response with full backup documents.

6. Set up and revise access level for different participants to project data base: This may include different levels of access for PM, LE, engineering team, and client team. In some cases to avoid mistakes different servers or physically separated networks may be used for different teams. Normally each engineer prepares assigned documents in his or her own computer or in the dedicated section of server. Then he shares it with the checker or LE for approval. In each discipline only LE has access to all native files and issued documents of that discipline. PM also has access to all disciplines. Only DCC has the responsibility and therefore the ability to remove or copy a file to the assigned section. Others may only have "read" access to other disciplines' documents. It is even better to transfer required documents between disciplines through DCC personnel. To have a clear line of responsibility it is recommended to

assign someone for interaction with DCC in each discipline and give cut-and-paste authority only to one of DCC's personnel.

7. Corresponding with client to ensure documents are downloadable after transferring them in soft copy format (via e-mail, or Internet to download from an FTP site): In certain cases the Internet access may impose some difficulties. This problem is in third-world countries. Internet access speed is continuously increasing. This correspondence ensures that the client has received required documents. Immediately after uploading, client shall be informed that certain documents are available for access.

8. Scanning and uploading issued documents on the server and if required on the Website: Some documents may contain several hundred pages of A3 or A4 size papers. Some may have been printed on both sides of paper. Some pages may not be legible or may have been put in the wrong direction or page number order. Before scanning, all these issues shall be reviewed very quickly. With automatic high-speed, high-resolution scanners the above checks will produce fast and neat results. Almost all software now has the ability to directly print to imaging files (like PDFs). The produced document is searchable, is smaller, and is clearer. However, since it can be altered, normally it is avoided and scanning is done from hard print. Security limitations can also be installed on the files. This depends on the contract. One solution is to send one set of hard copy in addition to the image file from software.

9. Preparing/issuing necessary transmittals and correspondences to client: All correspondence with client including letters, MOMs, engineering documents, invoices, change requests and orders, technical queries, etc. shall pass through DCC. For some, DCC has to prepare transmission document and for others shall only archive or list it. In any case it is DCC's duty to ensure correspondence is received by the client. If an action/response is expected, DCC shall expedite to receive it.

10. Preparing necessary number of hard copy documents: Although gradually decreasing, even now a majority of clients prefer to receive a necessary number of hard copies from consultants. Commenting on soft copies still has some restrictions. This will reduce client headaches on preparing additional copies and distributing. In project peak progress normally some documents are issued daily. Later they may reduce the number of documents per week. In some projects parts of engineering activity may have been subcontracted. This means specialist

staff may be stationed in a separate building. Some specialist studies like HAZOP or computational fluid dynamics (CFD) studies are of this category. Although in a majority of projects client members may be stationed in the engineering office, they may insist to deliver a certain number of copies to their main office. Each project will have a printing policy that shall be followed by DCC. This is specially recommended if personnel handling different projects are stationed in the same department.

11. Dispatching printed documents to client on a daily/weekly basis as per contract requirement: DCC shall find a procedure to receive a receipt for each transmission. Even for electronic transmissions a read receipt can be obtained and shall be archived.

12. Coordinating with client and the dispatch organization to make sure documents are received and signed transmittals are returned/archived for future reference: This is specially required when a third-party organization out of consultant personnel is hired/contracted to perform this task.

13. Printing, faxing, scanning, and uploading project correspondence: As explained in item nine herein, all correspondence shall pass through DCC. Some of them, like invoices, need preparation of a certain number of original copies. They shall all be signed and stamped. While others, like letters, need only the PM's signature. The contract defines the authorized representative signature for commercial and technical documents. Normally PM is the fully authorized representative.

14. Receiving client comments on project documents, listing them, and dispatching to discipline engineers: Client comments are the main issues that have to be discussed, responded to, and resolved. Accurate tracing considering comments themselves, their receiving dates, dispatching to discipline, and receiving response is a major necessity. In some contracts clients may insist to receive responses, discuss, and resolve them before the next-stage document is issued. This may be applicable for some main documents. Following the same procedure for all may take a long time.

15. Keeping track, expediting, and receiving consolidated discipline engineers' reply to client comments on issued documents: This may vary from accepting, responding back with additional backup documents, to rejecting or accepting a compromise solution. DCC shall issue response letters/transmittals. In some projects client may instruct to receive a response sheet and close it before issuing the next revision

of a document. They may want to have a meeting to discuss/agree on the responses. This is only possible if client members are stationed at the same office. Meetings will consume considerable man-hours of a project's budget, unless they agree on quick informal discussions. Too many meetings/discussions may also detract consultant engineers from their intended plan. On the other hand, if client intention is not fulfilled they may continue endless commenting. Therefore it is preferred to avoid this as a general approach for all documents if there is no contractual obligation. For some specific key documents like design basis, philosophies, main specifications, and data sheets, following this approach of closing comments in a meeting before issuing a revised document may be beneficial to all sides. Meeting agreements shall be carefully listed in MOM to know each party's undertakings. In a majority of cases a meeting with no MOM will be forgotten by all, or at least majority of, attendants.

16. Uploading revised documents by discipline engineers onto the server: To avoid introducing viruses to the server and losing some documents, only DCC staff shall have write permission on the server. Other team members have only read permit. Inside DCC they may assign a single username or different usernames.

17. Preparing necessary transmittals to client: Based on contract requirements, consultant may need to issue documents to several users like EPCIC contractor main office, fabrication yard, commissioning subcontractor, installation subcontractor, company, third party, etc. Some of these recipients may not receive all documents. A distribution matrix shall be prepared at the beginning of the project to inform DCC staff who shall receive what.

18. Distributing copied documents: Normally clients will receive as many hard copies as needed but will only send one set of a hard copy. In general only the discipline that has prepared and is responsible for issuing a document shall receive comments on project documents. Vendor quotations and documents shall be distributed to several disciplines for review and comment. The discipline responsible for that package determines which other disciplines shall receive what portion. They may mark on the hard copy or the soft copy of transmittal. This will inform the DCC of the required number of hard copies or recipients of related e-mails. In each case DCC staff shall trace issued documents to enable them to receive responses on time and transfer them to the originating discipline.

19. Preparing transmittals and issuing client comments with consolidated discipline engineers' response to third party for endorsement: This may not be required in all contracts. However, in some projects a third party may wish to know client comments on IFC documents for their own review. If possible, consultant shall not accept this term. The point is that the third party shall be independent and act unbiased. Their decision and approval shall be based on general engineering practice and applicable standards. Even if needed they can perform their own independent analysis.

5.3.3 Document Control

Parts of tasks listed under the document control topic are project control department tasks. Here it is assumed the project control engineer is stationed in the DCC office. To avoid duplicated effort, they are listed under DCC tasks. Assignment of tasks listed in Section 5.3 to different groups is no problem. Care shall be exercised that tasks required by the project are not neglected.

20. Prepare and update on a weekly basis the list of issued letters, correspondences, etc. showing what has been sent in which parcel, who has issued it, when it was issued, when the client received it, who in the client organization received it, if and when we expect to receive any response for that, keep track of its response, when has it been uploaded onto the Website, etc. This information is very vital for project control requirements. This shows why in some cases project control personnel are stationed in the DCC or have direct access to some of their documents and data base.

21. Expediting discipline engineers for response to IDC documents, keeping track of interdisciplinary transmittals deadlines, and highlighting delays if any: As explained, each discipline gives priority to its own documents. Therefore IDC documents may wait in the queue or even be neglected. This may have adverse impact on other disciplines' progress. Some documents (especially those directly used for procurement or fabrication) may have cascading impacts, causing delays on total project progress.

22. Updating discipline documents registered in cooperation with discipline lead engineers: Sometimes disciplines may have their own personnel for archiving and updating MDR and sometimes DCC personnel may do it for them. Especially if a company is using an early plan for internal usage and late plan for reporting to the client, both shall be updated—the first one in LE computer and the second one in DCC server.

23. Keeping a list of project correspondence and their closing status on a weekly basis: Some of the letters may not need a direct response, but other correspondence like technical queries, invoices, change requests, and change orders require a clear response. At each stage the PM shall know which items still need expediting and which may impact project progress. It is important to note that some queries that need definite instruction may delay project documents. An accurate report not only helps to have fruitful meetings but also may be used later in claims and counterclaims.

24. Keeping track of documents issued for endorsement and highlighting if there is any delay: In some projects client may instruct another certifying authority (CA) to either endorse or approve consultant documents. CA will not review all documents. They may not even look at certain documents like MTOs, lists. However, some others like philosophies, basis of designs, PIDs, etc. will be carefully reviewed and commented upon. Client may also request to receive a copy of the CA comments and consultant responses. They may even prevent a document to be issued in AFC if CA is not satisfied with the responses. Closing CA concerns may play a major role in project progress.

25. Keeping track of issued documents to client for approval, commenting, and expediting on a weekly basis: In almost all projects in the contract it has been stipulated that any document not responded to within a certain period (normally 14 days) will be deemed approved. But again in a majority of projects the client has not followed this rule. They use a very well-known excuse (which is again stipulated in all contracts) that a technical fault shall be rectified any time. From the author's experience, if out of 100 late client comments the consultant manages to prove 50 are only personal preference and not a technical requirement, he has had good luck! In many cases time and effort needed to justify this is more than the time required to redo the job. This report will be used in claiming against delays more to save the consultant from liquidated damages and not to receive compensation.

26. Keeping track of final AFC documents, preparing transmittals, and issuing to client: Before issuing AFC, consultant may add any number of changes. After that any change shall be carefully investigated. Normally the contractor is allowed to fabricate based on AFC drawings

unless some portions of it are HOLD. It is possible to upgrade a document/drawing in AFC but HOLD an entire sheet. It is clear that consultant shall have a proper reason to HOLD a document. Otherwise stopping the contractor will have considerable cost impact for yard man power and delay in project.

27. Keeping track of and listing document revision permits from different disciplines: This permit may be defined in the engineering execution procedure. LE or an authorized SE shall prepare it to allow the respective engineer working on a document to spend man-hours and upgrade it. This permit may contain brief information describing the reason for and the extent of revision. Majority of LEs don't like this permit due to considerable paperwork. However, it is a very good source to prevent unimportant changes, to track implemented changes, and keep man-hours within budget.

28. Keeping track of project audit dates, findings, and remedial action status: Other than technical audits, HAZID, HAZOP, and design reviews there are certain nontechnical audits that only focus on the formats and checks if adapted procedures have been properly followed or not. These are defined in consultant quality plan. Audit results are consultant internal reports and need not be issued to client.

29. Updating list and revision number of issued documents with weekly report: Preparing detailed and accurate weekly report is an important task. It is to be done by project control department/group or representative. I have named all three to show it is possible for project control to be performed in a separate department. This may have the advantage that their reports are free from influence of PMs or discipline leader and can therefore evaluate project progress from an unbiased point of view. In addition their representative may be assigned as project task force and be directly connected to PM. Since PM is responsible to perform his tasks within approved man power and allocated time frame, he may use project control help to this benefit.

30. Keeping track of documents issued to yards and coordinating to receive list of documents necessary for their fabrication: This is part of EPCIC contractor DCC team. However, in some contracts they may not mobilize the necessary employees in their project team and instruct consultant to perform it on their behalf.

In this section many times we have referred to preparing a transmittal. Although it may have different formats, the information included is more or less similar. A transmittal is a covering page accompanying each batch of documents issued to a receiver. It may contain the following information:

- Project name and number
- Name and logo of parties involved in project
- Name and reference number of the originator
- Name and reference of the receiver
- Document(s) title and number
- Document(s) revision number, status, and class
- Page size and total pages
  Each transmittal shall have a location for signature of the receiving body.

5.4 VENDOR DATA CONTROLLER

Vendor data controller (VDC) has almost the same function as DCC with respect to vendor data. There are some differences both in originating and handling of documents:

- Main difference is that related documents in DCC tasks are generated by engineering team and commented on by others, while in VDC tasks documents are generated by vendors and commented on by engineering team.
- As per engineering contract, consultant is responsible to rectify any engineering mistakes any time. Although normally client commenting cycle has a definite time limit, generally they continue commenting any time it suits them. In contrast, if consultant starts a new comment on vendor documents, they will claim delayed time.
- Before purchase order (PO), vendor may accept comments leading to changes in technical specifications. They will calculate its financial impact in their final quotation. Even if they ignore cost impact it may be a calculated action to improve their chance to win the tender. After PO vendor normally doesn't accept any change without prior approval of its time/cost impact. In some cases in which they plan to introduce themselves in a new market or want to improve their reputation (expecting future contracts), vendors may ignore small cost impacts. In others they may use any small opportunity for a claim.
- DCC has direct communication with producers and recipients of documents while VDC normally receives documents from and passes comments to EPCIC contractor. There is no direct contact between consultant and vendor. This indirect contact is essential during vendor quotation review. The main intention is for consultant to concentrate

on technical issues and disregard commercial items. However, after PO direct contact to some extent may be preferable.

- In majority of documents distribution matrix defined for VDC and DCC is completely different. EPCIC contractor may not want some documents to be issued to client. Selected topics and number of required hard copies differ considerably in the two groups.
- Client/EPCIC contractor actions to project documents other than class category are more or less the same. While vendor documents during quotation evaluation/ technical bid evaluation (TBE), after placing, PO and final data book have different status. For example, client may not need to review comments on vendor quotations but requires reviewing comments on specific vendor documents after PO. However, EPCIC contractor is interested in all comments having time/cost impact.
- Engineering team action and consequently VDC action on vendor quotation and vendor data are different. Their format and extent also differ. In majority of MRs specific list is given for inclusion of necessary documents in quotation. This is intended to facilitate one round of clarification and possibility of comparison between different quotations. Some vendors don't follow it. After PO majority of vendors follow approved supplier documentation requirements schedule (SDRS).
- VDC Transmittal format is completely different from DCC. It shall enable related disciplines to identify which part of quotation or vendor data shall be handed over to whom. Normally a table is placed at top of the transmittal. Below it all received documents are listed. The horizontal row of the table has numbers corresponding to the list below it. The vertical column of the table shows discipline representative names. The discipline responsible for that package shall receive a whole pack of documents. In this table and in front of other disciplines' rows they mark the cells related to document numbers as given in the list.

An example is given in Table 5.5. Assume Oily Water Package vendor has issued five documents. Mechanical discipline is responsible for this package. Documents include:

1. Degassing skid GA drawing
2. Degassing vessel nozzle location and orientation
3. Degassing skid PID
4. Degassing skid control panel
5. Inspection and test plan (ITP)

Table 5.5 Typical vendor data distribution

Discipline	1	2	3	4	5	6	7	8	9	10
Architectural	—	—	—	—	—					
Electrical	—	—	—	x	—					
Instrument	—	x	x	x	—					
Piping	x	x	x	—	—					
Process	—	x	x	—	—					
Mechanical	x	x	x	x	x					
Safety	—	—	—	—	—					
Structure	x	—	—	—	—					
Telecom	—	—	—	—	—					
QA/QC	—	—	—	—	x					

As can be seen in Table 5.5, AR, SA, and TL are not interested in any issued documents. EL and ST only review one. QA/QC group is normally out of consultant team. IN and PI each review three documents. This pattern shall be clear for documents included in each transmittal.

It is understood that there are more similarities than differences between the tasks undertaken by each group. Main activities of archiving, expediting, checking, etc. remain the same.

In this section first different types of vendor data (after PO) are described, and then VDC action on their handling and progress control during quotation review and after PO are described.

5.4.1 Types of Vendor Data

Documents issued by vendors during bid stage are very different from those issued after PO. Documents issued by vendor after contract is assigned follow SDRS list. Of course vendor may not follow SDRS included in MR. It is mainly intended to serve as a sample and guideline. At first stage after PO, vendor will issue its own SDRS, which shall be discussed and agreed upon. Later a vendor document list will follow agreed SDRS. Broadly, vendor data may be categorized in four types including contractual, technical engineering, QA/QC, and operational documents.

- Contractual

  These are documents that define manufacturer schedule (engineering, procurement, fabrication, testing, and dispatch), organization (man power, facilities/equipment, subcontractors), subvendors for raw material or package items, testing, progress measurement, delivery schedule, progress measurement procedures (physical, invoicing, weekly or monthly reports, areas of concern), etc.

- Technical Engineering
These are documents showing vendor design, tie-in flanges/interfaces with other platform equipment and utilities like compressed air system, sea water, fuel oil, blanketing gas, power system, consumption, general arrangement, technical catalogs, PIDs, BDs, control system, package engineering calculations like process, structural, piping, electrical, package 3D modeling (if applicable), data sheets, specifications, etc. In engineering documents vendor shall give calculations for package capacity, function, mechanical design, power consumption, performance, etc. In some cases vendor technical catalogs may be accepted. For proprietary items vendor may not give their design details. Instead, some certificates from third parties may be accepted.
- Manufacturing procedures
In these documents vendor needs to give relevant bending, welding, forming, painting, etc. procedures. The list shall be extensive enough to show vendor has completely understood the steps involved in manufacturing this package.
- Quality Assurance/Quality Control
These documents show vendor organization for quality management. QA documents may include procedures for material selection/procuring, performing engineering design, welder qualifications and evaluation, welding procedure qualification, equipment suitability, worker knowledge of the job, tools calibration, nondestructive tests, dimensional checking, quality plan, inspection test procedures, etc. QC reports may include material composition/mechanical property test reports, third-party certificates, nondestructive (or if applicable destructive) test reports on the selected specimens, ingress protection/hazardous classification test reports, parts or full package operation tests, etc.
- Operational and maintenance
These documents show vendor instructions for package handling, installation, operation, maintenance, and repair. Vendor shall give related procedures to enable safe and sound usage of supplied equipment by operators during platform life time.

Each group of documents may be needed by a certain party. For example, EPCIC contractor procurement department is interested in contractual documents. These documents allow them to check manufacturing status, arrange package testing, transport, and handover to yard personnel. Engineering team is interested in technical documents. This enables them to provide required utilities in the correct location and in sufficient capacity. Client/EPCIC contractor QC team are interested in QA/QC procedures/reports. EPCIC contractor yard and client operators are interested in operational and maintenance documents. Therefore not only consultant but also EPCIC contractor and client shall comment on vendor documents. Regardless of who is commenting on what, vendor final dossier shall contain a complete package for future usage.

5.4.2 Vendor Data Handling

Similar to DCC, vendor data handling covers all stages from receiving a document to gathering consolidated comments and dispatching to EPCIC contractor. At each stage general status shall be clear. Before start of actual manufacturing, responsible discipline shall give "go ahead" signal for fabrication to vendor. This doesn't mean that back-and-forth comments can continue endlessly. Other than teleconferences, face-to-face meetings with vendors may also be arranged. This shall exclude major events like KOM, HAZOP (if applicable), factory acceptance test (FAT), etc. VDC tasks from start of quotation review to final dossier can be listed as follows:

Quotation Stage

1. Establishing project filing system (including hard and soft copies) and continuously updating it: Quotations, technical queries, and vendor documents shall be archived in different locations. Responsible discipline shall have close cooperation with VDC to determine how shall documents be distributed. For example, assume a quotation includes GA, preliminary PID, and BD. Structural discipline may only need to review GA. PID and BD are not required for them. PID shall be distributed to process discipline. Piping may require both GA and PID. Instrument requires BD. If it is a vessel then electrical discipline may not want to comment on any part of this quotation or vendor data.

2. Receiving the quotations, listing their status, and dispatching to responsible discipline for technical clarification (TC): Expediting discipline engineers to issue TC may be in VDC or project control scope of work. VDC may keep quotations in its office and dispatch to discipline after sufficient number of quotations become available, or send to discipline to keep and maybe start a preliminary review if they find spare time. Later in this chapter a first-stage control checklist will be explained. First-stage checking may be done by a junior engineer to ensure

submitted quotations are complete. This will save senior engineers time. If anything is missing, they can inform EPCIC contractor to provide missing information.

3. Downloading vendor/client procurement responses to TCs, printing and issuing them to discipline engineers responsible for the package and distributing IDC transmittals: Vendor response may range from revising quotation, adding new data to it, changing parts of it as per requirements, providing backup documents, etc. Client may normally not interfere in TC stage. Only after issuing TBE they may comment on parts of it.

Although TBE and MR for purchase are related to procurement and vendor data activities, they are issued in DCC team. Only quotations and clarifications are kept in VDC.

After PO and at Vendor Data Review Stage

4. Coordinating with EPCIC contractor procurement department to issue the introduction letters: This may not be required in all projects. In some projects EPCIC contractor may accept or instruct consultant to directly contact vendors after issuing PO. Before TBE, communication with vendor is only through EPCIC contractor. After placing PO, package cost is fixed. Therefore to expedite information circulation, consultant may directly contact vendor. In this case EPCIC contractor procurement manager shall officially introduce consultant representative to vendor. In this introduction letter (which may be done in vendor KOM) contact points/representatives from each side and the extent of their authorities shall be clearly defined.

5. The first step in VDC is to receive vendor documents. VDC staff shall check if they are accompanied by a proper transmittal and are following project requirements. Quality of the document to enable engineers to review them like visibility of numbers, legible sketches, etc. shall be checked at this stage and remedial action taken before reproduction and distribution. VDC shall receive vendor data, download, and print them. If vendor data are transferred via e-mail or access to vendor FTP site, they shall print necessary numbers and pack if hard copy is received. A detailed list of received items shall be prepared to enable accurate tracking of the received data.

6. Comparing vendor data with their issued SDRS and sending expediting letters to vendors: VDC staff shall check received data with SDRS issued by vendor or document list attached as an appendix to

MR issued for purchase. Their issuing sequence, dates shall match vendor SDRS.

7. Vendor shall follow ISO requirements in issuing documents. Review of vendor documents format to ensure they follow project quality system is part of VDC tasks. Proper project format (logos, numbering, title, etc.) shall be used, revision numbers shall be correct, changes should have been clouded, response sheet should accompany issued documents, etc. It is very difficult to instruct vendors to change their document format. Normally the vendor engineering department is producing several documents for the same equipment for different projects. Other than small changes in logo, title, and project name they will not accept any other changes. They normally resist instructions for adding new sections to each document, performing new analysis or simulations.

8. Distribution of received vendor data to relevant disciplines (commissioning, yards, client, QA, etc.): VDC staff related to detail design engineering team are only concerned with discipline requests. However, VDC staff in EPCIC contractor team may need to dispatch related documents to others like commissioning team, yard engineering, and client. In some projects EPCIC contractor may request consultant to perform this task on behalf of their team.

9. Receiving engineering group comments on vendor documents: Detail design team shall check vendor data for two main reasons:
   • Engineering team shall ensure project specifications are exactly followed. If there is any deviation it shall be discussed and reviewed/approved properly. All deviations (minor/major like material of construction, utilized power, capacity, etc.) shall be highlighted to enable a proper decision.
   • Engineering team shall ensure all tie-in points are properly taken care of. For example, if there are any water, fuel, electrical power, instrument cable connections they should have been properly reviewed to have the exact location and value on the platform.

10. Preparing necessary transmittals and correspondences to EPCIC contractor/client with a complete list of commented documents (scanning, copying, uploading, and packing as necessary): This is to ensure vendor has received full response on time to avoid any delay in package manufacturing. Again transmittal format may be different from DCC. Logos, project title, etc. are similar. To avoid mistakes, different

numbering can also be used. For example, if DCC transmittals start from 1, VDC can start from 5000.

11. Scanning and uploading commented documents by engineering group on the server and if required on the Website: Normally documents uploaded on server are for internal LAN users and those on the Website are either for client or vendor downloading. In this case proper username and passwords with correct authorization should have been defined beforehand. They shall be revised as suits each project.

12. When documents are transferred in soft copy format (via e-mail, or Internet to download from FTP site) correspondence with client to ensure documents are downloadable and files are not damaged: In downloading vendor quotation or data, they shall be legible and printable. As explained before, correspondence channels may differ based on contract requirements. VDC may directly correspond with vendor to send a damaged file or ask EPCIC contractor representative to communicate and receive required file. If communication is indirect it is to the EPCIC contractor's benefit to do this checking before transferring to consultant. This will save project time. The additional consumed man-hours in EPCIC contractor procurement or engineering office are much better than delay in fabrication yard.

13. Reproducing necessary number of hard copy documents: With paperless offices this may not be required. However, the originating discipline may want to keep one set of consolidated comments for easy access. They may prepare it themselves. It is too time-consuming to circulate one set of hard copy between all involved disciplines to write down their comments. All disciplines shall receive their assignments simultaneously. Only in this way can project deadlines be met.

14. Dispatching copied documents to client on a daily/weekly basis as per contract requirement: Although in distributing documents to disciplines, EPCIC contractor and vendor's VDC may need to work much more than DCC, here the workload is normally much less.

15. Coordinating with EPCIC contractor, client, and the dispatch organization to make sure documents are received and signed transmittals are returned/archived for future reference: Importance of receipts in VDC may be more than DCC, because vendors tend to claim more if there is any delay. Although this may seem pessimistic, sometimes vendors plan a production time schedule with their present load factor. After a first contract they may receive others. For manufacturers it is very difficult, if not impossible, to immediately mobilize additional man power.

They will try to keep the original plan with overtime and even maybe some subcontracting, but at the same time they will try to find some legitimate excuses for delay on vendor data comments. Therefore, project control team and VDC shall be very careful on project deadlines.

16. Printing, faxing, scanning, and uploading project correspondence: Other than technical queries by vendors, which may need correspondence, normally VDC has less work in this category.

17. Scanning, preparing the transmittals, and uploading the engineering comments/clarifications to vendors or client procurement team: Again, in paperless offices comments may be given on the soft copy document. Document management systems provide this facility to enable engineers to directly give their comments on the soft copy. Other than saving the environment by preventing many hard copy printouts, this has a very good advantage in allowing all related parties to view others' comments at the same time. Of course at any time only one person is able to comment and others will have read-only access.

5.4.3 Vendor Data Control

Controlling progress and status of vendor quotations and documents may have same importance as project documents. Parts of these activities may be assigned to project control team. However, information providing evaluation basis is generated in VDC. Therefore, similar to project documents, project control engineer can be stationed in VDC. Physically DCC and VDC may be stationed at the same office. In addition when vendor data starts, project document workload is reduced. Therefore the same project control engineer with the help of some VDC staff can undertake these tasks.

Quotation Stage

18. Keeping track of MRs issued to client procurement and expediting in issuing quotations: EPCIC contractors are very eager to receive long lead items MR at the first stages of the project. Long manufacturing duration, exotic material, sophisticated design, proprietary design, new technologies, etc. characterize long lead items. High-power turbo-generators and compressors, high-pressure pumps, dehydration and regeneration packages, Class 1500 and above ESD valves, etc. may fall in this category. In some projects client may decide to free issue them to save time. Even in this case vendor data review and matching the

input and output to project requirements will be done by consultant. EPCIC contractor shall also follow deadlines in submission of required quotations/vendor data. Engineering PM shall have a specific time limit. Delays in submitting quotations that exceed allowable values shall be documented and informed to EPCIC contractor PM. This letter will both act as a warning and later may be used as delay justification.

From another point of view as per engineering contract EPCIC contractor may have a certain deadline in obtaining the required number of quotations for engineering review. If quotations arrive late, their technical evaluation may not be possible within the expected time frame of the contract. Normally EPCIC contractors avoid issuing PO in favor of any vendor unless all technical issues are completed. This will cover complete scope of work as per MR and required package specifications. After receiving PO and gaining some manufacturing progress, vendors get the upper hand. They know EPCIC contractor is vulnerable to their requests. Therefore some of them may misuse this status to claim cost or time impact if they get any legitimate excuse for changes.

19. Keeping list of issued MRs and informing PM of delayed item(s) on weekly basis: Project control and VDC have no means to persuade discipline leaders to expedite issuing any document. They shall provide updated reports to PM to enable him to arrange necessary catch-up plans. This item is similar to the previous one with the difference that it was external to EPCIC contractor and this is internal to engineering disciplines.

20. Keeping track and expediting client procurement team to issue the TC/TBE starting letter: Issuing this letter again depends on contract conditions. Maybe after receiving a certain number of quotations consultant has to start automatically. Maybe client wants a specific time span to see if better quotations (cost/time-wise) could be obtained. It is to the consultant's benefit to start evaluation of all bids in one package simultaneously. First of all this will save time. The engineer responsible for that package will work more efficiently if has all packages on the table. He can search for a specific topic on all quotations and comment on them equally. More important, since he is reviewing similar items at the same time his evaluation between different quotations will be fair. Again, if EPCIC contractor delays in issuing TC/TBE start instructions, and exceeds limits set by engineering team, they shall officially document it.

21. Issuing an early status report on received quotations: This may include an explanation whether they are complete according to MR checklist and compliance letter, whether they are from approved vendor list, whether they are suitable for technical clarification, and issuing TBE. This report needs responsible discipline engineering input. Based on this report, PM may issue a letter to EPCIC contractor and state that if on-time response is not issued, the quotation will be included in the TC and a conditional TBE may be issued or rejected.

22. Expediting discipline engineers for response to IDC vendor quotations/documents, keeping track of interdisciplinary transmittals deadline, and highlighting delays: Similar to project documents, disciplines have a tendency to give priority to their own packages. Continuous expediting will ensure no one has unintentionally forgotten a document in the rush of project workload.

23. Keeping track and expediting engineering group to issue TBE: TBE is an important project document that enables EPCIC contractor to prepare CBE. This is an important step in issuing MR for purchase and placing PO. TBE issue time shall be governed by project schedule. However, based on vendor response to a specific package some quotations may be clear and some others may need further clarification. Engineering PM shall coordinate with EPCIC contractor PM to find a solution. May be project procurement plan allows further correspondence with vendor, or they may ask consultant to issue a conditional TBE, or to expedite the job; a technical meeting is arranged between vendor and consultant to close the issue immediately.

24. Checking the issued TBEs and letters of acceptance with issued PO list: Each TBE shall lead to a PO, otherwise vendor data will not arrive. Vendor data review and implementation is an important part in completion of detail design. Vendors will not produce any data unless PO is issued in their favor and certain conditions like opening an effective LC, placing AP, etc. are fulfilled. Even after these prerequisites they need some time to develop documents to a certain extent to issue them. All these time constraints impact detail design time. Therefore, they shall continuously monitor PO status.

After PO and at Vendor Data Review Stage

At this stage it is assumed quotation review is finished and selected vendor has started to issue vendor data. Similar to any engineering project, at this stage vendor shall finalize related SDRS as first priority.

25. Issuing letters to vendors indicating issue dates of SDRS and deadlines: SDRS list and dates attached to MR for quotation is only indicative. Either during bid or during KOM, EPCIC contractor, vendor, and consultant may agree on certain list of documents and issue dates. It is understood that SDRS acts as vendor preliminary MDR. Therefore it may undergo some changes as per engineering progress. Consultant or EPCIC contractor in contact with vendors shall expedite them for producing related documents and highlight delays. Delays shall be carefully documented and expedited. Consultant has no leverage to enforce that vendor sends required documents. Therefore EPCIC contractor PM or procurement manager shall expedite vendors through means envisaged in PO. Placing PO is the honeymoon time. Sometimes during this joyful time both sides forget to put in necessary clauses to define mechanisms to expedite action from each side. In the absence of these mechanisms, continuous expediting will serve project goals.

26. Keeping track of vendor documents issued to responsible discipline for commenting and expediting on a weekly basis: Again this may be shared with project control group. As explained before, its importance may be more than project documents. This is basically a project control activity, but as explained for a single engineering project an independent project control group will not be assigned. In some cases consultant will assign someone to work under PM instruction or he or she will be stationed within VDC staff. When they issue an IDC transmittal he or she will immediately incorporate transmittal in the list and generate reports as per PM request.

27. Preparing the list and keeping track of expected reply date and expediting other disciplines' engineers: Originator discipline may not be able to track issued documents and expedite other disciplines to respond on time. Present software can easily provide required warnings. Setting warnings for 1 or 2 days ahead of actual deadline may enable disciplines that have forgotten a specific document to respond to it on time.

28. Reporting the vendor documents' status on a weekly basis: In this report VDC shall highlight concern points. They may include delays in receiving vendor data, mistakes in proper numbering, references, formats, quality of vendor data, etc. One general problem is that vendors may try to neglect consultant comments as much as possible. Or they don't bother to follow ISO rules and cloud revision points, etc.

This report may even consider internal problems like delays in commenting on vendor data, new comments that had been neglected in previous revisions by engineering team members. It is clear that this part of the report shall be strictly confidential and for internal use only.

29. Keeping status of vendor-issued documents and issuing updated status on a biweekly basis: Normally the engineering contract for review, comments, and implementing vendor data are for a fixed duration. If a huge number of documents arrive simultaneously, the engineering team may not be able to review and comment on all of them properly. In a rush to meet deadlines some important points may be neglected. On the other hand, if vendor documents arrive very slowly and with delay, although the engineering team may have plenty of time to respond the task will remain uncompleted. This list enables the PM to expedite receiving vendor data or request added time.

30. Keeping status of expected date for vendor reply and new revision of documents and issuing expediting letters to vendor: In many cases the engineering team may not be in direct contact with vendors; besides, they may not need to expedite vendors. However, keeping track of vendor delays may help in future claims.

5.5 ENGINEERING DISCIPLINES TASKS

Members of an engineering team may be assigned as one of task force members to a project or they may stay in their own department and provide specialist services to the project team. In any case their proper functioning and interface is an essential requirement for project progress. Lead engineers are in daily contact with each other and the project team. PM or discipline head may assign certain man-hours to each lead. The man-hours have a suitable margin below the actual man-hours estimated in the contract.

In fact consultants have two types of margins. The first is on the man-hours themselves and the second is on the unit rates. Normally unit rates not only include consultant overhead (please refer to chapter "Proposal Preparation") but also have some margin for time. Assume that for a specific task 200 MH with US$60 unit rate is assigned. Assume the engineer in charge is getting 30 US$/h and consultant overhead is estimated to be 60%. This leaves another 12 US$/h $(60 - 1.6 \times 30)$. Even using half of it provides additional 25 MH for this document with the same 30 US$/h $(6 \times 200/1.6/30)$.

This means that if there is a delay in performing this specific task, the consultant can assign extra man power as specialist study or additional part-time workforce within this limit without jeopardizing project profitability. Of course total profit is reduced.

Engineering team members have certain tasks that can be categorized as follows:

1. Preparing project documents to be issued by engineering group: In this assignment they will follow project MDR as prepared by discipline LE and approved by PM. It is important for PM to check interface documents' issue date with other related disciplines. For example, oily water separator PID (which is issued by process discipline) shall be at least in IFA status before oily water package MR (which is issued by mechanical discipline) is issued. EPCIC contractor may want this MR to be issued in early stages of the project as LLI. Process discipline can't say, "I am not concerned that mechanical discipline is responsible to issue this MR very early. I have some line sizing to do and that will get me occupied. Therefore, I will issue this PID very late." PM also can't say, "this PID has a small weight compared to other documents and to get project progress I will instruct process discipline to concentrate on other documents." Having internal meetings at the start of the project to finalize project MDR in a rational manner is very essential for smooth project progress. The same happens for other documents. As another example, note that electrical discipline is responsible to issue diesel generator sizing before its MR can be issued. However, diesel generator (DG) set sizing before issuing load list is impossible. To prepare load list, electrical discipline needs to get mechanical/process disciplines' evaluation of required mechanical power of pumps and motors. In this case, ME can't say, "I have to wait for vendor response to identify air compressor or sea water pump power."

2. Reviewing client comments, preparing response, backup documents, confirming, rejecting, etc.: For client it is very important to get a specific response to each comment. Even if it is a trivial comment, they expect to be treated with respect. This is natural as far as certain cutting-edge investigations have not been requested. PM and LEs shall always keep in mind that they have to respond to client comments and provide necessary backup documents as far as they are within normal engineering practice. PM and LE shall carefully distinguish between normal practice and cutting-edge academic studies. In the project there

is neither time nor SOW nor budget allowance to do this type of study. Unfortunately, in this respect there are no defined boundaries.

3. It is a fact that engineering knowledge is constantly growing. Hand calculation methods are being continuously replaced with more accurate FE and CFD analysis. Newly graduated engineers tend to perform complicated time-consuming analysis methods that in many cases may negligibly add accuracy. With computer hardware nowadays analysis time may not be the governing factor. On the other hand new software generate a large volume of output, which the engineer may become lost in due to their interpretation and usage. For example, in structural analysis for each single member software can calculate exact unity check ratio under any load combination. It is neither feasible nor practical to assign a certain member size to each member to have the most efficient unity check ratio. Similar groups of members will have uniform size even though some of them may be less stressed.

4. Issuing letters, transmittals to engineering group, yards, etc.: A majority of engineering letters contain technical data. As per contract, PM shall issue letters. He will have technical background only in one discipline. Even in that discipline he is not aware of all discussions and history of certain comments. Therefore, LEs shall prepare required draft. PM may add some contractual points to it. Or after discussion with other related disciplines, he may again request from the originator LE some alterations in the letter or some additional backup documents.

5. Issuing comments to engineering group that has issued documents: As explained previously each discipline shall take IDC tasks seriously. Discipline input may be crucial to originator discipline task. In many cases IDC is not only for comment but for filling in the missing data. It was said that in a project they may decide to combine all electrical, instrument, process, and piping data for each package/pump/tank/vessel, etc. in one data sheet issued by mechanical discipline. Although piping discipline may not gain any progress for the man-hours spent on this data sheet, their input is a main portion of this document. Therefore, priority shall be given to IDC documents.

6. Reviewing yard comments and preparing responses: Client comments are mainly concerned with minimum specifications as per contract. In addition, the consultant may receive comments from fabrication yard. Yard comments may cover:
   - Fabrication methods and whether the selected details are very complicated or not

- Clashes between different parts for example piping with structure, etc.
- Undefined details in any part
- Changing specified design with existing material or more familiar methods
- Rectifying design or fabrication mistakes

7. Liaison with yards to ensure necessary documents have become available to them: This is not a contractual responsibility for the engineering team. In some projects they may not be at all connected with fabrication yard. EPCIC contractor may have its own engineering team in the head office. In that case project team may require them to perform this task.

8. Coordination with procurement to evaluate bulk material proposals: Bulk materials are those that are normally manufactured and don't require specific engineering documentation. Structural steel including plates, rolled section and tubular, pipes, fittings, flanges, cable trays/ladders, and cables are listed in this category. In many projects, for these items only MTO is issued. MTO contains a list of required items and their standard designation. For example, stating ST355 J0 is sufficient to know material composition, yield stress, tensile strength, energy level in impact test, etc. Quantity of material will be given in the MTO. Additional contingency may be selected based on EPCIC contractor's previous experience or recommendation. Therefore, quotation evaluation for bulk items is a fairly straightforward step. To expedite procurement, majority of EPCIC contractors may not require consultant involvement in bulk material bid evaluation. However, some small amount of help may be requested. Discipline engineers may provide assistance in this regard.

9. Reviewing vendor quotations/preparing TCs is the first step in quotation evaluation. As explained in VDC SOW, when EPCIC contractor issues sufficient number of quotations or as per project deadlines, each discipline starts preparing TC for the submitted quotations. Each stage TC shall be complete. As much as possible new clarification shall not be given unless quotation is revised from the beginning. All disciplines shall submit their queries to the discipline that is responsible for that MR. Clarification may start from general queries to ensure vendor has completely understood project SOW, specifications/requirements, and has received latest revision documents to detailed questions on functionality of the proposed equipment, material of construction, required utility, etc.

10. After receiving vendor response to the first stage TC, a second TC may be required. Related discipline engineers shall review vendor response to previous TCs and second TC is prepared by engineering group. Consultant is only concerned with technical issues. If a commercial issue is brought up by vendor, consultant only has to bring it to EPCIC contractor's attention.

11. Issuing TCs to vendors: Final TC shall be properly signed and handed over to VDC as per project correspondence procedures. As explained before, normally the engineering team is not in direct contact with vendors. Discipline engineers hand over TCs to VDC, which in turn submits them with proper transmittals to EPCIC contractor procurement department.

12. Expediting vendor response to TCs is also normally out of consultant SOW. However, in some cases EPCIC contractor may ask consultant to do so. In this case the best method will be to arrange technical meetings. With new facilities telephone/video conferencing is easily possible. As per the author's experience, even these technical meetings may not completely close the issue. Almost all vendors prefer to have sufficient time to review meeting discussions, consult their fabrication shop, and issue a commercial/technical proposal. There is no doubt that these technical meetings will reduce vendor response time considerably. It is a fact that a simple e-mail or fax issued to a vendor doesn't show how serious the requesting body is. Therefore, they may not be so interested in responding in full detail. Each year vendors may participate in hundreds of bids. Only a portion of them are really serious and from those only from a few might a vendor succeed in getting an order. Therefore, most naturally they try to minimize their nonyielding efforts.

13. Responsible discipline for MR will receive vendor response to TC. Before instructing for distribution they have to review vendor response to TCs. In this preliminary review they will determine if requested data has been provided, and which disciplines shall receive and respond to this information. Shall VDC request for supplementary data, etc.

14. Keeping track of quotations and the response to TCs issued to the engineering group and issuing expediting letters is in VDC SOW. However, discipline engineer assigned to a specific MR shall always have a complete record of his own package status. At any time he or she shall know how many quotations have been reviewed, what are

the major/minor deviations of each quotation, which quotations are prominent, etc.

15. Reviewing vendor documents and commenting on them is the main step in vendor data review/implementation. This stage is critical in several ways. It may have cost and time impacts. This may delay manufacturing and delivery to site. Daily running cost of EPCIC contractor yard is very high. Delay in yard work schedule or rework has a considerable cost that EPCIC contractors shall try to avoid as much as possible. Therefore, consultants shall have sufficient backup documentation for each comment. This may vary from a vendor's previous affirmative response to TC or their direct statement in the quotation or approved deviation either in TC or during KOM. For each action the consultant shall have suitable backup.

16. HOLD points in the drawings hinder yard work. Consultant shall clearly define HOLD points. More than anything else this is required to protect consultant against yard rework claims. In spite of that, consultant shall help EPCIC contractor in proceeding with work. In some cases EPCIC contractors prefer to have small reworks instead of demobilizing their workforce. In structural discipline it is possible to tack weld certain areas and proceed with the remaining work. In case future vendor data necessitates some changes, breaking tack welds is fairly easy. Consultant PM shall coordinate with discipline engineers to keep track of HOLD points. Keeping track of HOLD points and checking the possibility of removing them is to be done in conjunction with discipline engineers and PM. From one point of view, the engineering team's responsibility is only to issue documents. Whether fabricating yard can start the job or not is the headache of EPCIC contractor PM. But this is not the whole story. It is to the EPCIC contractor's benefit to start a mutual interaction with the consultant. This will ensure the consultant will continuously check and remove HOLD points as much as possible. In a majority of cases removing HOLD is only related to contractual liabilities. Consultants don't want to be liable for any claims from fabrication yard rework due to implementation of new vendor data. It is to EPCIC contractors' benefit to keep their yard progress continuous. Joint meetings can be arranged on a biweekly or monthly basis between EPCIC contractor, manufacturing yard, and consultant PMs or representatives. Such a meeting is not required in the first half of the project. Its duty will be to identify project HOLD points and the steps to

be taken to remove each one. At any time the engineering team shall know the following:

- Where are the HOLD points?
- Why are they kept HOLD?
- How, when, and by whom can they be removed?

17. It is clear that removing HOLD points is not a single-person task. Client and vendors shall participate actively in removing HOLDs. Expediting client and vendors to facilitate removing of HOLD points is another task that discipline engineers can actively participate in to help yard fabrication. This is done by clarifying what are the implications? In some cases even before receiving vendor data consultants can send existing limitations to vendors and get their confirmation. This shall be completely done via EPCIC contractor. Since vendors have to guarantee their package functioning after installation if they find anything wrong with the proposal at any stage, even if they had previously given the go-ahead signal, they will keep their right to change. In this case EPCIC contractor shall accept responsibility and doesn't blame the consultant. Although removing HOLD points shows consultant good will to expedite project progress, it shall be very carefully controlled via EPCIC contractor's project team. DCC team reports to PM can be helpful in expediting discipline engineers (and vendors) to facilitate removing HOLD points. Based on the previous meeting decisions/action list, consultant PM may instruct certain activities to engineering task force. This may range from repeating some analysis with certain assumptions or revisions in some documents. Interaction with vendors to provide certain documents on time and keep the issued data unchanged is EPCIC contractor's duty. Manufacturing yard team may need to fit some sections and weld later after receiving a necessary permit signal.

5.6 GENERAL ITEMS TO BE CHECKED IN QUOTATION

Receiving a sufficient number of quotations as per contract is the first step in preparing TBE. Highly reputed/experienced vendors and newcomer vendors (which want to have a good impression on first introduction) normally prepare complete quotations. However, there are some vendors that may not send a complete quotation. They may have either had some bad previous experience with this EPCIC contractor, or think this bid is not so genuine, or simply be overloaded and don't bother for a new project, etc. Incomplete quotations may waste technical staff time. Normally EPCIC contractor doesn't have either the man power or the technical expertise to distinguish qualified quotations.

Before the actual start of TC/TBE process, a junior engineer from the consultant team may be assigned to review received quotations with a checklist. This checklist shall enable him to get a true image of quotation quality. He shall understand if the quotation is in a condition to pass TBE after rounds of technical clarification or not. If not, even before the official start of TC he shall return it to EPCIC contractor or ask for supplementary information.

A majority of, if not all, packages shall be reviewed by several disciplines. Each discipline may be interested in specific parts of the quotation. The following tries to explain what items generally shall be checked by engineers before the actual start of TC process. In addition, the interested discipline is also identified. This helps preparation of IDC transmittal. These items may be broadly categorized in contractual and technical points as follows.

Although technical clarification of a quotation is a multidisciplinary task, at this stage, only general factors that can be viewed by common sense engineering are checked. Detailed information of these items may be reviewed at vendor data stage. For quotation review, brief information or in some cases only a statement or addressing the issue may be sufficient. The responsible engineer shall decide on a case-by-case basis.

Based on the engineering contract parts of the information may be checked by EPCIC contractor or consultant. In any case, issuing a checklist to be filled in by the vendor will improve quotation quality. For small packages, requesting a large number of documents will be useless.

5.6.1 Contractual Points

Contractual points cover items needed by EPCIC contractor or client or help consultant in the evaluation process. Some EPCIC contractors may check their own requirements and some may assign this task to the consultant. Contractual points may range from cost-wise points to document items. In this section all nontechnical items are covered under contractual points.

1. Has vendor given the experience or reference list? This list will ensure client that vendor package is not first tested on his plant. Many clients have a requirement that accepts only well-proven designs in their platform. They are not so enthusiastic about new inventions.

2. Has vendor given the ISO or other regulating body certificates? ISO certificate is a first step assurance to client. It means proper QA and QC procedures have been established in vendor premises.

3. Majority of projects have 25–30 years of design life. Most vendors without even a track record of this long for their equipment confirm it. It shall also be kept in mind that technologies evolve very fast. Using 25-year-old technology is not recommended. Consultant shall check whether vendor has confirmed project design life or has left it unanswered. Not all packages can serve this long. However, with good maintenance and sturdy design some equipment may exceed it. The author has seen 1000 KW diesel generators that after more than 35 years' service still could produce more than 800 KW in the worst environmental conditions. The operators credited this to sturdy design and their strict maintenance procedures set by the first area manager in that field.

4. Has vendor filled in the table of compliance? Vendor shall accept to comply with project requirements. Normally standards define the minimum requirements. In each project the client quality system or simply project interconnections may dictate specific requirements that in some aspects may exceed standard requirements. Some vendors (especially those with a production line manufacturing) may not comply completely with all project specifications. To confirm all project requirements they may need to implement some changes in their production line, which have cost and time impacts for them. They may have some deviations that shall clearly be highlighted during bid evaluation. Major deviations shall be approved by client. This table may be used by EPCIC contractor commercial team in probable future claim/counterclaim negotiations.

5. Has vendor confirmed to give documentation as per requested list? Although during bidding all documents are not required, vendor shall ensure EPCIC contractor that they will provide all documents after receiving PO. The extent of necessary documents is governed by package size, complexity, function, etc. It is quite clear no vendor will provide the same number of documents for US$10,000 and US$10,000,000 packages. Requested documents may serve different purposes. As described in Section 5.4.1, vendor documents may be divided in several groups like contractual, engineering, manufacturing procedures, QC/QA, and operation/maintenance groups.

6. Has vendor accepted the priority of documents in case of conflict? This may seem a trivial point but it is not. Many projects have their own specific requirements, which may be higher than minimum standard requirements. If document priority shows general standards to be higher than project specification, then defining more strict requirements in project specification becomes useless.

7. For some packages client may request special certificates. These may vary from design verification certificate (DVC) for engineering activities or certifying authority (CA) for applicable procedures or TPA approval for manufacturing. Consultant shall check which one of these has been requested in contract and implemented in the MR. Some vendors may accept TPA or CA but for DVC request additional cost as option. Consultant shall ensure which parts have been confirmed by vendor and highlight the remaining. In some cases maybe type-approved certificates can be replaced with some requirements.

8. Most vendors have their own numbering system. This covers both equipment part list numbers and document numbering. Changing it and adapting to project numbering system is time-consuming and they dislike it. Changing part list numbers is impossible because later they can't find that part if a new order is placed. Normally after a package is sold for a few years the vendor has to supply spare parts. In almost all projects clients give an additional order for spare parts recommended by the vendor for two years' duration. In addition, vendors have to guarantee several more years (at least 10 yr) to provide package services. Normally vendors have no objection to put project tag numbers on equipment nameplates. In documents usually they put both numbers. Consultant has to check if vendor is using his own tag number or is following project tag numbers. This will facilitate future document tracing or spare part ordering.

9. Some vendors may have specially developed in-house software for their technical calculations. This is okay. However, there is general purpose software for writing reports, preparing sketches and drawings, etc. Vendor documents shall follow project requirements on the type and version number of this software. Consultant has to check whether vendor has confirmed to issue documents in the requested software's format or not.

10. In many packages using a typical design will considerably reduce package cost and manufacturing duration. Consultant shall check whether

vendor has provided any typicality or not. If yes, then each typical item specification shall be checked with project requirements to ensure they fulfill project minimum requirements. In a majority of cases there will be some deviations. Consultant shall identify major and minor deviations and if necessary ask vendor to provide backup documents for relaxations.

11. Has vendor accepted the guarantee period for his package? If not specifically mentioned in the quotation, then vendors' normal terms and conditions are applicable. This is not a technical issue but is an important contractual item. After dispatching the package and before its actual installation and commissioning on the platform there is always a time lag. Normally vendors can accept a period of 18 months after dispatch from their yard to12 months after installation on the platform (anyone that occurs first). Some clients may request longer duration. Vendors may accept it, especially if they want to promote their production.

12. Some vendors may develop special welding procedures for some materials. Based on material type, hardness, thickness, and grade they may mobilize certain equipment. It is very unlikely that only for one project a vendor mobilizes new equipment. Consultant shall check whether vendor has confirmed material of construction according to project requirements. The following items shall be checked: skid, vessels, piping, instrumentation, cables, trays, electromotor, mechanical equipment, etc. For example, a rolling machine suitable for a diameter up to 800 mm can't be used for vessels with a diameter of 2300 mm. In that case the vendor has to nominate subvendors eligible for the required tasks. Normally in the quotation subvendors will not be listed. However, during quotation review stage EPCIC contractor or consultant shall check whether subvendors are needed, and during vendor data (if PO is placed in favor of this vendor) shall ask for subvendor list and their capability definition. In some projects or for certain unimportant equipment client may relieve itself from subvendor checking. In this case since the vendor has full responsibility for the whole package the client may accept vendor confirmation of subvendor capabilities. For important packages availability of necessary machinery can be verified at KOM during vendor fabrication shop visit before the start of any activity.

13. Attending factory acceptance tests (FAT) is very helpful to ensure package will function properly during commissioning and later in

platform operation. In third world countries bureaucratic routines to get visas and arrange for tickets, hotels, etc. need some time. In some cases higher management permission for each trip shall be obtained. Even with a competent project team the coordination takes considerable time. In the majority of projects an inspection test plan (ITP) requires that client representatives at least take part during FAT. Few clients may grant vendors the permission to proceed with TPA in client absence. Some may HOLD test, which leads to delay in package delivery. To avoid this problem consultant has to check whether vendor has confirmed to give sufficient advance notice for attending the FAT.

14. Before installing the package in its final location, it has to be lifted and transported from vendor workshop to transit truck, from truck to shipping vessel, then to customs area, from there to the next transit truck, from truck to storage area, and finally to final location on the platform. In all these movements package has to be lifted and transported in a safe and proper manner. Some packages may need special lifting tools. Others may use normal items like nylon straps to avoid damage to paint, spreader bars to transfer sling lateral force to vertical load, etc. It may require securing specific lashes or removing them at later stages. Vendor has to specify any special lifting device, procedure and undertake its cost/responsibility from its own manufacturing yard to shipping vessel. Consultant has to check whether vendor has provided for temporary lifting, transportation, etc. in its quotation.

15. Has vendor confirmed to supply the special tools, insurance, and capital spares? Inside some packages there are some remote points to which access is impossible unless special tools are used. They may also be used for lifting and handling purposes. In addition some major items like valve seats may need replacement after several years' usage. These are capital spares. EPCIC contractor has to ensure that vendor has included these items in their proposal and as part of their SOW.

16. Has vendor given the list of commissioning and start-up spare parts and confirmed to supply them? This is again not a consultant headache. However, they are very important during package commissioning. Without them considerable delay and cost impact may be anticipated. Similar to 2-year spare parts, vendor is the best authority to determine it. Commissioning team of EPCIC contractor may have some minimum requirements but normally will not relieve vendor responsibility.

17. Has vendor given the list of 2-year spare parts? This is not a technical issue to be a consultant problem. However, the client always wants it. Operation and logistics engineers may comment on this document. Before their comment the related document shall be presented. Normally the vendor is the right one to propose the 2-year spare part list for their package. Client may add a few items or increase some of the requested numbers to be on the safe side during operation.

5.6.2 Technical Points

Technical points are those affecting consultant evaluation of vendor quotation. They may cover a variety of items that can be easily checked and have important impact on design.

18. Does proposed package fit to the area allocated by the design team? Package GA and footprint dimensions are very important in technical evaluation. In some cases it may affect structural design considerably. Normally packages are not rejected due to structural problems. As much as possible structure is adapted to accommodate them. In onshore plants there may be plenty of space to relocate large equipment. Normally a wide area is used for access and transportation. A small part of it will solve increases in package dimensions. In offshore platforms space has always been a main concern and is a precious commodity. Economical adequate space for accommodating all packages, providing sufficient access, maintenance area, and escape route are important design criteria. For all packages a tentative GA dimension is necessary. In each quotation consultant shall check if vendor has provided the preliminary GA or not. Later in the vendor data review stage layout team (PI) and support design team (ST) shall carefully review this information.

19. Has vendor confirmed receiving latest revision of documents? Sometimes to expedite project progress EPCIC contractor may receive quotations with IFA revisions. They may undergo some changes after upgrading to AFC. Consultant shall always ensure that vendor has received the latest project documents. This may impact material of construction, scope of supply, package input and output interfaces, etc.

20. Has vendor provided any data about center of gravity (COG)? Support skid info and COG location affect load distribution. For small packages it may affect local design and for large packages it may impact both local and global design (ST).

21. Has vendor filled in project data sheets? Project data sheets provide complete information for the package. If vendor fills them in, it means they are accepting responsibility for the given data. It will reduce considerable work later for consultant and will enable early approval of vendor documents. This will expedite production and finally reduce delivery time. Several disciplines like mechanical, process, electrical, etc. may review parts of data sheet info.

22. Has vendor provided mechanical or process guarantee? Some packages (especially process equipment) require process guarantee. At least on paper if a vessel designed to remove a certain percentage of water from crude doesn't perform this function, the vendor has to adopt certain corrective measures. If vendor design team or design subcontractor doesn't possess required qualifications then reviewing their proposal will be useless (PR, ME).

23. Has vendor provided PIDs for his package? For some single-item packages like vessels PID is not required. For others even a preliminary PID accompanying quotation will provide lot of information about package operation, isolation, and controlling. Most vendors are reluctant to do this and try to postpone it after firm PO. This may be for several reasons like not showing their proprietary design to competitors, not accepting liabilities on the design. The main reason they officially state is that they have not designed the package yet and therefore data may be misleading. This is not a good excuse and simply tagging "For Information Only" in the title block or in the background of the document will solve it. Reviewing PID is a multidisciplinary request and task.

24. Environmental conditions play a major role in each package operation. Consultant shall check whether vendor has confirmed environmental conditions like temperature/humidity, seismic levels, etc. as specified in MR. Equipment designed for arctic conditions may not perform well in tropical area. As an example, diesel engines have a high cooling demand. One common solution is to circulate cooling water, which absorbs heat from combustion cylinders and exchanges it with another absorbing media. One of the most common heat exchanger media is air. The heat dissipation is done in air-cooled radiators. Heat dissipation rate depends on change in air temperature. If outside air is warm (say 40°C) a large volume circulation is needed. Because at maximum its temperature may be raised to 50°C, this means larger fan diameter and rotation speed, which in turn requires

more power. The same generator with cooling air temperature of 20°C requires much less cooling air circulation (multidisciplinary request and task).

25. Most vendors have few painting specifications which have been suited to their paint shop. Project paint specifications may be different. Consultant shall check whether vendor has accepted project specifications or is providing his own surface preparation and painting. It shall either be according to project specifications or vendor shall confirm suitability of his own proposal for project environmental conditions. Some vendors may highlight their routine painting specifications in the main cost and request for optional additional payment to follow project instructions. Many vendors have special topcoats for their brand like yellow for Caterpillar and Liebherr. Code regulations specify red for safety equipment like deluge valves, or orange for life boats. These requirements can't be changed. It is to be noted that topcoat color is not the main concern for packages and consultant shall concentrate on thickness and material of primer and main coats (PI).

26. Each package has interface with others. They will use utilities provided by platform equipment like electric power, compressed air, etc. In calculating these utility demands and preparing their data sheets the related discipline depends on the data provided by other disciplines. If there is a gross mistake in calculating utility capacity the possibility of shutdown during platform operation will increase. Many packages (like high discharge or high head pumps) have high current demand during start-up. This may lead to generator shutdown due to high demand, which in turn may lead to platform shutdown. The same may happen to others. Generators have high cooling demand. If radiator is air cooled and located in a tropical area in hot conditions it may require a large air flow. Air fan increased dimensions not only may impact power demand but also have structural and architectural impacts. The same may happen to fuel demand. If a generator is consuming a large volume of diesel oil, proper storage and pumping capacity has to be foreseen. This is valid for all packages that require a utility line to be supplied from the platform. Consultant shall check whether vendor has stated the utility consumption including compressed air, electricity, fresh water, sea water, etc. Checking, providing, and design of necessary tie-in for interface items is also a multidisciplinary requirement and task.

27. Other than QA requirements all projects have their QC requirements. Standards define a minimum level of QC tests. Some projects may require more tests. In addition to that, if a vendor is on the client approved vendor list, minimum nondestructive testing (NDT) levels may be okay, but for vendors not on it additional tests may be specified. This may range from increasing NDT percentage (especially radiographic tests (RTs)) or even requesting destructive tests. Performing these tests increases manufacturing time and associated costs considerably. To keep their quotation costs at a reasonable and competitive level some vendors may just mention QC tests as per code or even refer to their own in-house procedures. To ensure project quality level has been maintained, consultant shall check whether vendor has confirmed compliance with NDT requirements, eg, postweld heat treatment, RT levels, etc.

28. Some platforms may be manned. Recent design trend is toward remote control systems. Recent technological achievements have provided a very good chance for this issue. At the same time suitable measures shall be considered to prevent unwanted problems, which may lead to cascading events leading to disasters. This may be achieved by proper logic system and dual control unit hardware. Remote control unit may be located in central control room or be implemented in the control software. Consultant has to check whether vendor has confirmed local and remote control philosophy including required hardware and software.

5.7 PLANNING AND PROJECT CONTROL DEPARTMENT

Planning department is a powerful help for PM to measure progress, identify delays/areas of concern/select mitigating measures, provide backup documents for claims and counter claims and revise plan. Their tasks can be divided into two broad groups, the first being project financial issues and the second project control issues. In a sense these are not separated from each other because project progress is controlled to ensure contractual obligations are followed. Contractual obligations are followed to ensure client payments are on time and future projects may be secured. The distribution given below is not unique. As explained before (based on project scale) planning personnel can be stationed in DCC or in planning department. Their duties can also be distributed in part to DCC personnel.

5.7.1 Financial Issues

As stated already these two sections are not isolated items that don't have any impact on each other. The ultimate goal of financial check is to ensure a project is profitable. In addition to that, project cash flow shall also be positive. Some of the checkpoints include:

1. Defining physical and invoicing progress measurement procedure and agreement with client on documents' weight factors is the first and main task of planning department. In many projects invoicing progress is only defined as per specific milestones. Before receiving a document in a certain stage, client will not accept to pay for the work in its production stage. The most customary milestones are IFC, IFA, and AFC. Normally 60%, 80%, and 100% invoicing progress is calculated for them. In few projects and in order to help consultant cash flow status, client may accept to pay for intermediate steps like at the start of a document or its issuance for IDC. Whether a certain document like weekly assignment or IDC transmittal shall be presented for client approval of the related cost or not remains for mutual agreement. In some projects client may require a work progress appraisal report (WPA) before issuing any invoice. This WPA has to be reviewed, commented on, and approved by client planning team before issuing the related invoice.

2. All contracts require progress monitoring. It may be on a weekly/biweekly or monthly basis. In certain projects based on project status different reporting duration may be selected. For example, at the end stages when all design documents have been issued and only a few documents remain to be revised (due to vendor data implementation) longer reporting duration may be selected, while in the mid-stages weekly reports are needed.

3. PM may need to compare the engineering team outcome with the acceptable invoices from client. This may be understood as a managerial tool to predict project loss or profit beforehand. To do this, the planning department compares the consumed man-hours with the performed jobs in each week. Since all documents undertaken by disciplines in a week may have not reached invoicing stages, physical progress rates shall be used. Of course it is understood that physical progress will equal invoicing progress at the predefined stages. For example, if IFC invoicing is selected to be 60%, its physical progress after issuing the related document at this predefined stage can't be less or more than invoicing progress.

4. Although MDR determines documents to be worked upon at each week, PM has some freedom in defining weekly assignments. Each week the planning department reviews assigned jobs in disciplines and calculates physical/invoicing progress of each. Based on that, PM may reschedule MDR to replace new documents with higher weight with low invoicing-value documents. This is not a routine job. However, it allows some freedom to PM to control cash flow. In some extreme cases (mostly in IFC stage), PM may instruct discipline engineers to officially issue a document even before completing all design review stages.

5. Although latest status of issuing design documents may be continuously updated by DCC staff after issuing the related transmittals, the planning group also needs it for both project control and invoicing purposes. Therefore, in some projects maybe DCC SOW is limited to preparing transmittals. They will submit a copy of the issued transmittal to the planning department and updating MDR is assigned to planning.

6. Progress/invoice approval in client team may be done by two different teams or personnel. Progress is checked in WPA by project control team and invoice is checked based on approved WPA in financial department. For several reasons such as lack of project budget, contractual accepted time lag between progress approval and actual payment, client PM's intention to apply financial pressure on consultant or encourage its team, etc., actual payment may be different from approved progress. Planning department shall prepare payment status report to enable PM to highlight it in weekly or monthly reports for necessary client action.

7. Normally, consultant may not have subcontractors, but for special studies and activities like HAZID, HAZOP, etc. they may employ third-party services. Clarifying subcontractors' payment status is an important task that helps PM in project cash flow measurement.

8. Very few (if any) contracts may finish based on original instructions. In the majority, client will instruct changes orders (CO). Although it is preferable not to perform any CO without prior agreement on its time/cost impact, in majority of cases consultant has to proceed with CO before client approval. Planning shall keep a clear record of CO status including the claimed value, consumed man-hours and cost, percentage of performed job on each CO, their invoicing, and client payment status. If the gap between payment and performed job

increases tolerable values, PM shall be informed to take proper action. It may vary from a simple letter expediting payment to stopping work on some items.

5.7.2 Project Control

Project control intention is to give the real status of the project to PM. He or she shall understand, what are the shortcomings? Which items need further strengthening? What mitigation measures shall be applied? What catch-up plan shall be prepared? Some of the required information includes:

9. All weekly or monthly reports require next week or month look-ahead plan. This may include a brief explanation of general tasks to be undertaken by engineering group or a detailed list showing document number/tasks description to be assigned to each discipline for the coming period. To prepare it, the planning team uses MDR-planned column. However, they have to carefully examine the status of each document to check three issues:
   - Some documents may have been prepared ahead of schedule.
   - Some documents may lag behind schedule or be part of catch-up plan.
   - PM may wish to exclude some documents from next-period tasks and include some others. This may be due to EPCIC project requirement, yard request, procurement status, its higher weight to compensate for cash flow, etc.

10. Procuring long-lead items is an important factor for EPCIC contractors. Their procurement team requires some time for obtaining quotations. In addition, the engineering team itself requires some time for technical evaluation of quotations. Planning department shall continuously monitor MR issue status (with emphasis on long-lead items). This will enable PM to reschedule their plan or devote extra man power or resources to bring them in line with project requirements.

11. In some contracts client may not accept upgrading a document to next step before receiving/responding to client comments. At least this may be valid for documents in approval category. On the other hand, consultant can't wait endlessly for client comments. Planning group shall prepare the latest status of client comments on design documents and delays. If contractual terms allow it, after a certain period with no response from client the related documents may be deemed approved and consultant can proceed with next stage. If not, PM shall highlight delays impacting project progress and request for their immediate

response. This list may be the major source of backup document used for claims and counterclaims.

12. If engineering contract requires CA approval, planning department shall prepare and update status of issued documents for endorsement and CA response. It will serve two purposes. The first is to help in upgrading to AFC stage and the second is in delay calculations.

13. Almost all contracts have a limit on number of received/evaluated quotations. It is always more efficient to start technical evaluation of all quotations for a certain package simultaneously. Although in some contracts EPCIC contractor may be required to issue official bid evaluation letter, mostly they fail to do so. Therefore, planning shall prepare quotation status. This will enable PM to decide starting TC preparation for a certain package.

14. Issuing TBE by consultant is the first step to allow EPCIC contractor issue PO. Only after this can the consultant expect to receive vendor data, which is the start of the project end. Project control team shall continuously monitor TBE status. This allows PM to check with responsible disciplines to find delay reasons for TBE in each package and probably find a solution. Remedial action may range from technical meetings with vendors instead of issuing another TC round or technical meetings with client to close minor deviations, etc.

15. Although VDC team fills in the data base for vendor data status, it is project control responsibility to process it. Brief reports showing percentage of data received from each package in different disciplines help PM to understand how much of vendor data is covered. In spite of engineering documents, it is difficult to assign weight factors to vendor data. Weight factors enable accurate progress calculation. For the engineering team only those data that may affect platform design are important. In addition, there are numerous data about scheduling, organization, QC reports, etc. that are not reviewed. It is important to note that QC documents shall be carefully reviewed by EPCIC contractor team to ensure compliance with project minimum requirements. A simple approach may be to give zero weight to documents not having impact on engineering design. In this approach, to approve vendor data completeness first all documents should have been issued and, second, main documents shall receive approval.

16. It is a fact that no delay has a single cause. Each party may be responsible for a portion of the delay. At the same time that project control people are monitoring client delays to be ready for future

claims/counter claims, they have to monitor the engineering team's internal actions to be able to prepare catch-up plans and take remedial actions. In some cases PM may decide to have a stricter internal plan. As explained, PM may select to remove the contingency part of man-hours and assign actual calculated ones to each discipline. This may act as an internal plan. Progress and delay may be calculated on this basis. Project control department may use this plan to calculate discipline delays and prepare internal catch-up plan. If this approach is used, they must be very careful not to mix the two plans.

17. As explained for internal plans, external plan is according to consultant contractual obligations to client. PM will try to extend activities as much as possible. This will allow him to uniformly distribute man power. Catch-up plans formally issued to client are normally less strict than internal plans.

18. Calculating delays and preparing catch-up plan are similar to QC activities. They measure the performed action and recommend remedial action if needed. In addition to it, similar to QA activity, project control team shall expedite discipline engineers and keep track of promised issue dates of documents as per document register before their deadline arrives. This activity may help each discipline's lead engineer to reschedule tasks or put additional man power on a certain task that is delayed behind schedule, or accept overtime work on a certain activity, or even outsource it. Each LE has partial or full freedom in OT assignment, but prior to outsourcing shall get PM approval.

19. Updating discipline document register in cooperation with discipline lead engineers is mainly done to enable them to monitor their own activities.

20. At each stage, the project shall have a final list of latest-issued documents. This may be a separate list or may use MDR. Project control shall keep the final list and revised number of issued documents on a weekly basis.

21. Both for managerial information and to complete weekly/monthly reports the project control team shall prepare a report on document issue status, delays, expectations, etc. This may be in two levels. The first is for higher management information and is brief. It only shows a general list showing number of documents per discipline divided in each stage with their status. The second is more detailed. It separates different classes of documents per discipline and per document category. From this list the PM will understand how many data sheets,

specifications, MRs, etc. per each discipline have been issued and had to be issued. This shows number of delayed documents. It is possible that some documents intended to be issued are delayed but others are given priority. Based on discussion with PM, discipline LE may decide to reschedule some documents' order. Although PM may have some reasons other than progress in this decision, it is project control's duty to inform him of the net outcome of this decision.

22. Issuing some documents is related to client responses. Consultant may issue several queries to understand preferred course of action. It is recommended to set a deadline for response and inform the action that the consultant will continue if query remains unanswered until that time. Keeping a list of engineering queries and client expected response dates, their closing dates, results, etc. enables PM to decide with LEs.

23. Change orders may constitute a considerable portion. In a majority of contracts the client is allowed to add or deduct 25% of the original contract value without needing to make a new bid. Project control team shall update related change order list and prepare a summary report to enable PM to understand where the project stands. It may be that change orders detract project from its actual path.

CHAPTER 6

Proposal Preparation

As far as this book is concerned, the first step for all projects is when an invitation to bid (ITB) letter is issued. The steps before that and how a company or a government organization decides to start a project is out of our concern. In some cases before actual bidding, prequalification evaluation may be performed.

ITB letter normally asks participants to send with their proposal for the specific project some documents like:

- Company financial turn over and status accompanied by official audit report
- Company experience list including general description of work with reference contact points from client's organization
- Company's present workload and future forecast
- Key personnel CVs

In spite of this routine procedure, in many cases the client already knows the consultant. The oil and gas engineering business is a small world. In each region majority of clients have had one or two experiences with a certain consultant. Other than the bureaucratic paperwork, many clients have their own approved consultant list and may invite one without requesting these documents. Therefore, client may only send the scope of work (SOW) and other necessary documents and ask consultant to prepare a proposal. A major part of such a list is in financial department responsibility. The business development department participates in bid. Normal sized consultants can't afford to assign highly experienced engineers to stay idle and when a bid arrives estimate required man-hours. Even if consultant managers want, experienced engineers are not willing to do this. Job satisfaction obtained from a complex design can't be obtained from number crunching in a bid. It is understood that bid preparation is part of our job. Without successful bid proposal there won't be a successful project.

As the first steps, consultant shall:

A. Familiarize himself with the project

B. Find out project cost order of magnitude

Depending on the size of the engineering company, inside the consultant organization two separate departments may be dealing with the

Practical Engineering Management of Offshore Oil and Gas Platforms
ISBN 978-0-12-809331-3
http://dx.doi.org/10.1016/B978-0-12-809331-3.00006-5

project. In small companies everybody shall do everything. In big consultancies marketing and engineering have been assigned to specialist bodies. The nature of the approach to client may be different in these two departments. While a marketing manager may promise client his company "will do anything from A to Z" (to satisfy client and obtain a contract) an engineering manager will say we "will do everything in our SOW as per contract."

Preparing a successful proposal leading to a successful project depends on several factors. In general the marketing department and proposal manager may consider any contract won from a bid the result of a successful proposal. From the engineering manager's point of view, although this is a prerequisite "a successful proposal" may not necessarily lead to "a successful project." The author is coming from the engineering team. Although I want to be fair, I honestly admit that in some cases I am a little biased in favor of engineering. In a broad sense each proposal is prepared by a marketing team but it is performed by a project team. This is true even if physically the same person performs these two different tasks. Finally, different ideas/approaches of both groups shall be combined in the proposal. A won proposal with huge problems in its accomplishment and overrun costs/time is a failed entity. This may happen when SOW is not clearly defined and costs are very low or a very short duration is accepted. All of these happen if consultant higher management decides to win a bid *at any cost*. Engineering team has no problem with this policy. The only issue is that we want managers to be fair and if they have accepted some loss due to any reason *do not ask engineering team to compensate*.

Some engineering, procurement, construction, installation, and commissioning (EPCIC) contractors look at the project parts based on their share of the whole project. When they see engineering value is 5% of the total cost and procurement is 60%, they assign maximum 10% of project duration for engineering and minimum 50% for procurement. This may be correct from one point of view. The only thing is that it lacks the interface between the two. Procurement will not start until engineering has issued material requisitions and will not finish until engineering has received vendor data and commented/approved them and implemented their impact on related documents. Therefore, detail engineering duration shall be sufficient, which may be up to 50% of the project duration. Construction, installation, and commissioning engineering extend to almost the end of the project.

If proposal/marketing department works in a separate island they may not see these impacts and just to win the project may give all sorts of unrealistic promises to client. The best situation is that proposal/ marketing organization shall work in line and with engineering disciplines that in reality will perform the job to clearly define project SOW and duration.

6.1 FAMILIARIZING WITH PROJECT REQUIREMENTS

This may be divided into three steps:
- To know client organization as a whole
- To know project's major team members in client organization
- To know project specific requirements

In the first step, consultant shall know client behavior and approach, whether they follow bureaucratic procedures or not. Clients may be grouped into three categories:

A. Category I: Client has developed their own standards, has performed many projects globally, and wants uniformity and standardization in projects. This will reduce engineering duration, facilitate their procurement and future operation of the platform. This impact is much more than savings obtained by very accurate and sophisticated engineering software.

B. Category II: Client doesn't care how you design the platform. The only important item is that after handover the platform yields the intended export. This will mostly happen when client has given a lumped sum cost project. For cost-plus contracts, they will be concerned with each expenditure item.

C. Category III: Client is in between the two extreme categories defined herein.

These three categories may be applicable regardless of whether the project is awarded by government bodies, international/regional operators, or EPCIC contractors. Of course bureaucratic government organizations tend to prevent changes from their accustomed methods as much as possible.

In category I, in spite of any backup document and calculations, you are not allowed to deviate from prescribed specifications. This may be to your benefit if:
- You have already performed a project with this client
- You are familiar with client specifications/intentions/comments
- You have sufficient/suitable Goby documents from similar projects with this client

In this case you can consider some time savings. However, if this is the first time then you should put some extra contingency for the rework. Consultant engineers may follow other specifications or recommended practices. In case there are differences, some rework will become mandatory.

In category II, client is only interested in the final product. They have put certain criteria for production volume, environmental conditions, start of production, and total cost, which if met they are satisfied, otherwise you are in trouble.

Majority of clients are in between. While they don't want to take responsibility for engineering, they want their own comments to be taken into account and implemented. This may be to evade bureaucratic problems. At the same time there may be a time span between standards applicable to their contract and new regulations. In this case they want the engineering team to implement them without their direct instruction. This may be a means to avoid giving change orders to EPCIC contractor. They will argue that since consultant has implemented changes it is a "design requirement," therefore is part of EPCIC contractor undertaking. Consultant shall distinguish between "design requirement" and "design improvement."

A clear understanding of client approach helps consultant plan his work and mobilize for interfaces. In all cases a knowledgeable client is much to the benefit of the project and the consultant. This will help consultant improve his working procedures and engineering team.

In the second step, consultant shall know the behavior of major decision-making engineers in client organization. It is true that established procedures may govern a contract. However, individuals' interpretation of the same procedure may be different. This may not always be possible. It is very helpful if each discipline can adapt to the approaches of the respective engineer in the client office. Some may prefer meetings before issuing document. Some may prefer extensive narrative reports. Some may like short reports and extensive software outputs. The author has worked with a client that always liked to have several sheets of hand calculations to be included in the extensive software report. This hand calculation was a sort of verification for the software results. He was a knowledgeable and experienced engineer. The point was that although the software was verified, he doubted the input data.

In the third step, consultant shall thoroughly review project contractual requirements and SOW. In many of the contracts client has a comprehensive list of deliverables. Many of them may not be applicable for the specific period of the project. However, somebody from day one has put them in the list, and now due to bureaucratic measures nobody dares to remove anything from the list. This case is no problem with experienced client engineers. They know what they need and, more importantly, they know that getting unnecessary documents may not be to their benefit. Unnecessary documents may bring new comments and delay the project. In spite of this, the same extensive list is a nightmare if client representatives are inexperienced and meticulous. In that case the consultant may be forced to generate documents (with the same title in the list) even if the content is rubbish and may include rewording/compilation of parts from other reports.

For each case suitable contingency factors in estimating man-hours for each document shall be included. This doesn't mean estimated costs shall increase in all cases.

6.2 COST ORDER OF MAGNITUDE AND APPROXIMATE COST

Project engineering consists of several stages, including:
A. Conceptual Engineering
B. Basic Design (FEED)
C. Detail Design Engineering
D. Construction Engineering
E. Transportation and Installation Engineering
F. Commissioning and Hookup Engineering

Sometimes detail design is divided into detail design and procurement engineering. Detail design is the focal point in this book. As a rule of thumb estimate in different projects, detail design engineering cost may vary from 3% to 6% of the EPC project. Therefore, if total project cost is known, detail design order of magnitude can be estimated. Other than the normal estimate of man-hours required for documents, several factors shall be considered:
1. Client approach in document evaluation/approval.
2. Client approach in invoice payment.
3. Client representatives' knowledge of the job.
4. Typicality of the project and availability of Goby documents.
5. Market situation and competitors' approach.
6. Consultant need for the project and present workload.
7. Consultant plan for future expansion.

Each of these items may add a positive or negative contingency to the estimated cost. Estimated cost means the cost calculated by multiplying unit rates to man-hours estimated by discipline department managers.

EPC price is normally reported in the press release of the contract news or can be obtained from sources in owner, operator, developer or EPCIC contractor organizations. If none of these sources is available, we can guess the price from comparing the size of the project with previous ones.

Order of magnitude cost informs consultant managers how to deal with this bid. Shall it be taken seriously? Shall they assign a complete bidding team to prepare proposal and estimate costs? Shall they prepare some preliminary engineering documents? In one of the bids, we had to prepare a semiconceptual study with cost comparison including rate of return.

At bid stage project final price shall only be known to a few including proposal/marketing manager, consultant managing director (MD), and the intended project manager. Inevitably each discipline will know their proposed cost. But combined costs including overheads and contingencies shall not be disclosed.

6.3 PROPOSAL PREPARATION ORGANIZATION

In some companies (mostly in big consultancies) in order to assign specific responsibilities to each team, the marketing department is separate from engineering disciplines. They have enough jobs to justify assigning dedicated engineers to proposal preparation. Small consultancies can't afford this luxury. After giving necessary input/data to proposal manager, engineers shall return to their respective department and start generating documents to enable paying their salaries.

In large consultancies for each year top management issues a company annual policy and sets some minimum limit achievements for different departments. For example, marketing and proposal department shall participate in at least X number of bids and win projects at least worth Y \$. Of course they put a limit on the total expenditure in the marketing department not to exceed a certain limit. At the same time they shall either win a certain percentage of the participated projects or bring a certain contract value as new jobs to the company.

The same is applicable for engineering disciplines. They shall perform instructed assignments. In this regard and to ensure consultant will not lose money (when assigning job to the departments), project manager applies some margin for company profit and contingencies. Disciplines also do the same. Therefore, the final assigned man-hours to each single person responsible for each item of the job may be up to 60% of the actual time calculated in the proposal.

The author believes that it is better to include middle-stage managers (discipline lead engineers) in the proposal preparation procedure. In fact, they are the platoon commanders fighting at the foremost front. They should recognize the project as their own undertaking and shall see some parts of their own contribution in it. If they are not included, at any failure they will bring the excuse that the project was underestimated by the proposal department just to win the project. This way they will evade their responsibility of carefully monitoring the document preparation stage by their fellow engineers.

The best condition will be that the intended lead engineers participate in proposal preparation. However, normally they have running projects in hand. They are busy and devoting some extra time for a proposal (which is normally requested in a rush) is very difficult if not impossible. To overcome this problem, even if proposal is prepared by others, a discussion meeting can be performed with future lead engineers and their opinions obtained.

To cater to all of the previous factors it is recommended to always have some margin in projects' schedule and cost. Using the concept of early curve for internal purposes and late curve for presentation to client may be useful. But it requires double man-hours in planning discipline and may cause confusion. Sometimes an internal report may be mistakenly issued to client.

The author has used this technique in one project. But it was a small one involving only structural discipline and had none of the complexities of interface with other disciplines and few interfaces with client. The concept was very simple. Late curve was presented to client. It was based on distributing man-hours in 8 h per day, 5 days per week schedule. This was 40 h per week. In total, this was 3840 man-hours. Project duration was 16 weeks and number of deliverables was limited. Each and every deliverable was assigned to a person. The early curve was based on using overtime work with an average of 50 h per week. The internal schedule was set for a 13-week project. In reality that project stopped in the middle. In spite of fulfilling two milestones, we exceeded total estimated man-hours.

6.4 MAN-HOUR WEIGHT DISTRIBUTION

Engineering man-hours is the basic driving force for consultants. It has the role of raw material in factories. Consultants have a specific number of personnel. It is not possible to immediately recruit or release qualified engineers. More important than recruiting is that the newcomer shall be able to fit in the team. Therefore, a certain team is assigned to a project and a few engineers may be added in between or taken from this project team and assigned to other projects. The actual man power requirement in different disciplines may be quite staggered. But the company and project task force has to proceed smoothly. An exaggerated example of a small project for a short duration is given in Example 6.1. The following observations are made:

1. Project duration ~4 month.
2. Involved disciplines include process, mechanical, piping, electrical, instrument, structural, and telecommunication (safety is not included here).
3. Approximately 19,700 MH are required to complete this job.

Based on preliminary evaluations, the following man power distribution in biweekly time spans are estimated as presented in Table 6.1:

Maybe the summation in some of the columns doesn't match given figures. This is due to roundup errors for personnel involved in each discipline. In some companies part-time (moonlighter) engineers are not used. The author only recommends it for specialized studies.

As can be seen, differences in number of required engineers may vary from as low as 10 to a maximum of 42. In addition, it may change from this time to the next 2 weeks. For a task force group it is very difficult (if not impossible) to increase and decrease number of involved personnel in a 2-week duration as per Table 6.1.

For organizations using department job distribution it may be possible to reduce idle time in a similar situation. Although this is a fictitious exaggerated example, this table clearly shows using task force organization or department distribution is not a fixed policy for a consultant. Based on project size and requirements, one of these methods can be selected.

For this case, the author may recommend distributing job to departments. In case task force assignment is mandatory, they can use seven engineers in process discipline from the start, and in the end of October or early November assign four of them to other projects. Mechanical may use four engineers from the beginning to the mid-plan and then change to

Table 6.1 Estimated man power distributions

Discipline	15-Aug	29-Aug	12-Sep	26-Sep	10-Oct	24-Oct	7-Nov	21-Nov
Process	4	7	8	8	7	4	2	0
Mechanical	2	4	5	3	5	3	3	3
Piping	0	2	5	5	5	5	5	3
Electrical	1	3	6	7	7	6	6	4
Instrument	1	2	9	9	9	11	11	5
Structural	1	1	3	5	6	5	4	3
Telecommunication	1	1	3	4	5	4	4	3
Sum	10	21	39	40	42	38	35	21

three engineers. Piping may use one engineer in the first month and then increase to five engineers from early September to the end. Electrical may use six engineers throughout the project duration. Instrument discipline may use eight engineers from early September. Structural may use five engineers from mid-August and then release one at the last month. And, telecommunication may use three engineers from early September and in October increase them to four. The selected man power will be as per Table 6.2.

Original estimated man power was 123 man months. Selected man power is about 136 man months. This distribution may increase total assigned man power about 10% but reduces malfunctions due to improper coordination/data transfer. Besides, it gives project manager a better hand in dealing with unexpected work. In all projects client will ask for small jobs here and there that may be additional to the contractual SOW but consultant can't claim any man-hours in the form of a change request for it. In addition, it reduces erratic man power distribution in each discipline, which increases discipline lead engineer efficiency in distributing assignments.

In general piping and structure are mobilized at the very beginning. They shall determine layout that is the basis of platform design. Mechanical shall start with specifications and data sheets to be issued for procurement. Safety shall define hazardous zone definition. This will enable other disciplines to determine instrument, electrical hazardous type. Although this issue is not related to proposal preparation, it has to be noted that in spite of heated arguments to accurately determine border between hazardous zones (like 1 and 2), in actual practice and to avoid mistakes in yard during installation, field instruments and electrical appliances (at least in process area) are procured for the worst-case zone.

Estimated man-hours for this fictitious project are given in Table 6.3. When calculating necessary man-hours for a document or activity, as a minimum the following topics shall be investigated:

- Document preparation in revision 0 including project requirements review, transfer of data from Goby document, adjusting basic project info and applicable standards.
- Interdiscipline checking and filling the related information.
- Document control center activity in transmittal preparation.
- Response sheet preparation and client comment implementation in new revisions or updating. Meeting with client may become required.
- Vendor data review and implementing their impact on document information.

Table 6.2 Selected man power distribution

Discipline	15-Aug	29-Aug	12-Sep	26-Sep	10-Oct	24-Oct	7-Nov	21-Nov
Process	7	7	7	7	7	7	3	3
Mechanical	4	4	4	4	3	3	3	3
Piping	1	1	5	5	5	5	5	5
Electrical	6	6	6	6	6	6	6	6
Instrument	1	8	8	8	8	8	8	8
Structural	1	5	5	5	5	5	4	4
Telecommunication	1	1	3	3	4	4	4	4
Selected sum	21	32	38	38	38	38	33	33

Table 6.3 Fictitious project man-hour distribution
Example 6.1: Man-hour distribution as per CTR list.

CTR no	CTR description	Man-hours		Duration			Man-hour biweekly distribution (period ending specified date)							
		Eng	Draft	From	to	Days	15-Aug	29-Aug	12-Sep	26-Sep	10-Oct	24-Oct	7-Nov	21-Nov
PR	**Process**													
PR-100	Philosophies (eg, process design, utility design, operation, safeguarding)	600		1-Aug	25-Oct	85	100	100	100	100	100	100		
PR-101	Study reports (eg, process simulations, process safety)	400	80	1-Aug	10-Oct	70	80	80	80	80	80			
PR-102	PFDs (process flow diagrams), with heat & material balance	200	80	1-Aug	12-Sep	42	93	93	93					
PR-103	UFDs	150	80	1-Aug	12-Sep	42	77	77	77					
PR-104	Process and utility P&IDs	500	210	15-Aug	15-Oct	61		178	178	178	178			
PR-105	Cause & effect diagrams	100	90	15-Aug	25-Sep	41		63	63	63				
PR-106	Data sheets for process equipment and long-lead items	400		20-Sep	7-Nov	48			80	80	80	80	80	
PR-107	Equipment list	60		12-Sep	14-Oct	32				30	30	30		
PR-108	Line list	90		12-Sep	24-Oct	42				30	30			
PR-109	Safe charts and ESD diagrams	120		12-Sep	24-Oct	42				40	40	40		
PR-110	Lists of chemical and utility users	90		17-Oct	10-Nov	24						45	45	
	Subtotal process	2710	460				350	591	671	601	538	295	125	

ME	**Mechanical**													
ME 100	Equipment list	160		1-Aug	25-Nov	116	20	20	20	20	20	20	20	20
ME 101	Mechanical data sheets	400		1-Sep	14-Nov	74			67	67	67	67	67	67
ME 102	Input to layouts	1040	320	1-Aug	15-Nov	106	170	170	170	170	170	170	170	170
ME 103	HVAC design philosophy	100	120	22-Aug	15-Sep	24		110	110					
	Subtotal mechanical	1700	440				190	300	367	257	257	257	257	257
PI	**Piping**													
PI 100	Block model review/PDMS	320	640	15-Aug	15-Nov	92		160	160	160	160	160	160	
PI 101	Plot plans & overall layouts	400	850	1-Sep	20-Nov	80			208	208	208	208	208	208
	Subtotal piping	720	1490					160	368	368	368	368	368	208
EL	**Electrical**													
EL 100	Electrical design philosophies	250		1-Aug	26-Sep	56	63	63	63	63				
EL 101	Study reports	280		1-Sep	24-Oct	53			70	70	70	70		
EL 102	Elec. load schedule & calc.	210		1-Sep	10-Oct	39			70	70	70			
EL 103	Specifications & data sheets	360		1-Sep	7-Nov	67			72	72	72	72	72	
EL 104	Single line diagram	140	100	15-Aug	25-Sep	41		80	80	80				
EL 105	Power, lighting, ETC layouts	150	470	15-Sep	15-Nov	61				124	124	124	124	124
EL 106	HAZ. area DWGS.	30	140	22-Aug	15-Sep	24		85	85					

Continued

Table 6.3 Fictitious project man-hour distribution
Example 6.1: Man-hour distribution as per CTR list.—Cont'd

		Man-hours		Duration			Man-hour biweekly distribution (period ending specified date)							
CTR no	CTR description	Eng	Draft	From	to	Days	15-Aug	29-Aug	12-Sep	26-Sep	10-Oct	24-Oct	7-Nov	21-Nov
EL 107	3D CAD input	40	200	1-Sep	1-Nov	61			48	48	48	48	48	60
EL 108	Final MTO	30	90	25-Oct	15-Nov	21							60	158
EL 109	Cable, JB, MCT ETC schedule	170	460	1-Oct	20-Nov	50					158	158	158	
	Subtotal electrical	**1660**	**1460**				**63**	**228**	**488**	**527**	**542**	**472**	**462**	**342**
IN	**Instrument**													
IN 100	Update basic design philosophies and reports	250		1-Aug	25-Sep	55	63	63	63	63				
IN 101	Instrument list and calculations	140	200	22-Aug	27-Sep	36		113	113	113				
IN 102	Specification and data sheets	280	440	1-Sep	7-Nov	67			144	144	144	144	144	
IN 103	Schedules (cable, junction boxes)	20	160	26-Sep	25-Nov	60					45	45	45	45
IN 104	Instr cable routings, LOC plan & cable ladder	160	400	29-Aug	4-Nov	67			112	112	112	112	112	
IN 105	INSTR installation details & hookups DWGS	80	560	29-Aug	4-Nov	67			128	128	128	128	128	
IN 106	Wiring diagrams & termination	170	460	27-Sep	21-Nov	55					158	158	158	158

Item	Description			Start	Finish									
IN 107	Logic diagrams, schematics, block diagram	280	300	1-Sep	18-Nov	78			97	97	97	97	97	97
IN 108	Loop diagram	260	200	17-Oct	15-Nov	29						100	100	
IN 109	F&G design	20	200	1-Sep	25-Nov	85			77	77	77	77	77	77
IN 110	Final MTO		100	12-Oct	25-Nov	44						40	40	40
	Subtotal instrument	**1660**	**3020**				**63**	**176**	**733**	**733**	**760**	**900**	**900**	**416**
ST	**Structural**													
ST 100	Design basis/specifications	300		1-Aug	5-Oct	65	60	60	60	60	60			
ST 101	Design reports	300		15-Sep	25-Nov	71				50	50	50	50	50
ST 102	Final bulk MTO	250		1-Oct	20-Nov	50				63	63	63	63	63
ST 103	3D CAD input		250	15-Sep	10-Nov	56		58	58		63	63	63	
ST 104	Site grading details	36	80	22-Aug	15-Sep	24								
ST 105	Road & drainage details	36	80	1-Sep	15-Oct	44			39	39	39			
ST 106	Piperack and pipe support details	36	80	1-Sep	25-Oct	54			29	29	29	29		
ST 107	Foundation and piling details	36	80	1-Sep	25-Oct	54			29	29	29	29		
ST 108	Building details	36	80	1-Sep	25-Oct	54			29	29	29	29		
ST 109	Manholes, doors, windows schedule	36	80	15-Sep	5-Nov	51				29	29	29	29	
ST 110	Temporary work details	45	113	1-Oct	20-Nov	50					40	40	40	40
ST 111	Standard drawings	101	253	15-Sep	20-Nov	66				71	71	71	71	71
	Subtotal structural	**1212**	**1096**				**60**	**118**	**244**	**398**	**500**	**401**	**314**	**223**
TL	**Telecom**													
TL 100	Update basic design philosophies and reports	80	130	1-Aug	5-Oct	65	42	42	42	42	42			

Continued

Table 6.3 Fictitious project man-hour distribution

Example 6.1: Man-hour distribution as per CTR list.—Cont'd

CTR no	CTR description	Man-hours		Duration			Man-hour biweekly distribution (period ending specified date)							
		Eng	Draft	From	to	Days	15-Aug	29-Aug	12-Sep	26-Sep	10-Oct	24-Oct	7-Nov	21-Nov
TL 101	Specification and data sheets	100	120	10-Aug	10-Oct	61		55	55	55	55			
TL 102	Telecom equipment schedules	80	120	1-Sep	4-Nov	64			40	40	40	40	40	
TL 103	Termination schedules	60	260	26-Sep	21-Nov	56					80	80	80	80
TL 104	Layouts and routings	60	260	29-Aug	10-Nov	73			64	64	64	64	64	
TL 105	Wiring diagrams, block diagrams	80	240	1-Sep	15-Nov	75			53	53	53	53	53	53
TL 106	Telecom equipment mounting details	50	240	15-Sep	20-Nov	66				58	58	58	58	58
TL 107	Final MTO	40	120	17-Oct	20-Nov	34	42	97	254	312	392	349	349	53
	Subtotal telecom	550	1490				42	97	254	312	392	349	349	245
	Total	10,212	9456				767	1669	3125	3196	3356	3041	2774	1690
	Required man power based on 40 h/W	40					10	21	39	40	42	38	35	21

6.5 UNIT COST CALCULATION

Each person's cost includes direct and indirect payments. The salary paid to each individual is not his or her only cost. There are some additional direct costs including tax, social security organization, health insurance, stationery/consumables, office equipment, leave payments, sick payments, etc. that add considerably to the unit rate. In addition to these direct costs it is necessary to add the indirect costs of payments to others who don't contribute directly to the projects but their activities are needed for the projects. An example of a group of important personnel in this category are the company's MD and all members of the board.

As much as number of personnel increases, the common costs distributed between all reduces, but at the same time the general expenditures increase. Employees may have different contracts. Therefore a common rule can't be applied. For each specific case, based on the governing statuary requirements and the actual case, the unit rates shall be calculated. It is possible to calculate the unit rates for all personnel of a department and then combine and use an average value for those at the same salary level. The overhead for each employee shall be carefully calculated. Some people don't contribute directly to bring cash into the company but their services are required for others bringing cash in.

The author is not a financial expert. What follows is only a method that has been used in dividing company additional costs to calculate overhead rates of employees. Examples 6.2 and 6.3 describe two cases. For both cases, 44 h/W and 192 h/month are assumed.

Example 6.2: Lead Engineer.
Base salary = 7000 US$/month

Net unit rate received by this engineer = 7000/192 = 36.5 US$/h.

This is the gross company payment. But the actual cost for the company is higher. In fact in calculating a project cost based on the calculated man–hours, the actual cost has to be used. Each case shall be calculated separately. The next example gives an overview of what items have to be considered.

A Statuary Costs
Several items fall in this category. For each country specific items have to be carefully reviewed. Some consultants may provide basic statuary requirements. Any additional services can be procured by employee. Some

other consultants provide better health and retirement services. These two may have different salary payments.

1. Direct tax on the salary

   Tax system varies from country to country and contract to contract. In some contracts tax may be paid by employee and in some by employer. In any case, it has a stepwise structure and the tax rate increases with increase in salary. As an example, assume following steps are used:

   0 ~ 1000 US$ = 5% Pay 1 = 50 US$
   1001 ~ 2500 US$ = 10% Pay 2 = 150 US$
   2501 ~ 4500 US$ = 15% Pay 3 = 300 US$
   4501 ~ 7000 US$ = 20% Pay 4 = 500 US$
   Total tax payment = 1000 US$ ~14.3%

2. Tax on the contracts

   Some countries give some encouragement help to export engineering services in the form of tax exemption. In others the tax for engineering services is much less than business or fabrication contracts. Here assume a 5% tax is applied for engineering services. The employee in an engineering company has to serve in the contracts. Therefore, in addition to direct tax on salary, a tax on his or her activities is also applied. In calculating contract tax share, its duration has to be considered. A short duration contract has the same tax rate as the long duration. Assuming for a 1-year contract, the share will be 5%. In a 2-year contract the share will be 2.5% each year. In this calculation, 1 year is investigated.

   This makes 350 US$/month.

3. Leave payment

   Each employee may have up to 2 days per month vacation. In some cases this may be only 2 weeks per year. To facilitate calculation, assume 1-month salary, which in this case is 7000 US$/year and makes 583 US$/month.

4. Contract stoppage payment

   In some contracts between consultant and employee each side has to give 1 or 2 month's advance notice before stopping the contract. This duration can be easily planned not to waste it. However, as per some statuary requirements, each company has to pay 1 month's salary per each year that an employee has served there. This may not be applicable in some locations. Assume here it is applicable and add some 583 US$/month.

5. Health insurance
This is a mandatory payment. Insurance schemes vary from country to country. Most government regulations instruct each company to provide a minimum insurance and then employee can buy additional insurance by himself. Here assume 8% to be paid by company, which makes a payment of 560 US$/month.
6. Hospitalization insurance
This may not be a mandatory cost or may be partially paid by employee and partially by employer. Here assume a value of 4% by employer, which makes a payment of 280 US$/month.
7. Training during job
This may vary from actual training outside company premises in an institute or other company to inviting lecturers to the company for part-time drills, etc. Normally a high-salary employee is already experienced enough and has the required qualifications. Assume only 1% here, which makes a payment of 70 US$/month.
8. Retirement deductions
This is not mandatory in all countries. In some it may be up to 7% of monthly salary. But in a majority it is left to individuals. In this chapter it is not included.
In this case sum of statuary payments makes 3426 US$/month.

B Consumables/Indirect Costs
1. Reproduction/stationery
Papers, pen, pencils, normal office stationery may not be so much in a month. But some projects may require several copies of a document and then including the check prints, saved prints, etc. it may become up to 2%. In this case it is 140 US$/month.
2. Lunch, refreshments, and housekeeping
In many companies the employee brings his or her own lunch, but in some cases company provides its own restaurant or mess room with meals. Or they may even bring meals in from an outside restaurant for employees. Maybe a small value is charged to the employee. Normally a company has to subsidize and this service is not zero cost to the company. Here assume company spends 10 US$/day and has 22 working days per month, which makes 220 US$/month.
Refreshment normally includes tea/coffee services and the tea lady. This may be included in lunch service and here we don't add a separate value for it.

Office housekeeping services may also be included under this topic. Again, a new value is not added.

3. Team building

In some companies team building activities are regularly attended. This function may be with or without client personnel and may include dinners with all employees, sightseeing, sport exercises, etc. In a majority of them employee families are also actively engaged. This activity may have a very good impact on personnel motivation. Here assume 0.5% for it, which makes 35 US$/month.

4. Bonuses

Consultants may have different policies. Some of them may pay higher salaries but then insurance services are less. Some of them may pay lower salaries but provide better services. In some, employees share in annual profits or a bonus may be given after each project is finished. This bonus may vary from year to year or project to project. It may be a little difficult to accurately distribute bonuses based on each employee's contribution in a project. Bonus effect on the employee may also be different. If consultant is an "easy come, easy go" company, then employees may prefer to have a guaranteed pay at end of each month. If consultant is trying to keep employees in all market conditions, then bonuses may be preferred. This condition provides a "shareholder feeling" for employees. They think, "I am a partner, not a tool." In this sample calculation, 2.5% is considered for this portion, which is 175 US/month.

5. Indirect costs

The product of an engineering company is its generated documents. There are many employees who are not directly producing a document. The best example is the MD himself. Of course nobody dares to tell MD "after signing a contract you are useless." It is a fact that only those directly listed under a project task force are taking part in the actual document production. The salary of the rest shall be obtained through the overhead applied to task force personnel salary.

The management, office services like IT, human resources, financial department, etc. costs may vary from location to location. These people do not generate documents but their services are required for other employees and the company has to pay their salaries form the contracts in which they are not directly involved. Middle-range and higher-ranking managers receive a considerable salary, which in fact is being paid by the documents generated by engineers/draftsmen, etc.

In addition there are costs for managers' travels to find new projects or giving presentations or advertisements, etc., which shall be paid by the projects. This ratio depends to some extent on consultant size. For a big company it is much less and for small ones a considerable part of each employee's salary is spent in paying "nonproductive employees" salaries. Refer to Example 6.4 and Table 6.4 for detailed explanation. Here assume 35% for which it is 2450 US$/month.

In this case the sum of consumables/indirect payments is 3020 US$/month.

C Mobilization Costs

1. Office rent

Even if a company owns the office this cost has to be included because instead of putting their money on the office premises they can use it in some other business. Renting costs vary considerably from location to location. A rate of 10 US$ per square meter is assumed here. Although each employee may only need a 2.4 × 2.4 square meter space, additional space for meeting rooms, passages, common services etc. shall also be calculated. Normally a manager's room has more space than low-ranking employees. On average 8 square meters is used in this calculation, which makes a payment of 80 US$/month.

2. Office utilities like electricity, telephone, etc.

Nowadays Internet service provides a much cheaper access for international dialing. Electricity and other energy sources costs also vary from location to location. On average take about 2%, which in this case is 140 US$/month.

3. Desk, chair, filing

Desk, chairs, filing, etc. have a fixed cost at purchase and then they will be used for a considerably long duration before being renewed. Each financial department has their own rules to calculate depreciation, which have to be used in tax calculations. Here assume a value of 1%, which makes an additional payment of 70 US$/month.

4. Computer

Computers are rapidly evolving. The depreciation rate is higher than for other office facilities. However, the good news is that their cost is also continuously decreasing. I vaguely remember a paper stating that if the home building industry had evolved with the same speed as computers and at the same time experienced the same reduction in unit costs nowadays we should have been able to buy houses with the cost of say "1 US\$/M^2." To be honest I don't remember the actual stated figure, but am sure it was more than several hundred times less than present normal prices. Here a value of 0.5% is assumed, which is 35 US\$/month.

5. Printer, fax, scanner
   These machines are shared by many employees. As much as the number of people sharing them increases, the individual cost will become lower. But then the downtime for waiting period will be more. Also it will require shorter maintenance durations. A value of 0.3% is assumed here, which is 21 US\$/month.

6. Software
   Software undergoes daily development. With expansion of networks and widespread need for firewalls and antivirus programs, the costs may become considerable. Here it is assumed that the company already has the required software for its job and only needs to upgrade them. Cost of new software with its training and mastering duration may be more than this. Here a value of 0.5% is used, which is 35 US\$/month.

   In this case the sum of mobilization payments is 381 US\$/month.

D Profit

This depends on company policy. In each project it may vary based on market situation, client, company status, etc. Here a value of 15% is assumed, which is 1050 US\$/month.

Sum of the costs for this particular employee may be summarized as following:

Base salary	= 7000 US\$/month
Statuary payments	= 3426 US\$/month
Consumables/Services	= 3020 US\$/month
Mobilization costs	= 381 US\$/month
Profit	= 1050 US\$/month

This makes a sum of 14877 US\$/month. Therefore the unit rate in project cost calculation based on consumed man-hours by this employee shall be calculated as $14877/192 = 77.5$ US\$/h. This is about 112% above the base salary.

The type of employee contract and statuary requirements has a considerable impact on this ratio. Next, another condition is examined for a lower-ranking employee and a different contract.

Example 6.3: Senior Engineer.
Base salary $= 5000$ US\$/month
Net unit rate received by engineer $= 5000/192 = 26.0$ US\$/h

A Statuary Costs
1. Direct tax on the salary
   Here assume company has no obligation for employee tax and only reduces scheduled deduction. This is directly taken from employee earnings and is no burden to company.
2. Tax on the contracts
   In this contract it is assumed that company receives tax exemption from authorities. Therefore, no tax allowance is assumed.
3. Leave payment
   Here half a month's salary is assumed, which in this case is 2500 US\$/year and is 208 US\$/month.
4. Contract stoppage payment
   Here the only obligation is 1-month advance notice. Since during this period the employee is performing normal duties, therefore no burden to the company is assumed.
5. Health insurance
   This is a mandatory payment. Insurance schemes vary from country to country. In some the company has to provide a minimum insurance and then the employee can buy additional insurance by himself. Here assume 5% to be paid by company, which is a payment of 250 US\$/month.
6. Hospitalization insurance
   This may not be a mandatory cost or may be partially paid by employee and partially by employer. Here assume a value of 1.5% by employer, which is a payment of 100 US\$/month.
7. Training during job
   This may vary from actual training outside company premises in an institute or other company to inviting lecturers to the company to part-time drills, etc. Normally a high-salary employee is already

experienced enough and has the required qualifications. Assume only 0.5% here, which is a payment of 25 US$/month.

In this case the sum of statuary payments is 583 US$/month.

B Consumables/Indirect Costs

1. Reproduction/stationery

   Papers, pen, pencils, and normal office stationery may not be so much in a month. But some projects may require several hard copies of a document and then including the check prints, saved prints etc. it may add up to 1%. In this case, this is 50 US$/month.

2. Lunch, refreshments, and housekeeping

   Here it is assumed that the company is not providing any lunch services. Therefore the cost is zero. Refreshment and housekeeping costs are assumed to be 0.2%. This is 10 US$/month.

3. Team building

   Here no team building activity is performed. Of course this is a unique consultant. Even if not for personnel but for client members some functions are anticipated.

4. Bonus

   Again this consultant doesn't pay any bonus to employee.

5. Indirect costs

   Same as previous case, so assume 35% for which it is 1750 US$/month. Refer to Example 6.4 and Table 6.4 for detailed explanation.

   In this case the sum of consumables/indirect payments is 1810 US$/month.

C Mobilization Costs

1. Office rent

   Here it is assumed the office is in a cheap area and a rate of 5 US$ per square meter is assumed. The area occupied by each employee on average is assumed to be 6 square meters, which is a payment of 30 US$/month.

2. Office utilities like electricity, telephone, etc.

   Nowadays Internet service provides a much cheaper access for international dialing. Electricity and other energy sources costs also vary from location to location. On average take about 1.5%, which in this case is 75 US$/month.

3. Desk, chair, filing

   Here longer duration for depreciation is used with a value of 0.3%, which is an additional payment of 15 US$/month.

4. Computer

   Here a value of 0.3% is assumed, which is 15 US$/month.
5. Printer, fax, scanner

   A value of 0.3% is assumed here, which is 15 US$/month.
6. Software

   Here a value of 0.3% is used which makes it 15 US$/month.

   In this case the sum of mobilization payments is 165 US$/month.

D Profit

This depends on company policy. In each project it may vary based on market situation, client, company status, etc. Here a value of 15% is assumed, which is 750 US$/month.

The sum of the costs for this particular employee may be summarized as follows:

Base salary	= 5000 US$/month
Statuary payments	= 583 US$/month
Consumables/services	= 1810 US$/month
Mobilization costs	= 165 US$/month
Profit	= 750 US$/month

This makes a sum of 8308 US$/month. Therefore the unit rate in project cost calculation based on consumed man-hours by this employee shall be calculated as 8308/192 = 43.3 US$/h. This rate is about 66% more than the base salary.

It is seen with different contractual conditions that the overhead varies from ~66% to 112%.

Calculation of the just discussed factors is straightforward. The financial department shall continuously update their data base to enable the latest factors to be used for proposal cost estimates.

In this calculation the average unit rate is used for all discipline members. In any discipline there are newcomers who earn much less than the experienced ones. Some of the engineers' salaries may be higher than their direct manager. Specialists get a very high salary. But they are employed temporarily and their cost will be directly charged to the project as a subcontractor.

In some cases a company may outsource portions of the job. These subcontracts are also directly charged to the project with appropriate overhead.

Other than the project manager who is directly involved in the ongoing project, and therefore his man-hours are included in project cost calculation, other managers don't contribute directly to the project. A majority of managers don't like the idea that their work is not producing any direct benefit to the project. In order to justify their high salaries they claim to bring direct benefit to each and every project of the company. The author is not against high payment to managers. He also doesn't intend to imply that managers don't benefit the company. This is far from the idea presented in this book and the actual belief of the author. The only thing the author wishes to emphasize is a specific approach/procedure to project cost calculation to enable precise overhead definition. Otherwise, the 35% factor used in the previous calculations becomes vague and its definition will be left to individual interpretation.

Please note the items and factors used herein are given as an example. The number of factors is reduced to avoid lengthy descriptions and keep only the important items. Each company shall develop costs applicable to its own conditions. A proposed method for overhead calculation is given in Example 6.4. Table 6.4 shows detailed calculations. Again, the author warns against any direct usage of these tables. In each case they have to be expanded and adapted as appropriate.

The tables are only given to justify use of the 35% factor for overhead estimation.

The author wishes to point out and emphasize that project cost calculation is like all other engineering activities.

- First of all, there are many ways to solve an engineering problem. Therefore, you may use several approaches/methods to calculate project costs.
- Secondly, is that one can calculate up to the fourth digit of accuracy, but is this necessary? Does it add to our understanding of the phenomenon?
- Thirdly, the point is to use only important factors. It is understood that each company shall list all applicable factors for a particular project. That will be a long list. Then someone (MD with the help of financial manager) shall review it and cross out the unimportant factors or combine them under one topic. Only after this exercise will the financial department be in a position to calculate the correct overhead.

Table 6.4 Sample overhead calculation

Employee position	No	Unit salary ($)	Salary		Tax		Leave
			Indirect	Direct	Indirect	Direct	0.500
Managing director (MD)	1	15,000	15,000	0	3020		625
Marketing manager (MM)	1	10,000	10,000	0	1690		417
Human resources manager (HR)	1	10,000	10,000	0	1690		417
Financial manager (FM)	1	10,000	10,000	0	1690		417
Engineering manager (EM)	1	12,000	12,000	0	2210		417
Secretaries	5	2,000	10,000	0	750		500
Architectural manager	1	9000	9000	0	1460		417
Electrical manager	1	9000	9000	0	1460		375
HVAC manager	1	9000	9000	0	1460		375
Instrument/control manager	1	9000	9000	0	1460		375
Mechanical manager	1	9000	9000	0	1460		375
Piping manager	1	9000	9000	0	1460		375
Planning manager	1	9000	9000	0	1460		375
Process manager	1	9000	9000	0	1460		375
Safety manager	1	9000	9000	0	1460		375
Structural manager	1	9000	9000	0	1460		375
Telecommunication manager	1	9000	9000	0	1460		375
Architectural lead engineers	1	7000	0	7000		1000	292
Electrical lead engineers	3	7000	0	21,000		3000	875
HVAC lead engineers	2	7000	0	14,000		2000	583
Instrument/control lead engineers	5	7000	0	35,000		5000	1458

Continued

Table 6.4 Sample overhead calculation—Cont'd

Employee position	No	Unit salary ($)	Salary Indirect	Salary Direct	Tax Indirect	Tax Direct	Leave 0.500
Mechanical lead engineers	4	7000	0	28,000		4000	1167
Piping lead engineers	6	7000	0	42,000		6000	1750
Process lead engineers	4	7000	0	28,000		4000	1167
Safety lead engineers	3	7000	0	21,000		3000	875
Structural lead engineers	6	7000	0	42,000		6000	1750
Telecommunication lead engineers	2	7000	0	14,000		2000	583
Architectural engineers	2	5000	0	10,000		1200	417
Electrical engineers	6	5000	0	30,000		3600	1250
HVAC engineers	4	5000	0	20,000		2400	833
Instrument/control engineers	10	5000	0	50,000		6000	2083
Mechanical engineers	8	5000	0	40,000		4800	1667
Piping engineers	12	5000	0	60,000		7200	2500
Planning engineers	5	5000	0	25,000		3000	1042
Process engineers	8	5000	0	40,000		4800	1667
Safety engineers	6	5000	0	30,000		3600	1250
Structural engineers	12	5000	0	60,000		7200	2500
Telecommunication engineers	4	5000	0	20,000		2400	833
Architectural drafting	2	3000	0	6000		550	250
Electrical drafting	4	3000	0	12,000		1100	500
HVAC drafting	3	3000	0	9000		825	375
Instrument/control drafting	8	3000	0	24,000		2200	1000
Mechanical drafting	3	3000	0	9000		825	375
Piping modeling/drafting	20	3000	0	60,000		5500	2500
Process drafting	5	3000	0	15,000		1375	625

Employee position	No	Unit salary ($)	Salary		Tax		Leave
			Indirect	Direct	Indirect	Direct	0.500
Safety drafting	5	3000	0	15,000		1375	625
Structural drafting	10	3000	0	30,000		2750	1250
Telecommunication drafting	3	3000	0	9000		825	375
HR clerk	3	2000	6000	0	450		250
Financial department clerks	3	2000	6000	0	450		250
IT department	4	4,000	16,000	0	1700		667
Office services	6	1000	6000	0	300		250

Employee position	No	Unit salary ($)	Stoppage	Health	Hospital
			1.000	0.080	0.040
Managing director (MD)	1	15,000	1250	1200	600
Marketing manager (MM)	1	10,000	833	800	400
Human resources manager (HR)	1	10,000	833	800	400
Financial manager (FM)	1	10,000	833	800	400
Engineering manager (EM)	1	12,000	1000	960	480
Secretaries	5	2000	833	800	400
Architectural manager	1	9000	750	720	360
Electrical manager	1	9000	750	720	360
HVAC manager	1	9000	750	720	360
Instrument/control manager	1	9000	750	720	360
Mechanical manager	1	9000	750	720	360
Piping manager	1	9000	750	720	360
Planning manager	1	9000	750	720	360
Process manager	1	9000	750	720	360

Continued

Table 6.4 Sample overhead calculation—Cont'd

Employee position	No	Unit salary ($)	Stoppage 1.000	Stoppage	Health 0.080	Health	Hospital 0.040	Hospital
Safety manager	1	9000	750		720		360	
Structural manager	1	9000	750		720		360	
Telecommunication manager	1	9000	750		720		360	
Architectural lead engineers	1	7000		583		560		280
Electrical lead engineers	3	7000		1750		1680		840
HVAC lead engineers	2	7000		1167		1120		560
Instrument/control lead engineers	5	7000		2917		2800		1400
Mechanical lead engineers	4	7000		2333		2240		1120
Piping lead engineers	6	7000		3500		3360		1680
Process lead engineers	4	7000		2333		2240		1120
Safety lead engineers	3	7000		1750		1680		840
Structural lead engineers	6	7000		3500		3360		1680
Telecommunication lead engineers	2	7000		1167		1120		560
Architectural engineers	2	5000		833		800		400
Electrical engineers	6	5000		2500		2400		1200
HVAC engineers	4	5000		1667		1600		800
Instrument/control engineers	10	5000		4167		4000		2000
Mechanical engineers	8	5000		3333		3200		1600
Piping engineers	12	5000		5000		4800		2400
Planning engineers	5	5000		2083		2000		1000
Process engineers	8	5000		3333		3200		1600
Safety engineers	6	5000		2500		2400		1200
Structural engineers	12	5000		5000		4800		2400
Telecommunication engineers	4	5000		1667		1600		800
Architectural drafting	2	3000		500		480		240
Electrical drafting	4	3000		1000		960		480
HVAC drafting	3	3000		750		720		360

Employee position	No	Unit salary ($)	Stoppage 1.000	Health 0.080	Hospital 0.040
Instrument/control drafting	8	3000	2000	1920	960
Mechanical drafting	3	3000	750	720	360
Piping modeling/drafting	20	3000	5000	4800	2400
Process drafting	5	3000	1250	1200	600
Safety drafting	5	3000	1250	1200	600
Structural drafting	10	3000	2500	2400	1200
Telecommunication drafting	3	3000	750	720	360
HR clerk	3	2000	500	480	240
Financial department clerks	3	2000	500	480	240
IT department	4	4000	1333	1280	640
Office services	6	1000	500	480	240

Employee position	No	Unit salary ($)	Training 0.010	Stationery 0.020	Lunch 10.000
Managing director (MD)	1	15,000	150	300	220
Marketing manager (MM)	1	10,000	100	200	220
Human resources manager (HR)	1	10,000	100	200	220
Financial manager (FM)	1	10,000	100	200	220
Engineering manager (EM)	1	12,000	120	240	220
Secretaries	5	2000	100	200	1100
Architectural manager	1	9000	90	180	220
Electrical manager	1	9000	90	180	220
HVAC manager	1	9000	90	180	220
Instrument/control manager	1	9000	90	180	220

Continued

Table 6.4 Sample overhead calculation—Cont'd

Employee position	No	Unit salary ($)	Training		Stationery		Lunch	
			0.010		0.020		10.000	
Mechanical manager	1	9000	90		180		220	
Piping manager	1	9000	90		180		220	
Planning manager	1	9000	90		180		220	
Process manager	1	9000	90		180		220	
Safety manager	1	9000	90		180		220	
Structural manager	1	9000	90		180		220	
Telecommunication manager	1	9000	90		180		220	
Architectural lead engineers	1	7000		70		140		220
Electrical lead engineers	3	7000		210		420		660
HVAC lead engineers	2	7000		140		280		440
Instrument/control lead engineers	5	7000		350		700		1100
Mechanical lead engineers	4	7000		280		560		880
Piping lead engineers	6	7000		420		840		1320
Process lead engineers	4	7000		280		560		880
Safety lead engineers	3	7000		210		420		660
Structural lead engineers	6	7000		420		840		1320
Telecommunication lead engineers	2	7000		140		280		440
Architectural engineers	2	5000		100		200		440
Electrical engineers	6	5000		300		600		1320
HVAC engineers	4	5000		200		400		880
Instrument/control engineers	10	5000		500		1000		2200
Mechanical engineers	8	5000		400		800		1760
Piping engineers	12	5000		600		1200		2640
Planning engineers	5	5000		250		500		1100
Process engineers	8	5000		400		800		1760
Safety engineers	6	5000		300		600		1320

Employee position	No	Unit salary ($)	Training	Stationery	Lunch
			0.010	0.020	10.000
Structural engineers	12	5000	600	1200	2640
Telecommunication engineers	4	5000	200	400	880
Architectural drafting	2	3000	60	120	440
Electrical drafting	4	3000	120	240	880
HVAC drafting	3	3000	90	180	660
Instrument/control drafting	8	3000	240	480	1760
Mechanical drafting	3	3000	90	180	660
Piping modeling/drafting	20	3000	600	1200	4400
Process drafting	5	3000	150	300	1100
Safety drafting	5	3000	150	300	1100
Structural drafting	10	3000	300	600	2200
Telecommunication drafting	3	3000	90	180	660
HR clerk	3	2000	60	120	660
Financial department clerks	3	2000	60	120	660
IT department	4	4000	160	320	880
Office services	6	1000	60	120	1320

Employee position	No	Unit salary ($)	Team building	Office rent	Office utilities
			0.005	10.000	0.020
Managing director (MD)	1	15,000	75	200	300
Marketing manager (MM)	1	10,000	50	160	200
Human resources manager (HR)	1	10,000	50	160	200
Financial manager (FM)	1	10,000	50	160	200

Continued

Table 6.4 Sample overhead calculation—Cont'd

Employee position	No	Unit salary ($)	Team building		Office rent		Office utilities	
			0.005		10.000		0.020	
Engineering manager (EM)	1	12,000	60		160		240	
Secretaries	5	2000	50		600		200	
Architectural manager	1	9000	45		120		180	
Electrical manager	1	9000	45		120		180	
HVAC manager	1	9000	45		120		180	
Instrument/control manager	1	9000	45		120		180	
Mechanical manager	1	9000	45		120		180	
Piping manager	1	9000	45		120		180	
Planning manager	1	9000	45		120		180	
Process manager	1	9000	45		120		180	
Safety manager	1	9000	45		120		180	
Structural manager	1	9000	45		120		180	
Telecommunication manager	1	9000	45		120		180	
Architectural lead engineers	1	7000		35		90		140
Electrical lead engineers	3	7000		105		270		420
HVAC lead engineers	2	7000		70		180		280
Instrument/control lead engineers	5	7000		175		450		700
Mechanical lead engineers	4	7000		140		360		560

Employee position	No	Unit salary ($)	Team building 0.005	Office rent 10.000	Office utilities 0.020
Piping lead engineers	6	7000	210	540	840
Process lead engineers	4	7000	140	360	560
Safety lead engineers	3	7000	105	270	420
Structural lead engineers	6	7000	210	540	840
Telecommunication lead engineers	2	7000	70	180	280
Architectural engineers	2	5000	50	140	200
Electrical engineers	6	5000	150	420	600
HVAC engineers	4	5000	100	280	400
Instrument/control engineers	10	5000	250	700	1000
Mechanical engineers	8	5000	200	560	800
Piping engineers	12	5000	300	840	1200
Planning engineers	5	5000	125	350	500
Process engineers	8	5000	200	560	800
Safety engineers	6	5000	150	420	600
Structural engineers	12	5000	300	840	1200
Telecommunication engineers	4	5000	100	280	400
Architectural drafting	2	3000	30	140	120
Electrical drafting	4	3000	60	280	240
HVAC drafting	3	3000	45	210	180
Instrument/control drafting	8	3000	120	560	480

Continued

Table 6.4 Sample overhead calculation—Cont'd

Employee position	No	Unit salary ($)	Team building 0.005	Office rent 10.000	Office utilities 0.020
Mechanical drafting	3	3000	45	210	180
Piping modeling/drafting	20	3000	300	1400	1200
Process drafting	5	3000	75	350	300
Safety drafting	5	3000	75	350	300
Structural drafting	10	3000	150	700	600
Telecommunication drafting	3	3000	45	210	180
HR clerk	3	2000	30	210	120
Financial department clerks	3	2000	30	210	120
IT department	4	4000	80	280	320
Office services	6	1000	30	420	120

Employee position	No	Unit salary ($)	Desk 0.010	Computer 0.005	Printer 0.003	Software 0.005
Managing director (MD)	1	15,000	150	75	45	75
Marketing manager (MM)	1	10,000	100	50	30	50
Human resources manager (HR)	1	10,000	100	50	30	50
Financial manager (FM)	1	10,000	100	50	30	50
Engineering manager (EM)	1	12,000	120	60	36	60
Secretaries	5	2000	100	50	30	50
Architectural manager	1	9000	90	45	27	45
Electrical manager	1	9000	90	45	27	45
HVAC manager	1	9000	90	45	27	45
Instrument/control manager	1	9000	90	45	27	45
Mechanical manager	1	9000	90	45	27	45

Employee position	No	Unit salary ($)	Desk 0.010	Desk	Computer 0.005	Computer	Printer 0.003	Printer	Software 0.005	Software
Piping manager	1	9000	90		45		27		45	
Planning manager	1	9000	90		45		27		45	
Process manager	1	9000	90		45		27		45	
Safety manager	1	9000	90		45		27		45	
Structural manager	1	9000	90		45		27		45	
Telecommunication manager	1	9000	90		45		27		45	
Architectural lead engineers	1	7000		70		35		21		35
Electrical lead engineers	3	7000		210		105		63		105
HVAC lead engineers	2	7000		140		70		42		70
Instrument/control lead engineers	5	7000		350		175		105		175
Mechanical lead engineers	4	7000		280		140		84		140
Piping lead engineers	6	7000		420		210		126		210
Process lead engineers	4	7000		280		140		84		140
Safety lead engineers	3	7000		210		105		63		105
Structural lead engineers	6	7000		420		210		126		210
Telecommunication lead engineers	2	7000		140		70		42		70
Architectural engineers	2	5000		100		50		30		50
Electrical engineers	6	5000		300		150		90		150
HVAC engineers	4	5000		200		100		60		100
Instrument/control engineers	10	5000		500		250		150		250
Mechanical engineers	8	5000		400		200		120		200
Piping engineers	12	5000		600		300		180		300
Planning engineers	5	5000		250		125		75		125

Continued

Table 6.4 Sample overhead calculation—Cont'd

Employee position	No	Unit salary ($)	Desk 0.010	Computer 0.005	Printer 0.003	Software 0.005
Process engineers	8	5000	400	200	120	200
Safety engineers	6	5000	300	150	90	150
Structural engineers	12	5000	600	300	180	300
Telecommunication engineers	4	5000	200	100	60	100
Architectural drafting	2	3000	60	30	18	30
Electrical drafting	4	3000	120	60	36	60
HVAC drafting	3	3000	90	45	27	45
Instrument/control drafting	8	3000	240	120	72	120
Mechanical drafting	3	3000	90	45	27	45
Piping modeling/drafting	20	3000	600	300	180	300
Process drafting	5	3000	150	75	45	75
Safety drafting	5	3000	150	75	45	75
Structural drafting	10	3000	300	150	90	150
Telecommunication drafting	3	3000	90	45	27	45
HR clerk	3	2000	60	30	18	30
Financial department clerks	3	2000	60	30	18	30
IT department	4	4000	160	80	48	80
Office services	6	1000	60	30	18	30

Example 6.4: Sample calculation is performed for a consultant with following conditions:

- Managerial group including: MD, marketing manager, human resources manager, financial manager, and engineering manager. These are also members of the board and each of them has one secretary. This makes 10 personnel.
- In addition to those in the managerial group and engineering team there are three HR staff, three financial department clerks, four IT engineers, and six services staff. This makes 16 additional personnel.
- Each engineering discipline has one head of department, one or more lead engineers, several engineers, and modeling/drafting people.
- Engineering disciplines in alphabetical order include: Architectural 6, Electrical 14, HVAC 10, Instrument/control 24, Mechanical 16, Piping 39, Planning 6, Process 18, Safety 15, Structure 29, and Telecommunication 10. This is total of 187 personnel.
- Project managers are selected from the engineering team. They are not a separate department. When a project finishes, the selected project manager returns to his department.
- In general this consultant has about 213 personnel and is a company above medium size.

All items explained herein are calculated for this fictitious consultant. To facilitate calculations some main assumptions are made:

- Equal salary is assumed for similar ranks. This is not true. Even within a discipline two engineers working with the same experience may receive different salaries.
- Tax rate as explained in "Case 1 — Lead Engineer" is used. For each category applicable tax is calculated separately.
- Half-month of annual leave is assumed for all regardless of position.
- One-month advance notice for contract termination and yearly bonus are assumed for all.
- Costs are calculated at 8% for health and 4% for hospitalization.
- Costs are calculated at 1% for training and 2% for stationery.
- 10 US$/day for lunch, refreshment, and housekeeping plus 0.5% team building are calculated for all regardless of position. It is understood that part of team building is related to employees and part is related to clients. Normally managers spend much more to buy gifts and presents for clients and in advertising.
- Meeting rooms, public areas, and services are all included in the area dedicated to each employee. Office rent is calculated to be 10 US$/M^2.

Occupied area for MD is 20 M^2, other senior managers 16 M^2, their secretaries and HODs 12 M^2, lead engineers 9 M^2, and all others is assumed to be 7 M^2. This consultant with 213 employees occupies about 1692 M^2 building. On the average this is a little less than 8 M^2 per person, which is okay.

- Office utilities cost is calculated to be 2% monthly salary, desk and furniture cost is 1%, computers 0.5%, printing facilities 0.3%, and software is calculated to be 0.5% monthly salaries.

The following figures may be deduced:

- Total Salary payment: Direct 826,000 US$, Indirect 200,000 US$.
- Total Tax payment: Direct 99,525 US$, Indirect 30,010 US$.
- Total Leave payment: Direct 34,417 US$, Indirect 8333 US$.
- Total Stoppage/yearly bonus payment: Direct 68,833 US$, Indirect 16,667 US$.
- Total Health payment: Direct 66,080 US$, Indirect 16,000 US$.
- Total Hospitalization payment: Direct 33,040 US$, Indirect 8000 US$.
- Total Training payment: Direct 8260 US$, Indirect 2000 US$.
- Total Stationery payment: Direct 16,520 US$, Indirect 4000 US$.
- Total Lunch, Refreshment and Housekeeping payment: Direct 38,720 US$, Indirect 8140 US$.
- Total Team Building payment: Direct 4130 US$, Indirect 1000 US$.
- Total Office Rent payment: Direct 13,040 US$, Indirect 3880 US$.
- Total Office Utilities payment: Direct 16,520 US$, Indirect 4000 US$.
- Total Desk and Furniture payment: Direct 8260 US$, Indirect 2000 US$.
- Total Computer payment: Direct 4130 US$, Indirect 1000 US$.
- Total Printing Facilities payment: Direct 2478 US$, Indirect 600 US$.
- Total Software payment: Direct 4130 US$, Indirect 1000 US$.

Sum of direct payments (salary only) is 826,000 US$ and indirect payments is 306,630 US$. This makes about 37% overhead (indirect/direct) costs. Table 6.4 shows detailed representation of calculations.

6.6 MISCELLANEOUS COSTS

Under this topic some small invisible costs are explained. Although they may be small, the sum of two or three item costs may be considerable. The engineering market is a low profit margin business. Therefore neglecting several of these hidden costs may adversely affect consultants.

In a majority of projects client accepts to open a letter of credit. This guarantees payment. In the contract a deadline is put for review/approval of invoices. The time lag between consultant expenditure and income is the basis for developing the cash-in/cash-out curve. Very few clients (if ever) meet this deadline. With positive cash flow the consultant receives credit from its bank account. The minimum impact of payment delay beyond contractual limits is reducing this credit. On a purely commercial basis, this may be equal to the bank interest rate for the delayed duration of the unpaid invoice.

For each project the consultant may receive advance payment (AP). To receive AP the consultant shall hand over some sort of guarantee (APG). Normally it is a bank guarantee. All banks require a certain amount of deposit plus some task fee to issue the guarantee. It depends on bank regulations and consultant credit in that bank. AP portion is deducted from each invoice on a pro rata basis. Normally the guarantee shall also be released with the same amount. In many cases clients don't care to release the amount until AP is fully compensated. In some cases even if client permits the pro rata release of the APG, bank authorities tack on so many costs and conditions that it is easier for the consultant to keep it and only release when fully redeemed.

The same is applicable for good performance guarantee (GPG). Normally it shall be released after a certain period from issuing the final dossier. But again it takes more time than originally stipulated in the contract. Normally the client wants further clarification and improvements or additional documents. This delays issuing the final acceptance certificate (if ever they bother to issue it).

Therefore, during cost estimate and proposal preparation the consultant has to decide on these issues and define the hidden costs attributed to delays in invoice payment and APG/GPG release. The author has seen up to 3% total project cost for these issues.

6.7 CHANGES/VARIATION ORDERS

In all projects some changes may occur during the engineering phase compared to the original SOW. Changes may happen due to internal and external factors.

Some of the internal factors include:
- New findings in reservoir study and behavior, which may revise master development plan (MDP)
- Changes in client's operating system, which may revise required equipment
- Changes in export conditions, which may revise plant layout
- Changes in platform location
  Some of the external factors include:
- New standard or statuary regulations

- Introduction of new equipment to market
- Value-added proposals by engineering team

These are relatively major changes. There may be some minor changes as well, such as changing a vendor, rerouting some piping, etc.

6.7.1 Change Order Cost Estimate

Change order cost estimate has to be based on unit rates included in engineering contract. A method for calculating applicable unit rates has already been explained. Project manager shall not recalculate it. In addition, during contract effective period constant unit rates will be used. Normally escalation is not applicable unless clearly specified in the contract.

The contractual procedure may be different in projects. In a majority the first request for change shall be officially issued by client. This is correct because the actual scope shall be clearly defined by the originator. In spite of that, many times the client is reluctant to do so. Or the issued request is vague and doesn't clarify all aspects. In this case the consultant shall not waste time and shall issue the complete description himself. It has to be noted that finally the EPCIC contractor will place all blame either for missing target milestones or document delays on the consultant. It is essential to always provide a complete set of backup documents. Minutes of meetings shall be carefully written. As much as possible all major participants' signatures shall be obtained. Newly established targets and their impact on project timetable shall be defined.

In many cases client accepts the cost but puts in a condition that the contract's original milestones shall not be changed. Since consultant can't add new engineers for a very short period and neglect the impact of new documents on the rest, this is almost impossible. Consultant shall not accept unreasonable requests or instructions. Many clients may promise to compensate in the next project or in later change orders. Consultant shall either not accept this promise or consider that he is doing this change free of charge.

After change details become clear, consultant shall evaluate necessary man-hours to perform instructed activities and issue a change proposal. Some clients have an approach to neglect change proposal letters. They neither confirm nor reject it. As time passes they press on consultant for project milestones' fulfillment. Then consultant in spite of several requests has to either contractually or reluctantly perform the necessary change.

In each change it is important to thoroughly check the interdisciplinary impacts. Some previously issued documents may need to be changed. Some procured material may need to be revised. It is essential to mention in all correspondence that consultant has only evaluated engineering impacts. All rework in yard, procurement team, payments, construction, and installation plans shall be carefully evaluated by EPCIC contractor team.

6.7.2 Contractual Obligations for Change

In some contracts several items may be clearly mentioned like:
- If consultant forgets to highlight changes within a certain period then he loses the right to claim. Therefore, all newly issued documents shall be carefully reviewed with priority.
- Unless a complete agreement on changes is not reached, consultant is allowed not to perform it. Of course they put so many conditions on this article that it is better for consultant to perform small change orders even before receiving their official approval.
- EPCIC contractor may like to connect change order payments as back to back to main contract payments. Consultant shall be very careful here because some of engineering change orders are so small compared to EPCIC contract values that may never be proposed to the main client.
- In some contracts there may be a distinction between change orders issued by owner/operator and EPCIC contractor.
- If parts of change order that impact procured materials, EPCIC contractor may instruct consultant to use procured material as much as possible. This will place an additional burden on consultant analysis. The reason is in some cases, to fulfill this undertaking the consultant has to perform much trial and error, which will impact both document issuing timetable and consumed man-hours.

6.7.3 Change Order Payment

Although in all contracts client accepts to calculate and pay change orders, a majority of them dislike the idea. As far as the author has seen, in almost all contracts a limit (25%) is specified for the maximum value of change orders. This means that if more than one-fourth the contract value is instructed as a change order, a new bid shall be performed. It can be a limited bid. But in any case the financial department shall get other proposals in addition to the consultant.

In bureaucratic organizations approving a change request requires a long and tedious procedure that may take much longer than engineering duration. From another point of view accepting any change order in addition to original lump sum cost may be considered as client failure to foresee all project requirements and to impose them from the beginning. The majority of companies start with a fixed budget. In this book the author is not referring to main oil companies but to EPCIC contractors.

Engineering cost is estimated between 4% and 6% of total EPCIC contract. EPCIC contractor doesn't estimate engineering costs as consultants do in detail and based on preliminary generated master document register. They just put a value in the main proposal during original bid to get the main project. After that they will initiate a new bid to select consultant. For sure they will select proposals below their estimated cost. The difference may be understood as their profit in this section of the total project. Therefore, they are reluctant to accept any change order that reduces this profit margin. However, there may be several factors that may necessitate change orders. Some of them are listed following:

- Changes introduced by client
- Value added proposals
- Changes introduced by EPCIC contractor procurement department
- Changes introduced by EPCIC contractor fabrication yard
- Changes introduced by offshore installation contractor (OIC)

Any time during a project client may find new information about the reservoir and its yield to require a revisit and change in the original scheme. This change may normally happen during a revisit/study of the MDP.

During value–added studies/proposals some major changes from project basic requirements may be proposed. This change may come from either the client itself or consultant. However, before accepting any proposal a comprehensive backup document will be needed to get client approval. This may have considerable cost/time impacts and may only be acceptable if it is suitably proved that the gain for the project during its intended design life is more than its costs.

In the engineering contract a specific duration is assigned for procurement engineering. To accomplish this schedule the consultant may limit the duration for which quotations can be received or limit the number of quotations or technical clarifications. Procuring packages have much more cost than engineering. During this process EPCIC contractor procurement department may find cases in which by introducing a new vendor can save in the package cost.

In addition to the procurement department, the EPCIC contractor yard may change fabrication methods. This may impact engineering design and consequently necessitate a change.

In some projects OIC is different from EPCIC contractor. The final selected methods for loadout, transportation, and installation have a considerable impact on engineering design. Any change in transportation barge, lifting crane, etc. may require reanalysis or even rearrangement of structural members that have time/cost impacts on engineering.

Each party has certain reasons for proposing a change. The impact will not only be limited to engineering but will extend to other stages of EPCIC contract. EPCIC contractor project manager shall foresee whether introducing this new change will save total project time/cost or not. Any savings in a specific part of the project may lead to time delay/additional cost in another part. This shall be fully investigated by project members including client/EPCIC contractor/OIC and consultant before instructing/accepting a change.

CHAPTER 7

Planning the Project Budget

Similar to projects involving procurement, construction and installation those that only include engineering activities shall also have a budget plan. Since the total value of engineering projects is much less than PCIC projects, normally budget planning is limited to monthly reports prepared by financial departments. The author has the idea that they shall be treated in more detail. Financial reports may have three main topics: received money, spent costs, and expected money.

For each project a budget plan has to be prepared to address all topics in more detail. This will include expected budget topics and its cost distribution throughout the project duration, plus procedures to rectify short-comings and mishaps.

A sample is presented in this chapter. In fact, this report was prepared for two projects by the author. In the first project based on project progress it was updated twice to reflect actual conditions. Unfortunately in both cases the consultant executive manager didn't show any interest in this report. Instead, the report prepared by the financial department was used. I still think this format is better. To be able to present it in this book without information property violation, figures, salaries, and some other information are changed.

Consultants are invited to review the proposed methodology and try it in their internal procedures. The reader shall note this is only a sample out of several ways that financial evaluation of a project can be done. Each consultant may have its own practices and preferred methods that shall be considered. It has to be noted that if a procedure is continuously followed by a consultant and every team member is familiar with it changing it will still always have some difficulties. With the so-called "old method" everybody knows what a term means and what is expected from each person in each stage. To implement the new method, first the terminology shall be defined. A learning curve needs its own time span to demonstrate results.

A first revision of this report may be prepared by the assigned project manager (PM) based on information received from the bid proposal manager. In some consultancies the person assigned to prepare the final bid

Practical Engineering Management of Offshore Oil and Gas Platforms
ISBN 978-0-12-809331-3
http://dx.doi.org/10.1016/B978-0-12-809331-3.00007-7

proposal and negotiate with the potential client may be selected as project manager. In this case the two positions may be held by the same person.

This document shall not be distributed openly. Its restricted distribution may cover only engineering manager, financial manager, and company executive manager or board of directors. A summary of this report has to be updated on a monthly basis (or other intervals suited to project duration and consultant financial status) to show project status for quick information at the top company managerial level. The following is a proposed table of contents:

- Introduction
- Budget Proposal
- Project Costs and Planning
- Expected Income and Deficits
- Expected Deviations
- Proposals and Incentives

The topics in this sample report are taken from an actual project prepared by the author. As explained before the figures have been altered for confidentiality reasons. Therefore, in some cases the figures may not seem so realistic. Each company's management shall assign policies and discuss the required topics to be addressed in the project budget plan. The topics proposed here are only based on the author's experience in the selected case with some modifications. The same report in another project may have different topics. However, the major items of how the budget/costs have been estimated, what the expected invoicing and expenditures are, what the major setbacks in the project course are, and what are the mitigation measures that may be used for each case shall be addressed in any manner appropriate for the company's system.

7.1 INTRODUCTION

The introduction section can be further subdivided into subsections. It has to give a brief explanation of what the project is and what parties are involved. These may be taken from the general explanation given in the project contractual documents. They may include main contract, amendments, bid circulars, minutes of meeting (MOM) and kick off meeting (KOM) notes, etc. The introduction may contain the following subsections:

- Project definition
- Purpose of document
- Objectives
- Abbreviations

7.1.1 Project Definition

This section gives a brief definition of project main items like structures, geographical location, systems, and intention. It may contain number and type of platforms, their elevations, coordinates, distances from each other and onshore facilities, available systems on the platform, intended purpose including a very brief description of their facilities, interfaces with each other and onshore facilities, etc. Although a general list was given, for the purpose of this document, this part can be very brief.

As this report is not intended to be a comprehensive design basis document, the description can be limited only to major items.

In addition to the general topics mentioned, some basic information will be used in this document. They may be explained in more detail. For example:

- Involved parties and their interface with each other like owner, main contractor, subcontractor, etc. shall be clearly defined. How will each of the concerned parties impact consultant work?

  These parties have different interests. Their contractual roles will inevitably impact consultant work. As much as possible the consultant shall try to maintain his independence and remain out of conflicts between parties. In theory it is understood that engineering design shall try to balance costs with intended scope of work (SOW). Cost studies are performed in general terms during conceptual study for the whole offshore complex. But it is impossible to perform an individual cost study for each package. In each case the consultant will refer to standards, recommended practices, and similar successful designs. In most cases a compromise between different interests (without jeopardizing minimum standard regulations) is the best that the consultant can achieve.

- Document approval process and the effect of conflicting interests: Let's be frank. The owner wants engineering documents to describe a quality much more than standard obligatory requirements. The contractor wants specifications to cover minimum requirements. This is for lumped sum contracts. For cost plus contracts each side intention may be different. The common, unchanged intention is to get maximum profit from the performed job. In some conditions reputation, environment, safety of personnel, operational simplicity, etc. may also mean profit. Unfortunately it is very difficult (if not impossible) to translate them in terms of comparable numbers.

- Progress evaluation (physical and financial) procedure: They may be different from each other. The planning and project control team may find it better to give weight to each document based on original estimated man-hours (lead engineer, senior engineer, engineer, drafting, etc.) multiplied to related unit rates. This weight will be used for both physical and financial progress calculation. Even with this condition, figures quoted for the two progresses may be different. Contractually invoicing can only be done at predetermined intervals (normally monthly). At invoicing deadline, only completed stages of documents (issued for comment, IFC; issued for approval, IFA; and approved for construction, AFC) can be listed. At the same time, a discipline may have worked on several documents. But their completion status may not match agreed stages. This means that even under the best conditions when the measurement tool is the same, physical progress may be a few percentages higher than invoicing progress. This is important because "cash in" is governed by invoicing (financial progress), but remaining SOW is governed by actual performed work (physical progress).

- Invoice payment procedure is also very important. Is it directly paid after project team approval? Shall the financial department get approved invoice value from the project team and include it in a plan to be approved by the company's executive manager? Is there any time delay between approving an invoice and preparing necessary funds for it? Does the project team control project cash or is it part of general company cash flow? Is procedure and time span between approving an invoice to its actual payment automatic and always the same or may it change in each case? After all, money settled in the consultant bank account is "real" cash flow while the approved invoice is only "potential."
- Document stage progress (IFC, IFA, and AFC) is very important in invoice calculation. Clarify if each issue of the document will increase its stage one step or if it needs approval. For example, an owner may state that unless all my comments are implemented I will not accept a document stage increased from IFC to IFA or AFC. To follow this intention, the consultant may immediately implement all comments without considering their impact on project cost. This may be against engineering, procurement, construction, installation, and commissioning (EPCIC) contractor interests. They may instruct the consultant to argue for each comment if it may eventually impact cost. This means more manpower consumed on each document and longer duration to raise document status. Owners may use this contractual definition as an effective tool and leverage to force the consultant to accept and implement

their comments on each document. It has to be noted that from consultant arguments, the contractor is benefiting the most. Therefore, it is not fair to leave the consultant to bear financial impacts alone since the owner has not approved document progress to the next stage.

These five bullet points are understood to impact consultant invoicing and consequently its cash flow. In some contracts major affecting parties may be different, which have to be adjusted as necessary.

7.1.2 Purpose of Document

In oil and gas projects each contract may have a specific structure. A government regulatory body may assign a field or part of it to a developer. They may subdivide a very large development plan to several areas and/or stages. Each stage may be further divided between several general contractors. Each general subcontractor may hire a consultant.

Only for clarification sake, assume an oil field is able to produce under normal conditions for 5 years. After that some sort of enhanced oil recovery (EOR) becomes necessary. In the approved master development plan (MDP) several satellites, one main and one living quarter are envisaged for the first stage, which includes normal production.

For satellites it may be more economical to fabricate and install risers and space needed for future EOR equipment in addition to present export line riser. This will increase consultant SOW. Another approach is to ignore future risers and only install a jacket and topside that are able to tolerate loads and have additional space suitable for future development. Unfortunately, this may not be so economical. Because installing external risers and necessary equipment after several years of production requires mobilizing an offshore fleet, which inevitably will be more costly. But this will reduce consultant SOW.

In the same project, for main and living quarter platforms future bridge connection to new platforms and tie-in points may show to be more economical. For maximum utilization of fabrication yard free space, a developer may divide a project between one or two main contractors. They may establish a consortium between themselves and hire a common consultant, or each one may have their own consultant. These cases may impact consultant SOW and therefore engineering project flow in a different manner.

For the purposes of this discussion, the scope of consultant involvement shall be defined in two or three paragraphs as per engineering contract between EPCIC contractor and consultant.

The scope of other parties' involvement in the project also needs to be defined in two or three paragraphs as per main contract or consortium agreement (if any).

In addition to the previous requirements, the scope of subcontractors' involvement in the project also needs to be defined in two or three paragraphs as per related subcontracts (if any).

It is understood that the project budget plan is a live document and will be updated from time to time to reflect changes, additions, or new attachments necessary or introduced during the execution of the project. Updates can be made monthly or bimonthly or as other selected periods found suitable. A project budget plan is not intended to act as a catch-up plan. In addition, it is not intended to act as a pure financial report. But it has to specify project cost effectiveness and highlight whenever a specific managerial decision is needed. This decision may vary from allocating additional resources/budgets, discussion or amendments to contract with client, change in task force organization or personnel, etc.

The purpose of the project budget plan is primarily to control the cash in—cash out flow, forecast a required budget for the next period with an estimate of budget topics, control milestones, and propose mitigating steps (if necessary).

Responsibility for implementation of the project budget plan recommendations and proposals rests with the project manager. But its approval needs joint meeting of PM, consultant executive manager, and financial manager. In order to get a view of the present situation and plan for the next period, all direct/indirect costs shall be issued to the PM to plan for the future.

Suggestions and requests for amendments or supplements to this procedure should be addressed to the PM who will, when appropriate, issue revised and/or add new sections. Similar to all other documents, the revisions will be identified in the revision matrix at the front of the document and underlined in the text.

In some consultancies the financial manager may take this role and PM role may be limited to daily handling of project technical issues or implementation of the report suggestions. This may happen sometimes due to company executive managers' reluctance to issue cost data to a technical group. They think this is company confidential data and shall be restricted. Although the author doesn't approve of this system, it is left to each company's norms. Each consultant is established with an organization chart and job definition. It is not recommended to change all established procedures that have been running for a long time just for the sake of one project or the PM.

7.1.3 Objective

The objective of the project budget plan document is to establish a tolerable cash flow and ensure reasonable profit is generated for the project. It has to be noted that all profitable projects don't have tolerable cash flows. In addition, a nonprofit project may be accepted only to help consultant cash flow. In the preliminary stages of assessment, the project may face some negative cash flow due to the following reasons:

- Normally EPCIC contractors are reluctant to pay directly and out of their own pocket to a consultant. They prefer to first receive payment from owner/client and then pay their subcontractors. It has to be noted that a consultant is also one of the many subcontractors selected to finish a project. Other than delay in progress approval between consultant to EPCIC contractor and EPCIC contractor to client, there is a delay in the relevant payments as well. Sometimes the accumulative delay may exceed 2 months.
- Consultant personnel shall be paid at the end of each month while the cost of documents produced by them in this duration may be received with 2 months' delay.
- For some specialist studies, a consultant may hire subcontractors. They may not have the financial ability to wait for their payments for a long time. Therefore, although unlike to consultants' personnel, payments may not be at end of each month, but the delay may be less than "cash in" delay for the consultant.
- Consultant payment milestones may be different from EPCIC contractor payment milestones for the same engineering SOW. For example, clients are normally reluctant to pay for start and interdisciplinary check (IDC) distribution of a document between related disciplines. For consultants, IDC completion is a major portion of the job.
- At the beginning of the engineering phase very few documents are expected to be issued. This means very little invoicing. However, exactly at this stage very large man-hours shall be consumed. These man-hours are needed for preparing design basis, checking different options, clarification meetings for discipline interfaces, etc. This is similar to the famous progress S-curve in construction projects.

Although in some cases mobilization fee/advance payment are paid, in some cases it is not so effective. In some cases the author has seen companies that differentiate between the mobilization fee and advance payment.

Continuous updating and monitoring of a project plan enables to minimize the gap between cash-in and cash-out curves and ensures a positive profit at the end of the project.

7.1.4 Abbreviations

This section defines the abbreviations used in the specific project. Normally each document will have a list of abbreviations that are used. In the project budget plan the needed abbreviations include:

- Parties involved including owner, client, EPCIC contractor, consultant, subcontractors, management's inside consultant organization, etc.
- Stages of document issue/project progress
- Task Force Disciplines

Abbreviations are generally defined when first used. Similar or slightly different abbreviations have been used in different projects. The use and extent of an abbreviations list depends on each company's practice. However, some of them have been many times and very widely used. Some of the most commonly used abbreviations are listed at the beginning of this book. For each specific report, applicable abbreviations shall be used.

7.2 BUDGET PROPOSAL

This section gives a brief description of bidding history. The invitation to bid (ITB) stages include steps taken in estimating the price, man–hour estimate, cost breakdown, planned organization chart and intended task force, major team players, main items affecting final calculated cost, contingencies implemented in cost estimate (man–hour and unit costs), etc.

Documents are engineering design output. An accurate document list at the bidding stage will provide an accurate man–hour estimation. A list of required documents is prepared based on project SOW. Major engineering assumptions including those related to platform systems/packages, structural assumptions to comply with EPCIC contractor yard facilities or offshore installation contractor fleet like transporting barges, lifting cranes, geotechnical/geophysical data, etc. will all affect the document list.

Use of highly qualified engineers, availability of goby documents from similar projects, previous work experience with the same client or in the same region, etc. will facilitate document issue or approval stages.

Contractual agreements on advance payment, guarantees like good performance guarantee (GPG) and advance payment guarantee (APG), letter of credit opening and its becoming effective, definition of commencement date/project start, project effective duration, etc. will impact cash flow.

7.2.1 Bidding History

This section gives a brief history of the bidding process. Some of the items to be covered are:

- When and how the ITB was received?
- Which organization/company has invited the consultant for bidding?
- Is this the first encounter with this client? What do we know about how they treat their subcontractors? Do they play fair or should the consultant always be on guard?
- What are the consultant's priorities (establishing itself as a competent engineering company, reputation, gaining profit as much as possible, saving through a financial crisis, etc.)?
- What is the present workload? Shall we dive in price calculation?
- What is the status of subcontractors? Can they absorb part of the commercial risk? Shall we help them? Have they been evaluated both technically and financially? What is their rating? What portions of the job are being subcontracted?
- What are the cost revisions? Why have there been cost revisions? Are they due to competitors? Are they due to changes in SOW? Are they due to more accurate evaluation received from disciplines?

The oil and gas offshore world is relatively small. Major players already know each other or have heard something from others. In each region there are only few engineers with consultant, contractor, operator, and client experience. A majority of them work only in one field/party.

7.2.2 Reasons for the Selected Price

Project man-hours are calculated through a normal procedure of sending bid documents showing SOW and document list (or instructing discipline leaders to prepare their own list) to discipline leaders. After review, each discipline will come up with an estimated list of topics and required man-hours. It has to be noted that at this stage detailed document register

(MDR) is not required. A list of topics in each discipline is sufficient. Then the proposal manager shall review it from an interface point of view. In a joint meeting with all disciplines, man-hours may be reevaluated. Man-hours shall include the following shares:

- First issue preparation and internal reviews.
- Response to client comments and other revisions updating.
- Time consumed for each discipline to review and input necessary data in related sections of documents belonging to other disciplines.
- Vendor data review and implementation.
- Possible amendments due to vendor data changes and impacts.

Based on consultant unit rates and multiplying to the calculated man-hours, the actual cost is estimated. This is the first step. Then a committee including the company executive manager, financial manager, proposal manager, and the administration manager or their representatives shall review the proposal and determine what mobilizations are needed. This may vary from renting new offices, changing existing arrangements, and procuring additional equipment like computers, printers, faxes, telephones, scanners, furniture, etc. New recruitment in each discipline and the duration for it shall also be investigated and determined. All these items shall be added to the original cost calculation to find out actual project cost. Overhead ratios as determined by the financial department shall be implemented. After this stage, company long-term policies shall determine the final fine-tuning factors.

All these items shall be clearly listed in the project budget plan report. This is not only needed to save the skin of a manager if the project fails but also when trying to find mitigation measures. It will indicate where the contingency points are and how much savings can be expected from each budget topic.

In a multidiscipline project when several parties are involved interfaces between disciplines and related parties like client, EPCIC contractor, offshore installation contractor, etc. play a major role in project progress. The same team of engineers may fail if there is a conflict between client parties. Some items that may be addressed are listed briefly following:

- If a project is the first multidisciplinary undertaking of this size for the consultant or its track record is poor, then individual task force members shall have several years of experience in oil and gas projects to persuade a client to award the project. This issue of task force members' experience shall be highlighted in the organization chart and CV section of the proposal and in the presentation meetings. Consultant shall confirm they will be dedicated throughout the project.

- Some competitors may have advantages in their offer. This may be due to previous experience with this client and therefore receiving their support, having an ongoing similar project that provides suitable goby and leads to saving in man-hours, etc.
- Some competitors may have a ready team that has just finished or is in the last stages of another project. Therefore, their mobilization process may be very rapid.
- In any condition client shall be convinced that this proposal is attractive. This may be cost-wise, engineering/procurement, or construction schedule-wise, etc.
- The previous experience of consultant and client key personnel in the same company or project is also an advantage.
- If client/EPCIC contractor has delayed project start due to any sort of inactivity, then including some contingency in engineering schedule to compensate a portion of this delay will be very advantageous.
- If parts of project basic data have been delayed (for example, geophysical/geotechnical study, design basis, process assumptions, etc.) then it is advantageous to propose starting the study with presently available data from a nearby location or similar projects trying to accommodate the worst condition. However, this is a two-sided blade. It may save the client some time before actual site-specific data becomes available, but the data may vary from the assumptions and require considerable rework for the consultant. This risk factor shall be suitably reviewed.
- When offshore projects are booming, some mills may be booked for a long duration. In this case they will give long production durations. It may be advantageous to provide a preliminary list based on previous experiences to book mill fabrication. Consultants shall be very careful to follow deadlines for producing correct material take off (MTO). Otherwise, large changes in the ordered material may cost heavily to the client, which in turn may claim liquidated damages from consultant. This is only applicable for bulk items and not for packages.
- In order to start yard fabrication with an early schedule, client may assign predefined milestones or consultant may voluntarily undertake to issue AFC drawings for some parts of the platforms in an early stage of the job. This is also a two-sided blade. Sometimes client may insist on their comments and their resolution may take considerable man-hours.

In this case if milestones are missed, EPCIC contractor may face heavy yard mobilization costs.

Stages explained herein may seem too complex. Bringing a majority of the consultant managerial team together to share responsibility for every bid may not be possible. Circulation of letters and estimates between all departments also doesn't seem to be possible within a short period in a majority of bids. In some cases the bid organizing committee wastes a lot of time on small items and at the same time wants all bidders to respond within a very short time.

One of my most respected colleagues (who taught me a lot) used to say: "The biggest problem with our clients is that they plan a project to start from time X and end at time Y. They waste a lot of time on negotiations and bureaucratic measures and X is continuously dragged forward. This will override inter stage milestones. But nobody neither cares nor dares to move Y a little (not even proportionally) forward." Contractors also say "yes" to all requests before signing a contract. After signing a contract with an impossible time span they start to find means to evade these incorrect obligations. This way much energy and effort is lost for useless and unnecessary arguments that could have been avoided if from the beginning the client had taken a more rational approach.

7.2.3 Budget Topics

Cost of small and routine projects may be calculated by rule-of-thumb comparison with previous ones. Major or new project costs shall be calculated by two methods to countercheck each other.

The first method may be based on listing cost, time, resources (CTRs) in each discipline and assigning man–hours for each item. Man–hours may be distributed in several categories, for example, lead engineer, senior/junior engineer, lead designer, and drafting personnel. Each group's share in different type of documents varies. For example, in design basis documents may only have LE and SE man–hours. While in a general list, only drafting staff may consume man–hours. The man–hours spent in interfacing, IDC checks, vendor data review, etc. may be calculated separately under "tasks" topic equal to a predefined percentage of discipline total man–hours or added individually to each document. However, in the first case since the client does not receive any hard copy or document for performing the interface activities, the task's total weight is again distributed between discipline CTRs listed as a deliverable based on their weight factor.

The second method is calculating/estimating the number of engineers mobilized in each discipline and summing up other nonlabor costs. Some trial and error may be needed so that calculated costs from the two methods converge. A proposed list of budget topics may include:

- Personnel salary (Budget topic 01)
- Office mobilization and utility (Budget topic 02)
- Overhead (Tax, Social Security Organization (SSO), Statutory) (Budget topic 03)
- Engineering Subcontractor/s (Budget topic 04)
- Contingencies (Budget topic 05)

Final discount/escalating may be granted by company executive manager or project sponsor in contract finalization negotiations. List of topics shall be prepared for each project. They are not universal. Each topic also may be divided into several subtopics.

7.2.3.1 Personnel Salary and Subcontracts (Budget Topics 01 and 04)

Due to their somewhat similar nature these two topics are gathered in one section. They all refer to monthly payments to individuals (person or a company) for the provided services. In a project organization chart some subcontractors may have been included. The contract with subcontractors is normally in a lump sum basis. In this case, a consultant shall not be concerned with the number of their involved personnel and only check the schedule. However, for some reason this is not so easy.

First, they may require some services from the consultant. Therefore, the consultant shall have an estimate of their personnel number to provide the required services.

Second, the average man-hours can't be very much different. Therefore, the consultant shall have an estimate of their mobilized manpower to evaluate the cost effectiveness of each proposal.

In addition, to calculate project price by the second method, an estimate is needed. Table 7.1 gives an example of the unit prices used in calculating a project budget. These are fictitious and shall be adjusted based on actual paid values.

Table 7.1 Unit costs

Position	Project manager	Lead engineer	Engineer	Modeler	Designer	Drafting/ Staff
Unit cost USD/ Month	6000	4500	3500	2200	1800	800

Table 7.2 gives an example of the original estimated values in each discipline for a fictitious project. Total sum of budget allocated for this topic is US$1,707,100.

The following notes shall be considered:

1. Some subcontractors may be selected for portions of engineering. This may happen for specialist studies. Their cost and the required double taxation/overhead shall be included in the estimated man-hours.

2. A brief history of discussions with subcontractors shall be given. Company managers may not be involved in the smallest details. However, a history showing the original prices of subcontractors, escalations, discounts, major assumptions, etc. shall be given in this report.

3. With some of the subcontractors consultant may have an agreement and undertake some of their responsibilities as per main contract. Or, to relieve itself it may transfer all the responsibilities toward EPCIC contractor and client mutatis mutandis to them. At the same time, consultant may prefer to have all the task force under the same roof. In this case they have to provide some services to subcontractors personnel. The different conditions of subcontracting shall be considered when calculating the charges.

4. Some subcontractors may require special facilities like sophisticated software or laboratory equipment, etc. This cost may be separated under budget topic 02. The cost may vary from leasing specific software to assigning a laboratory for a test, etc.

5. EPCIC contractor may request further services in addition to project requirements in order to facilitate their procurement or construction process. This may be included in related discipline man-hours.

6. In this table all personnel of a similar rank have been considered with a constant salary. This may not be true in the actual situation. Even within a discipline two engineers may receive different salaries. Some of the main factors with an impact on personnel salary have already been discussed in chapter "Balancing Between Client and Task Force Engineers." At first revision of this document, which assigned personnel to the project are not clear, this approach may be acceptable. However, in later revisions actual salaries have to be incorporated.

7. If personnel are recruited only for one project, their additional social benefits and costs, like 1-month advance notice, yearly bonus, insurance and tax payment, medical benefits, 1-year leave, etc. shall be included as considering additional months in the project or distributed in monthly salary.

Table 7.2 Personnel costs

Item no	Discipline name	Position	No	Month	Subtotal USD	Total USD
01–PM	Project management	Project manager	1	12	72,000	162,000
		Project planner	1	12	42,000	
		Senior document control center (DCC)	1	12	21,600	
		Office staff	3	11	26,400	
01–PR	Process	Lead process	1	12	54,000	139,800
		Engineer	2	11	77,000	
		Draft	1	11	8800	
04–SA	Safety	Lead safety	1	12	54,000	216,800
		Engineer	4	11	154,000	
		Draft	1	11	8800	
01–ME	Mechanical	Lead mechanical	1	12	54,000	178,300
		Engineer	3	11	115,500	
		Draft	1	11	8800	
01–PI	Piping	Lead piping	1	12	54,000	283,900
		Engineer	3	11	115,500	
		Modeler	2	11	48,400	
		Designer	2	11	39,600	
		Draft	3	11	26,400	
01–EL	Electrical	Lead electrical	1	12	54,000	159,600
		Engineer	2	11	77,000	
		Designer	1	11	19,800	
		Draft	1	11	8800	
01–IN	Instrument/control/ Telecom	Lead instrument	1	12	54,000	245,400
		Engineer	4	11	154,000	
		Designer	1	11	19,800	
		Draft	2	11	17,600	
01–ST	Structure/architecture	Lead structural	1	12	54,000	321,300
		Engineer	5	11	192,500	
		Designer	2	11	39,600	
		Draft	4	11	35,200	

7.2.3.2 Office Mobilization and Utility (Budget Topic 02)

Under this topic several issues are to be covered. They may include office rental costs, computer for each engineer, software costs, office furniture, printer/scanner, lunch, midday refreshment, reproduction costs, team-building activities, client costs in consultant office, hazard identification (HAZID)/hazard and operability (HAZOP) chairman and scribe in basic and detail design, and travel costs. Total sum of budget allocated for this topic is US$773,930. Table 7.3 gives the estimated costs.

The following notes shall be considered:

1. Office space for about 65 persons shall be much more than 300 m^2. If some spaces like meeting rooms may be shared with other projects, that will reduce the required area. Rental cost of such a building depends very much on the country, city, and location. In the same city, cost in two different locations may vary by half. However, consultants can't rent a place in any location of the city. Other than the total available area, factors like the surrounding location, vicinity to client, etc. shall be considered. Even if consultant owns a building, they shall include related rent cost. Procuring a building is considered a preinvestment. This cost may later be used as contingency, or management can neglect it to win the project.

2. One of the few commodities that their cost reduces continuously is computer. In fact with the same price paid today for a specific capability, next year can buy a better specification. The problem is due to rapid changes; computers, monitors, and similar equipment soon become outdated and need to be replaced. This means their depreciation rate is high.

3. A lump sum value of US$9000 is assumed for office network and connections. Similar to calculated rent for a building that is owned by a consultant, parts of this cost can be distributed between several projects. In addition, if for any project a partition arrangement is changed, total cost shall be included in that project.

4. An engineering office requires a huge number of printing/scanning pages per month. For large consultancies having a dedicated printing department in addition to normal printing machines in each discipline is cost-effective. The check prints can be taken in the discipline. But when the document becomes ready and needs to be reproduced in the number of hard copies as required by the contract, it may be delivered to the printing department. A normal-sized project may need around 200,000 A4 size equivalent pages to be printed per month.

Table 7.3 Office costs

Item no	Description	No	Month	Unit price USD	Subtotal USD	Total USD
02–Of	Office rent	1	12	8000	96,000	96,000
02–MO	Computer	65	—	1000	65,000	149,450
	Table, chair	65	—	650	42,250	
	Office preparation	—	—	Lump sum	9000	
	Printer/scanner	2	—	13,000	26,000	
	Printer	6	—	450	2700	
	Filing	30	—	150	4500	
02–LU	Lunch (22 days)	65	12	6	102,960	161,280
	Midday refreshment	65	12	2	34,320	
	Water/electricity/telephone	1	12	2,000	24,000	
02–RE	Reproduction	—	12	3,500	42,000	42,000
02–TE	Team building	—	—	Lump sum	30,000	30,000
02–TR	Travel	40	—	3,000	120,000	120,000
02–CL	Client in office	6	12	1,100	79,200	79,200
02–HA	HAZID/HAZOP	2 × 2 × 2	—	1,200	96,000	96,000

Maintenance of the print/copy machines is very essential. Nowadays paperless offices are becoming more popular. Even with this policy, a minimum amount of printing is necessary.

5. One year has 365 days with 52 weeks. Official national holidays in addition to two days per week shall be deducted from working days. However, in certain stages of the project heavy overtime may be required. In each project a percentage of overtime shall be considered. One reason may be delay in issuing documents. The main reason is that the consultant is not willing to hire a huge number of personnel. No company can build up or decrease suitable man power immediately. After a project finishes, there may be a gap until a new assignment. Therefore, consultants prefer to have limited, manageable man power.

6. Utility costs including electricity, water, gas, telephone, etc. vary from country to country. Normally within a country costs are almost the same or have negligible variation.

7. A lump sum price has been considered for client entertainment and a number of selected gifts or other items for the selected persons.

8. Contractually consultants may have to provide a specific number of working stations or offices for the client. Normally client personnel expect to get better services compared to their fellow engineers in EPCIC contractor or consultant offices. This may include larger working space, better chairs, desks, computers, etc.

9. Specific HAZID/HAZOP sessions in basic and detail design stages shall be foreseen. Normally the chairman shall be from the engineering group. Sometimes a client may require a separate company (instead of a specialist) to undertake HAZID/HAZOP. This may be treated as a subcontractor. Exact duration for each HAZID/HAZOP session can't be estimated. Some additional costs can be used for other underestimated items.

7.2.3.3 Overhead (Tax, SSO, Statuary) (Budget Topic 03)

Tax and SSO payment shall be included for the whole contract including subcontractors and consultant SOW portion. This can be calculated very simply as a specified factor multiplied to the total calculated cost. Subcontractors shall include their own tax separately. Although this is some sort of double taxation, in almost every project that the author has seen this has been the same. Of course subcontractors' tax and other statutory payments have already been included in their proposal. Some extra factors to calculate the additional expenses for late payments of each invoice and guarantee costs may be included.

Table 7.4 Overhead costs

Item no	Description	Total USD
03—TA	Tax 6% of total	225,548
03—SS	SSO 8% of total	300,731
03—PR	Profit ~ 10% of total	375,914

In spite of what is usually mentioned in the contract, in many projects APG and GPG are not released on time. This imposes additional unforeseen costs (like bank charges to extend guarantee) to the consultant, which can be included in the total cost. It is the executive manager's decision to delete these overhead charges so that the final proposal cost becomes attractive. But in general cash flow they shall be considered.

The profit may also be included here. This way profit is included as one of the cost topics. This is not a method normally followed by accountants and is only adopted here to ensure all lost money is in the contingency. Total sum of budget allocated for this topic is US$902,193. Table 7.4 gives the estimated values. Since constant percentages have been multiplied to total sum and the result is rounded up, in some cases last digit accuracy shall be checked. In this example no cost has been included for the additional costs related to late release of APG, GPG, and invoice payments.

7.2.3.4 Engineering Documents Certification
In some projects the client may require engineering documents to be reviewed and certified by a third party authority (TPA). Certification may be needed by:
• Insurer for legal purposes
• Owner to ensure minimum health, safety, environment (HSE) standards have been implemented
• Owner to ensure selected material match project requirements
• EPCIC contractor as guarantee for contractual obligations, etc.

A third party may be hired by client, EPCIC contractor, or consultant. In a project headed by the author a main consultant was responsible for the whole basic and detail design duration. In spite of that, basic design had to be done by a third party and the same team had to endorse detail design documents. Therefore, basic design was subcontracted. Client intention from endorsing was to ensure that in all aspects the detail design team followed basic design intentions. In practice many changes were implemented in design. The changes not only improved basic design considerably but also reflected new vendor data requirements.

This section may be applicable to some projects and not applicable to others. Endorsing detail design documents by basic design subcontractor in that specific project was in addition to TPA approval. Due to its importance (from documents approval point of view and the cost, which constituted a large portion of the engineering contract) this portion was given a specific budget topic. This topic is not used for the summarized report in this chapter.

7.2.3.5 Budget Summary
The total project budget is now divided in four categories as per Table 7.5. The fifth row is kept for contingency value.

In final contractual negotiations some discount may be given. This reduction shall be divided proportionally between budget topics, or as instructed by company executive manager only to a few of them. Another approach may be to include it totally in the contingency. Selection of each of these approaches may impact PM decisions in utilizing resources. For example, if a discount is only applied to "contingency" section, then PM can use originally selected number of engineers but increase routine check of progress delay.

If discount is applied on topic "01," then number of engaged engineers shall be reduced. It is understood that reduction in number of personnel can't be unlimited. If company executive manager wants to reduce costs as much as needed only to get the contract, then they shall accept its cost impact and possible loss of money without jeopardizing the engineering team structure.

In any case, since tax, SSO, and profit are calculated as a percentage of the total lump sum value automatically by reduction of the total price, their respective values will be reduced.

Table 7.5 Project bid cost summary

Item no	Description	Total US$
01	Personnel salaries	1,490,300
02	Office mobilization and utility	773,930
03	Overhead	902,193
04	Subcontractors	216,800
05	Contingency	375,913
Total		3,759,136

7.2.4 Areas of Concern and Mitigation Measures

Certain inaccuracies and unpredictable variables may affect each project cost estimate and progress. Identification of potential problems and measures to rectify them is essential to ensure that each project will end successfully. In this chapter only financial issues are discussed. Therefore, only in this chapter is the other name for a "successful project" a "project performed within budget with profit." It has to be highlighted and emphasized that to the authors a successful project is the one that "benefits people"—this means the environment, people of the involved country in particular and human nature in general, clients, EPCIC, consultants, etc.

7.2.4.1 Areas of Concern

There are several uncertainties and concerns in the estimated costs:

1. Costs may escalate from those predicted in any of the topics. This may happen due to the natural yearly escalation rate or any other unpredictable event. One of the examples is the guarantee costs. Based on consultant reputation and general economic condition, banks may be willing to issue a guarantee with a minimum charge or require considerable payments or certain amount of money deposited in a bank account before issuing or releasing the guarantee.

2. Client delays in invoice payment are added to the expected contractual time lags. Normally at least 30 days payment lag (after approving invoice) is anticipated in each contract. Delays referred here are in addition to the time lag anticipated in the contract. These delays may impose some additional costs to consultants. This will be more aggravated if a consultant depends on a bank loan for a portion of its daily expenditures. At least in the author's country you can never take your client to court and get payment for extra bank charges. On the other hand, banks can and will (as a routine basis) confiscate even your computers and the chairs you are sitting on for their money. In other countries also (I guess), even if you get your money for a project through court verdicts, you should not expect another project from the same client. Remember, oil and gas business at least in my country is a small world. Many people know many others.

3. Estimated man-hours may change due to external reasons. This may be due to increased/repeated rounds of comments from the client, delay in sending the comments, cosmetic comments, and client persistence to

implement them. This is in addition to internal factors that may affect estimated man-hours.

4. Estimated man-hours may change due to internal reasons. This may be due to reduced efficiency, due to employing unskilled man power, due to coordination/interface problems, inaccurate preliminary estimates, any type of rework, etc.
5. Estimated unit costs may change. This may be the office renting, procurement of new computers and other facilities, etc.
6. Personnel salary may change. Normally it is expected that an engineer with a higher salary is more efficient. But this is not always true. It is understood that each position requires a certain level of expertise, which normally requires higher payment for higher positions. It has to be understood that very high payments are not a guarantee for very high efficiency.

There may be some other unforeseen factors, which shall be evaluated case by case.

7.2.4.2 Mitigation Measures

Even within a consultancy mitigation measures may vary for different projects. In order to overcome these concerns, several approaches may be adapted. The project manager shall determine the problem. Is it total cost? Is it cash flow? Is it mobilization (personnel, equipment, etc.)? Is it specialist study?

1. Portions of the job may be subcontracted. This will divide the risk between engaged parties. However, at the same time it reduces a portion of the profit and if the selected subcontractor is not properly selected or managed or faces financial/technical problems, this may add burden to the consultant. This is a two-sided blade. It is important to note that always the consultant signing the contract with the EPCIC contractor has full responsibility toward the client. This is true unless a consortium is established. In that case, responsibility will be divided based on terms of agreement. In that type of contract perhaps everyone both joint and severally undertakes full responsibility to the client. This means that if one party fails, others shall undertake to complete the job under the same cost and time span.
2. Payment to subcontractor(s) may be connected as back-to-back payment from EPCIC contractor with suitable time lag. This will only help the consultant in the cash flow. It will not improve total cost. It has to be kept in mind that if the subcontractor's ability to work for a

certain period without payment is limited, then full risk can't be shared. In addition, they are also like us. In a certain period they may be in a grave financial situation and accept any project, but later improve their conditions. It is natural that they divert their main efforts toward more attractive projects. Regardless of all contractual conditions, all disputes, all expediting efforts, attractive projects (more profitable) get more attention from managers. Therefore, in assigning man power and other resources, they will be more favored.

3. Advance payment rate is very important in improving cash flow. This will cover the first portion of the job, in which the progress is very slow and consequently invoice values are also low. Normally advance payment is about 10–15% of the project lump sum price. For a 1-year project this may be equivalent to 2 months of expenditures. However, if large mobilization is required this money may be entirely consumed and only may last for 1 month's expenditures. Therefore, consultants shall immediately start issuing possible documents to gain some progress and issue invoices.

4. Invoice payment is based on document weight factor. Some documents such as design basis, data sheets, specifications, typical drawings, etc. will be issued in the early stages of the project. Their weight shall not be calculated exactly as per their consumed man-hours. First, these are main documents that are the basis or prerequisites for next-stage documents. Second, they are issued at a stage when only a few documents can be issued and other documents, which are more or less connected to the contents of these documents have to be issued later. Third, at the beginning of the project these documents shall keep the project going. Therefore, an importance factor shall be multiplied to their actual weight. The author has used up to 25% increasing factor for these documents that are issued at the first four months of the project. Of course, it is clear that this factor shall not be revealed to the client.

5. Some documents like MTOs and material requisitions are very much needed by the EPCIC contractor procurement department. Only in some cases is it possible to issue revisions of them with less accuracy. Later revisions with more accurate data are only possible after closing all parties' comments on prerequisite documents. First, inaccurate revision will help the procurement department identify their sources and start commercial negotiations. Therefore, a first revision can be issued with considerable weight to gain progress. The unfortunate issue is that you can't keep vendors waiting for a long time. They will

immediately ask for accurate data to close their contract. Especially in times of volatile economic conditions all quotations will have a limited validity period.

6. In many projects progress of topics like management can't be directly calculated. All the weight is either given to deliverables or at best first deliverables' progress is calculated and then an equivalent value is assigned to the management portion. However, many management procedures shall be closed at the start of the project. For example, all correspondence procedures, skeletons of the various reports, numbering systems, document handling procedures, document registrations, etc. shall be prepared at the early stages of the project. The fact that they are more or less constant from project to project will help in absorbing man-hours consumed for them. But in any case, some specific man-hours are needed to get client approval on them. Consultants can list management documents and assign weights for each. Their issuing milestones can be as per project normal deliverables milestones. Issuing these documents will help project progress at the starting phase of engineering.

7.3 PROJECT COSTS AND PLANNING

Project expenditures may be divided in two broad categories. They will have certain subdivisions. Some costs are on lump sum basis, which has to be paid for a specific requirement at a certain stage of the job. Others are continuous approximately equal installments throughout the project. For the first type, specific events expenditures like KOM, HAZID, HAZOP, mobilization costs, etc. can be listed. The main items under the second category are personnel salaries, office rent, utility costs, etc.

7.3.1 KOM Plan

KOM costs may include advance payments to subcontractors, gifts and celebration functions, reproduction of the huge number of documents used as the project basis, mobilization of highly qualified engineers only to review client submitted documents and familiarize with the project requirements, etc. All these will sum up to a considerable lump sum that has to be spent in a short duration of maybe 2 weeks. The conclusion will be summarized in KOM MOM. It is essential to clearly write down and obtain client confirmation before implementing these findings in project documents.

7.3.2 Office Mobilization Plan

Before the start of the engineering job, office mobilization shall be finished or completed to a certain extent. Therefore, based on the project schedule an early and late mobilization/demobilization plan shall be prepared. In some cases suppliers may be willing to hand over the equipment or parts of them at a certain date and receive payment in several installments. Company credit plays a major role in this respect. Even a specific consultant may have everything ready for a certain project while requiring a huge amount of mobilization for the next project. Some of the mobilization items may include:

- Renting an office with sufficient area.
- Providing utility connections like telephone, network, electricity, etc.
- Providing desks, chairs, files, cabinets, etc.
- Providing computers, monitors, etc.
- Providing required software.
- Providing printers, scanners, etc.
- Providing stationery.
- Providing refreshments and food.

Some of these items need payment in several installments and others may need continuous monthly or even weekly payments. Their impact in monthly expenditures is shown in the cash-out curve.

7.3.3 Personnel Mobilization Plan

Personnel mobilization is the most essential part of any engineering project. Key task force members shall have client approval. This will include their academic degree, total years of experience, year in the key point position, experience in similar projects, etc. For others minimum qualifications and number of required personnel shall be determined by the lead engineers. The process of selecting candidates, interviewing, and contracting has been discussed in chapter "Balancing Between Client and Task Force Engineers." Here the intention is to highlight that the salaries shall be calculated as part of the monthly cash-out curve. This includes direct and indirect payments to the personnel involved in the project. Salary of consultant managers and other personnel (like administration group) that are not directly involved in project document/drawing preparation are calculated by one of the following two methods:

- The best approach is to separate it from project costs and ask the financial department to report overburden costs separately. Therefore, consultant managers can compare those costs with the sum of profit and contingency. This way the PM is relieved from this burden. The example given in this chapter follows this method.

- In case the executive manger insists on including overburden costs in project expenditures, the first and easiest approach is to calculate a portion of overburden costs (indirect salaries/expenditures, etc.) equal to the ratio of the project lump sum cost divided by the total company projects. In this method each project receives a specific amount of surcharge. It can be added as a specific item in the report expenditure table.
- In the second approach indirect salaries for all persons are separately calculated. Each person (even the managing director) fills out a time sheet that shows his percentage involvement in each project. Based on this share the salaries are distributed between projects. In this method the indirect payment assigned to a specific project in two consecutive months may be different. However, the report table is simpler and doesn't include separate rows.
- Some portion of the manager's time is devoted to market development, which is not related to any project. In addition, there are certain costs that are not related to a project, for example, costs for advertisements, government fees for consultant certificates, etc. This portion is better to be divided based on each project portion to the whole consultant project's value.

The actual progress always has a lag to actual received invoices. Hence, each consultant shall adopt a method of using the latest received payment or issued invoice in this report.

It should be noted that although company managers instruct all personnel to fill in weekly time sheets and identify man–hours spent in each project or even group of documents, they themselves are very reluctant to do so. In many cases it may be impossible to get necessary data to prepare a report as per bullet point two. The author never had this chance. Therefore straight from the start follow procedure one and don't waste your time in useless negotiations with higher managers.

7.3.4 Estimated Monthly Expenditure Plan

Other than the subcontractors, who invoice according to document milestones, the normal monthly expenditures may include several items like rental, utility, reproduction, and salary costs. Normally reproduction and utility costs are divided also as per each project share compared to the total

Table 7.6 Estimated cash-out curve

Month	Sum USD
1	225,230
2	474,460
3	699,690
4	948,920
5	1,174,150
6	1,369,490
7	1,588,830
8	1,808,170
9	2,003,510
10	2,198,850
11	2,394,190
12	2,481,030

value of projects. The same is applicable for consultant costs in marketing, taking part in new bids, capabilities presentations to potential clients, etc. With these assumptions a first estimate of the monthly expenditures can be made. This estimate shall be reevaluated at the end of each month based on actual data and the cash-in/cash-out curve to be reproduced by the accountant. Table 7.6 gives the estimated cash-out curve.

The actual table may contain several columns. For brevity's sake, in this example only two columns have been kept. In each project consultant managers shall decide to what amount of detail this table shall be prepared. Some of these topics have a fairly or exactly constant monthly paid value, like rent, utility, salary, etc. Some have monthly payments but may not be constant like subcontractors invoices. The third group is those in which total amount is paid in several installments, like office mobilization costs.

Tax and SSO related to personnel salary will be continuously deducted and paid to government organizations. Company annual tax is paid in one installment and is not a monthly payment. Personnel salary tax and SSO are monthly installments.

In this sample expenditure it is assumed that tax and SSO are deducted monthly from consultant payments. At specific time intervals the consultant takes the proofs of deduction to related authorities and they issue a clearance certificate. This case may not be valid everywhere. For each case actual conditions shall be checked. Even if monthly deductions are not sufficient to cover full tax and SSO payments, they will cover a major portion of them.

In this example it is assumed after month 12 that there is no payment for this project. This assumption is not strictly correct. Although a project office (or at least major portion of it) may be demobilized, but there are still some payments for this specific project. For example, a project team may still expedite to get a final acceptance certificate. This may need several meetings and document handovers or material presentations, etc. Or, the accounting group may correspond with client, banks and other authorities to close project-related issues.

7.4 EXPECTED INCOME AND DEFICITS

The main intention of these calculations at budget planning is to help consultant managers foresee whether a certain project can fund itself and provide cash necessary for company routine/expansion activities or if it needs to rely on company financial support. In this way preventive/mitigation measures can be foreseen, areas and potentials for improvement identified and implemented.

This function is not possible unless a cash–in/cash–out curve is prepared. The previous section explained the cash–out curve. With some clients estimating a cash–out curve may be more straightforward than predicting a cash–in curve. Other than uncertainties in estimating project progress and potential invoices, client financial status and their adherence to contractual terms also affect payments.

Normally clients finally pay the approved invoices but with a time delay. Considerable delays impact consultant cash flow and in some cases may hinder/stop development plans. A development plan doesn't necessarily mean adding new personnel or buying a new office building, etc. In some cases it may only mean distributing some bonus money between employees to increase their loyalty toward the consultant or increasing key personnel salaries to prevent them from joining other companies.

7.4.1 Expected Invoices

Client payments are only made after receiving/evaluating consultant invoices. Even after proper invoicing and following all agreed/assigned progress calculation/invoicing procedures, there is always a delay in payment even compared to contractual deadlines. Clients tend to forget that the continuous payment of invoices enables consultants to improve their services to clients. The smooth progress of technical issues plus the invoicing are very important.

As a general rule, in all projects S-curve philosophy applies to the amount of progress. This means that at the first stages of the project, progress is very limited. In each engineering project, philosophies, specifications, and data sheets are issued in the first 3 months of the project. After issuing these documents, projects gain some momentum and considerable invoices can be made. Table 7.7 shows the expected cumulative progress and invoices. Just to show the impact of client late payment, two scenarios have been included. In the pessimistic scenario, client payments are made within a 3-month time span. In the optimistic scenario, payments are made within a 2-month time span. Normally, a 1-month time lag between approving the invoice and actual deposit of money in the consultant account is put in all contracts. The assumptions in this example include:

- At the start of the project the client may pay 15% of the total project value as advance payment. This payment has no deduction. It is paid against acceptable advance payment guarantee and the pro rata value is deducted from each consecutive progress payment, at the end of which or progressively the APG value is reduced or returned to the same ratio. However in many cases the client doesn't care to release the amount until AP is fully compensated. In some cases even if the client permits the bank to release the APG, there are so many bureaucratic obstacles that it is easier for the consultant to keep it and only release once when the AP value has been fully redeemed.

- TAX and SSO shall be deducted from each invoice payment. In this example they are taken as 14% of total project value. The details of invoice calculation are not repeated here. Progress percentages are fictitious. Total amount of statuary deductions depend on the country and the type of contract. It is the financial team's duty to find out which system is applicable and shall be used.

- Good performance guarantee (GPG) value is also not constant. It may vary in value and in the form. The author has normally faced 10% GPG. However, its form with even the same percentage may change. For example, 5% may be deducted from the invoice but another 5% may be requested in the form of a bank guarantee, or the whole 10% may be deducted from the invoice, or a 10% bank guarantee requested. Part of the deducted value may be released after issuing a provisional acceptance certificate. The bank guarantee will be released after issuing the final acceptance certificate. In this sample example, 5% is deducted and released 6 months after the last invoice.

Table 7.7 Estimated total cash in

End of month	Estimated progress	Gross $	Net $ approved	2-month delay + 1-month time lag		1-month delay + 1-month time lag	
				Net $ received	Net $ cumulative I	Net $ received	Net $ cumulative II
1	0.040	150,365	120,894	563,870	563,870	563,870	563,870
2	0.070	263,140	328,116	0	563,870	0	563,870
3	0.090	338,322	586,425	0	563,870	120,894	684,764
4	0.100	375,914	862,722	120,894	684,764	207,222	891,987
5	0.095	357,118	1,114,758	207,222	891,987	258,309	1,150,296
6	0.095	357,118	1,356,616	258,309	1,150,296	276,297	1,426,592
7	0.090	338,322	1,576,356	276,297	1,426,592	252,036	1,678,628
8	0.090	338,322	1,786,962	252,036	1,678,628	241,858	1,920,486
9	0.085	319,527	1,977,480	241,858	1,920,486	219,740	2,140,227
10	0.085	319,527	2,159,849	219,740	2,140,227	210,606	2,350,832
11	0.080	300,731	2,324,048	210,606	2,350,832	190,518	2,541,350
12	0.080	300,731	2,481,030	190,518	2,541,350	182,370	2,723,720
13	0.000	0	0	182,370	2,723,720	164,199	2,887,919
14	0.000	0	0	164,199	2,887,919	156,982	3,044,900
15	0.000	0	0	156,982	3,044,900	0	3,044,900
16	0.000	0	0	0	3,044,900	0	3,044,900
17	0.000	0	0	0	3,044,900	0	3,044,900
18	0.000	0	187,957	0	3,044,900	0	3,044,900
19	0.000	0	0	0	3,044,900	0	3,044,900
20	0.000	0	0	0	3,044,900	187,957	3,232,857
21	0.000	0	0	187,957	3,232,857	0	3,232,857

In order to show the importance and impact of client delays in invoice payment, two scenarios have been shown. In the first scenario a 1-month delay is added to the normal 1-month contractual time lag, and in the second scenario a 2-month delay is considered.

7.4.2 Cash in—Cash out Table and Curve

Even a "successful project" may face periods of negative cash flow. This is the nature of each project and is not a scare factor. The main factor is to know how long this deficit will continue and how much its maximum value will be. Very short duration deficits can be tolerated more easily than long duration deficits even if their absolute value is much larger (Table 7.8).

Negative numbers show deficit. AP is a very good help in keeping the consultant budget in positive side. However, as much as delays in addition to contractual time lag happen in invoice payment, the deficit increases. As can be seen in scenario 1, more than $282,163 deficit happens. In addition, the deficit continues for 9 months. In the second scenario, the maximum deficit is limited to $56,933 and only continues for 3 months.

Table 7.8 Expected deficit

Month	Deficit 1	Deficit 2
1	338,640	338,640
2	89,410	89,410
3	−135,820	−14,926
4	−264,156	−56,933
5	−282,163	−23,854
6	−219,194	57,102
7	−162,238	89,798
8	−129,542	112,316
9	−83,024	136,717
10	−58,623	151,982
11	−43,358	147,160
12	60,320	242,690
13	242,690	406,889
14	406,889	563,870
15	563,870	563,870
16	563,870	563,870
17	563,870	563,870
18	563,870	563,870
19	563,870	563,870
20	563,870	751,827
21	751,827	751,827

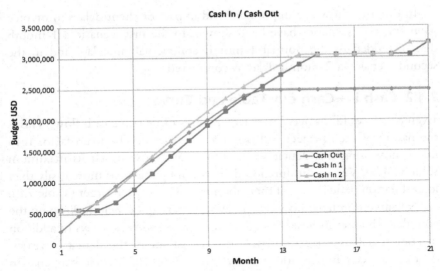

Figure 7.1 Cash in–cash out curves.

This difference occurs with only 1 month additional delay between the two schemes. In both cases if only 10% was paid as AP (instead of 15%), the situation would have become worth.

Cash in–cash out curves are shown in Fig. 7.1. The deficit curves are shown in Fig. 7.2. Figures below the reference zero line show deficit months. This is useful in showing both the month in which the deficit shall be expected in addition to its value. Project manager and consultant financial and executive managers shall cooperate closely to overcome the crisis duration.

This example was designed to have similar plus value at the end. It wanted to show cash flow impact regardless of whether at the end a profit is received or not.

Adding a table showing total payments at the end of each month, total approved invoices, and actual received money is helpful for managers. A sample brief table for the end of month eight is shown in Table 7.9.

Total actual payment for this month is US$1,808,170. Total net approved invoices are US$1,786,962. Actual payment including AP for scenario 1 is US$1,678,628, which shows US$129,542 negative cash flow and for scenario 2 is US$1,920,486, which shows US$112,316 positive cash flow.

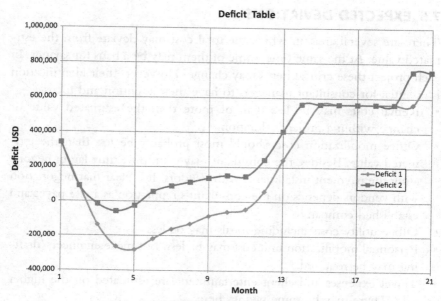

Figure 7.2 Deficit curves for two scenarios.

Table 7.9 Brief status end of month eight

Item	Name	Sub item	Description	Allocated budget USD	Sum USD
01	Personnel	01–PM	Project management	109,600	1,056,000
		01–PR	Process	98,400	
		01–ME	Mechanical	126,400	
		01–PI	Piping	203,200	
		01–EL	Electrical	112,800	
		01–In	Instrument	115,200	
		01–ST	Structure/ Architecture	230,400	
02	Office costs	02–Of	Office rent	64,000	597,770
		02–MO	Equipment	149,450	
		02–LU	Utility	107,520	
		02–RE	Reproduction	28,000	
		02–TE	Team building	20,000	
		02–TR	Travel	80,000	
		02–CL	Client in office	52,800	
		02–HA	HAZOP/ HAZID	96,000	
04	Subcontractor	04–SA	Safety	154,400	154,400

7.5 EXPECTED DEVIATIONS

There are several areas in which the total cost may deviate from the estimated value. At the same time, some of them may be a basis for savings. In each project these critical items may change. However, their identification is essential for consultant managers to have their attention and focus.

- Rental costs may be less than or more than the estimated value or change within a project's duration.
- Office mobilization costs should most probably be less than the estimated value. Besides, the consultant may gain some time lag in negotiating the payment installments with vendors. It is clear that negotiation with vendors depends on the consultant's reputation as a safe payer and established company.
- Office utility costs including meals may be less.
- Personnel mobilization unit cost may be less. But more engineers/drafting may be required.
- Travel expenses and client entertainment are estimated on the higher side. There may be some savings here.
- Project total duration may become longer.

Once more it is emphasized that the assumptions in this example are fictitious. They are just explained to give an idea of what factors may need further consideration. In each project the applicable factors shall be decided.

7.6 PROPOSALS AND INCENTIVES

It is natural that the project manager tries to save on each budget topic. First, he shall find which topics have better savings potential without adverse impact on general progress. In some topics that are based on lumped sum value, he may not have so much chance and has to try not to exceed the expected value. In other topics, the project team has to find methods for saving. Following are preliminary guidelines:

1. Procuring computers and other office facilities (office mobilization) is a great cost. It has to be directed from a company with suitable discounts and in several installments.

2. A more decentralized management system not only allows but also encourages the PM to give more freedom to each lead engineer. In this case they may have further authority in selecting the number and quality of their staff. The goal of this authority is to optimize discipline outcome. This allows the lead engineer to distribute or replan his or

her team performance to get better efficiency. A portion of this saving may be paid to each leader as incentive.

3. The lead engineer may be allowed to give incentives to his team fellow engineers in preparing/issuing some specific documents. This may range from overtime to bonus.

4. In addition, some specific documents can be distributed between discipline engineers on a lump sum basis. The logic is very much similar to the case in which the specific document or task is considered as a sort of further subcontracting on a small scale.

5. Improved interaction with client team members will help the approval process and prevent back and forth document and response sheet issuing. Of course this shall be carefully done because it may have the disadvantage of wasting a great deal of time implementing clients' various requests on studies out of project SOW. Lead engineers shall always be on the alert to stay out of general study and academic research.

REFERENCES

[1] Dawson Thomas H. Offshore Structural Engineering. Prentice Hall Inc.; 1983.
[2] Mather A. Offshore Engineering an Introduction. Witherby and Company Limited; 1995.
[3] Devold H. An Introduction to Oil and Gas Production. ABB Oil and Gas; 2009.
[4] Pope JE. Rules of Thumb for Mechanical Engineers. Gulf Publishing Company; 1997.
[5] Chakrabarti SK. Handbook of Offshore Engineering Vol. 1, 2. Elsevier; 2005.
[6] Patel MH. Dynamics of Offshore Structures. Butterworths; 1989.
[7] Wilson JF. Dynamics of Offshore Structures. John Wiley and Sons; 1984.
[8] Graph W. Introduction to Offshore Structures, Design, Fabrication and Installation. Gulf Publishing; 1981.
[9] Poulos HG. Marine Geotechnics. Unwin Hyman; 1988.
[10] Chakrabarti SK. Hydrodynamics of Offshore Structures. Springer-Verlag; 1987.
[11] Clauss G, Lehmann E, Ostergaard C. Offshore Structures. Vol. 1. Conceptual Design and Hydrodynamcis. Springer-Verlag; 1992.
[12] Clauss G, Lehmann E, Ostergaard C. Offshore Structures. Vol. 2. Strength and Safety for Structural Design. Springer-Verlag; 1992.
[13] Barltrop NDP, Adams AJ. Dynamics of Fixed Marine Structures. Butterworth–Heinemann; 1991.
[14] McClelland B, Reifel MD. Planning and Design of Fixed Offshore Platforms. Van Nostrand Reinhold Company; 1986.
[15] TOTAL. Exploration and Production, Design and Construction of Offshore Living Quarter. 2005. GS CIV 404.
[16] Lienhard IV JH, Lienhard V JH. A Heat Transfer Textbook. 3rd ed. Cambridge (Massachusets): Phlogiston Press; 2006.
[17] Det Norske Veritas. Offshore Standard, Electrical Installation. March 2001. DNV–OS–D201.
[18] American Petroleum Institute. Recommended Practice for Classification of Locations for Electrical installation at Petroleum Facilities Classified as Class I, Zone 0, Zone 1 and Zone 2. 1997. API–RP–505.
[19] International Standard. Electrical Apparatus for Explosive Gas Atmospheres, IEC–60079 Parts 0 to 20. 1998.
[20] British Standard. Code of Practice for Control of Condensation in Buildings. 2002. BS 5250.
[21] British Standard. Code of Practice for Ventilation Principles and Designing for Natural Ventilation. 1991. BS 5925.
[22] Quinn AD. Design and Construction of Ports and Marine Structures. McGraw Hill Book Company; 1972.
[23] Corps of Engineers. Shore Protection Manual, Vol. 2. Coastal Engineering Research Center, Department of the Army, Waterways Experiment Station; 1984.
[24] GLENN Report. Report on Meteo–Oceanographic Conditions Affecting Design and Operations. 1999. SP–2–GEN–1–A–TRG–502.
[25] TOTAL. Exploration and Production, General Specification Safety Layout. 2003. GS SAF 021.
[26] American Petroleum Institute. Recommended Practice for Design and Hazards Analysis for Offshore Production Facilities. 2001. API–RP–14J.
[27] Det Norske Veritas. Offshore Standard Safety Principles and Arrangements. 2001. DNV–OS–A101.

[28] AGIP. Offshore Installations, Design Layout and Safety Spacing. 1996. 20243. VONSAF.SDS Revision 0.

[29] Det Norske Veritas. Recommended Practice, Risk Based Inspection of Offshore Topsides Static Mechanical Equipment. 2002. DNV—RP—G101.

[30] NORSOK Standard. Risk and Emergency Preparedness Analysis. 2001. Z—013 Revision 2.

[31] American Petroleum Institute. Recommended Practice for Planning, Designing and Constructing Fixed Offshore Platforms — Working Stress Design. 2007. API—RP—2A.

[32] Young WC, Budynas RG. Roark's Formulas for Stress and Strain. 7th ed. MacGraw Hill; 2002.

[33] USGS. Reserve Growth Assessment Team, Assessment of Remaining Recoverable Oil in Selected Major Oil Fields of the Permian Basin, Texas and New Mexico. 2012.

[34] American Society of Mechanical Engineers. ASME, B16.5. 1996.

[35] Gas Processors Suppliers Association. Engineering Data Book. ed. 12 2004.

[36] European Standard. Safety Requirements for Secondary Batteries and Battery Installation. 2002. EN 50272.

[37] American Petroleum Institute. Recommended Practice for Fire Prevention and Control on Open Type Offshore Production Platforms. 1993. API—RP—14G.

[38] Shell Internationale Petroleum Maatschappij B.V. E and P Project Management Guidelines. The Hague; 1990.

INDEX

CTICM. *See* Centre Technique Industriel de la Construction Métallique (CTICM)
Cylindrical pigs, 273

D

DAF. *See* Dynamic amplification factor (DAF)
Data base, 390
Data sheet (DS), 65
Data sheets, 82–83
Davit, 319–320
DB. *See* Basis of Design (DB)
DCC. *See* Document control center (DCC)
DCS. *See* Distributed control system (DCS)
Deaeration tower, 271
Decibels, 207
Dedicated time/priority, 395
Deep waters fixed jackets, 5
Deflection, 462–463
Degassing unit, 250
Dehydration, 240–241
    contactor column internal, 242f
    internals, 244
    package, 241
        PFD, 241f
    PFD, 241
        items, 243
    rich glycol, 241–243
    TEG, 241–242
    theory water absorption and glycol regeneration, 243
Deluge valves, 314
Demulsification, 257
Demulsifier, 229, 257–258
Deoiling unit, 250
Desanding vessel, 240
Design verification certificate (DVC), 427
Deviations, review of, 211–212
Dew point, 68
    temperature, 218–219, 266–267, 283
DGF. *See* Dissolved gas flotation (DGF)
DGs. *See* Diesel generators (DGs)
DHSV. *See* Down hole safety valve (DHSV)
DID. *See* Duct and instrumentation drawing (DID)
Diesel engine, 289

Diesel generators (DGs), 289–290, 419
Diesel oil package, 286
Diesel oil system, 286–287
    diesel fuel consumption, 289t
Direct current consumers, 291
Direct tax, 458, 463
Disciplines in offshore platform design, 25. *See also* Systems and equipment for
Disciplines/engineers mobilization and requests, 358
    consultant expectations, 368–369
    engineering recruitment, 359–360
    human factors, 367–368
    required data, 365–366
    salary, 360–364
    work discipline, 366–367
    working conditions, 364–365
Dissolved gas flotation (DGF), 255, 256f
Distributed control system (DCS), 45
    typical signals in, 46–47
Distributed control system (DCS), 221, 303–304
Document
    approval process, 489
    class, 382
    control, 403–405
        AFC documents, 404–405
        CA, 404
        PM, 405
        transmittal, 405–406
    handling, 398–403
        client comments, 401–402
        distributing copied documents, 402
        engineering team, 399–400
        hard copy documents, 400–401
        transmittals preparation, 403
    importance, 379–380
    number and title, 380
    Numbering Procedure, 371–372, 380–382
    stage
        action dates, 385–387
        progress, 490–491
    status, 389–390
    type and class, 381–383
    weight in discipline, 383–385

T

Printed in the United States
By Bookmasters